D0083411

The Developing World

The Developing World

An Introduction

Second Edition

E. S. SIMPSON

Longman
Scientific &
Technical

Longman Scientific & Technical
Longman Group UK Limited
Longman House, Burnt Mill, Harlow
Essex CM20 2JE, England
and Associated Companies throughout the world

Copublished in the United States with John Wiley & Sons, Inc.,
605 Third Avenue, New York NY 10158

First published 1987
Second edition 1994

British Library Cataloguing in Publication Data
A catalogue entry for this title is available from the British Library.

ISBN 0-582-21888-8

Library of Congress Cataloging-in-Publication data
A catalog entry for this title is available from the Library of Congress.

ISBN 0-470-23387-7 (USA only)

Set by 13 in 10/12 Palatino
Produced by Longman Singapore Publishers (Pte) Ltd.
Printed in Singapore

For Julia, Alison and Ruth Claire
who growing up in the Developing World came to know mankind is one

Contents

List of figures

List of plates

(All photographs are by the author)

List of tables

Preface to the first edition

This book attempts to introduce its readers to the issues of development in the Developing World. It is designed as a beginning rather than an end and it inevitably leaves much unsaid. It is hoped that its bibliography will enable the reader to explore still further the issues raised. In it I have eschewed the use of the first person yet it is a book born of a long acquaintance with countries in the Developing World and my perception of it is inevitably a product of my experience. My first contact came as a result of my serving in the British navy during the Second World War. As a very young man I arrived in Ceylon, now Sri Lanka, and experienced the beauty, and the poverty, of a developing country. I had been at a school accustomed to see some of its pupils enter the Colonial Service and work in one of the patches of pink spread across the atlases of the day. A colonial empire had seemed a natural disposition of power and influence. I was shaken out of this unthinking acceptance when a young Sinhalese soldier told me that he had wished the Japanese had been victorious since this would have shown that the British were not all-powerful.

The intervening years have seen the collapse of that power as the British and other empires have been dismantled and scores of new independent nations have come into being. My eyes were opened further when, as a university don, I was seconded to Nigeria to help in the establishment of a new university. A minor incident gave me another glimpse into the development process. On the Jos plateau driving with half a dozen students on a field survey we passed a line of women from the local tribe. All were naked apart from a few leaves. Suddenly one of the students, a sophisticated young woman studying at this new university we had implanted in Nigeria, cried out 'There's my grandmother!' She was one of the line. It brought home to me the enormity of the shock which indigenous cultures experience as the material manifestations of development thrust their way into the heart of traditional values. But Nigeria taught me much more as it revealed to me the pain and trauma which many nations have experienced in the process of development. It put to the test many development theories in the searing heat of reality as I was witness to the tragedies of massacre and civil war.

In all this and in my present endeavour I am indebted to many individuals; indeed it is impossible to list them all but some I must mention. I owe a great debt of gratitude to Robert Steel who guided my scholarly interests into the Developing World and placed before me a challenge I could not resist. He has given me unfailing encouragement and support and I cannot thank him sufficiently. Second is the debt I owe to my friend Andrew Learmonth whose profound intellect I have always admired and envied. Third I wish to acknowledge the influence of Sam Richardson, former District Commissioner turned scholar, who showed me that the practice of pragmatism does not involve the abandonment of principle. All have been colleagues at one time or other. Their influence and friendship is always with me.

It has been my good fortune and privilege to have worked in the Developing World with and for men who were of rare distinction. I have abiding memories of the qualities of leadership and the unfailing courtesy of the late Ratu Sir Edward Cakobau, whose ancestors had ceded Fiji to Britain, and of the Prime Minister of Fiji, Ratu Sir Kamisese Mara, whose qualities of statesmanship are an example to everyone. But many are the men who each in his own way has revealed to me the nuances of meaning which are an essential to any attempt to understand societies other than one's own. Among them I must mention Savenaca Siwatibau and Isireli Lasaqa, Chor Pang Lo, Chek Lam So, Chi-Keung Leung and Tze-Nang Chiu, Kingsley Ologe, Samson Odingo and Justice Mlia. And there are the memories which remain vivid and indelible of the kindness and hospitality I have found throughout the Developing World. In Malawi I have enjoyed it both at the hands of those in high authority and of the poorest of peasants. In China I have been invited into the homes of people poor but dignified who with ancient courtesy have bid me share their limited resources. In the South Pacific I have enjoyed that open hospitality which the Fijians and Tongans combine with a gracious ceremonial. All shared kindnesses which made us one.

In the preparation of this book my particular thanks must go to three people; to Mrs Olive Teasdale for her great skill in the preparation of the maps and diagrams which illustrate it and to Miss Yvonne Lambord and Mrs Eunice Tubman who saw the conversion of a tangled manuscript into a typescript.

Most of all my indebtedness is to my wife, Christine, whose intellectual curiosity has been a constant delight and stimulus to me and whose ability to relate to, and be accepted by, peoples and societies new to her has done much to make us and our children a part of those societies we have known throughout the Developing World.

Preface to the second edition

The Developing World is far from static. In the past decade changes have been taking place which are of great significance for developing countries. This second edition not only updates the data and diagrams presented but also discusses these changes. Attitudes to development problems and views as to how best they may be resolved have themselves been modified by the evidence of the changing scene. Each chapter has been recast, where appropriate, to encompass these events.

This book, as an introduction to the Developing World, sets the framework for a consideration of its problems and the hopes of its people. I am very conscious that it does little to put the flesh on the bones of Third World reality and says little of the 'view from within'. In the greatly extended bibliography of scholarly works I have included five books which each in its own way helps to correct this omission. They are but a sample but each gives valuable insights and much food for thought. They are: *Wild Swans* by Jung Chang, telling of life in the People's Republic of China; Sembene Ousmane's *God's Bits of Wood*, a novel set in colonial French West Africa; Dominique Lappiere's account of life in a Calcutta slum, *The City of Joy*; Mark Tulley's collection of essays on India, *No Full Stop in India*; Ben Okri's magical novel, *The Famished Road*, set amidst the mysteries of the Yoruba people of southern Nigeria.

In the preparation of this revised edition my thanks must go to Mrs Lynne Martindale for preparing the typescript and to Mrs Ann Rooke who has drawn the revised diagrams.

Acknowledgements

We are grateful to the following for permission to reproduce copyright material: the Food and Agriculture Organization of the United Nations for fig. 2 from FAO World Report: State of Food and Agriculture, 1991 and for fig. 3 from FAO World Report: State of Food and Agriculture, 1986; the Editor, Journal of Developing Areas for fig. 30 (H C Weinand, 1973); Longman Group Ltd for fig. 20 from figs 7.3 and 7.4 in *Economic Development in the Third World* (P Todaro 1977); John Murray Ltd for six lines from 'The Planster's Vision' by John Betjeman from *Collected Poems*; The Society of Authors on behalf of the Bernard Shaw Estate for 'The golden rule is that there are no golden rules' by Bernard Shaw from *Maxims for Revolutionaries*; the World Bank, the International Bank for Reconstruction and Development, for fig. 7 from World Bank Development Report 1980. Whilst every effort has been made to trace the owners of copyright material, in a few cases this has proved impossible and we take this opportunity to offer our apologies to any copyright holders whose rights we may have unwittingly infringed.

Part 1 Dimensions and issues

'Now, here, you see, it takes all the running you can do, to keep in the same place. If you want to get somewhere else, you must run at least twice as fast as that.'

Lewis Carroll
Through the Looking-Glass

1 Worlds within worlds

Dos linages sólas hay en el mundo, como decia una abuela mia, que son el
tenir y el no tenir
(There are only two families in the world, as a grandmother of mine used to
say: the haves and the have nots)

Cervantes
Don Quixote: El Cabellero de la Trieste Figura

That the world is divided by levels of economic prosperity is well known. It is
possible to postulate that this has always been so. It is equally reasonable to
argue, however, that the contrasts have never been as great as those which now
exist and which had begun to emerge in the eighteenth and nineteenth centuries.
That great upsurge in productive capacity with its associated transformation of
not only the means of production but also the whole organization of societies
and their economies, that change we call the agrarian and industrial revolutions,
profoundly distinguished its participant nations from those not directly
involved. This distinction came to the eyes of merchants and missionaries who,
both knowingly and unknowingly, spun a web of interconnections which tied
the outer countries to the inner industrial nations. As empires grew, colonial
territories enhanced both the resource base and the potential markets of the
industrial economies of Europe. Whether it became colonial or not, the world
that was not part of the economic transformation of the nineteenth century
remained distinct. In terms of technological application and economic
organization it was undeveloped and, in consequence, poorer in material terms.
It was variously described as savage, primitive or backward by the newly-rich
world. In so much as these descriptions implied less developed in culture,
religion or political systems, they were the product of both ignorance and
arrogance; ignorance of the richness of the cultures of pre-industrial civilizations
and an arrogance in an under-estimation of them. Attitudes have changed; the
nature of the imperial powers and the political status of the non-industrialized
world have changed; the world of today is not that of 1885 or indeed 1945 and
the terminology has likewise changed. What remains are the differences,
differences of both a greater and lesser degree between the many independent
sovereign nations which now constitute the world community. There are still
two worlds and the difference turns upon comparative levels of wealth and
poverty.

The terms used to describe the poorer world are several. The most common
is Third World, an example of an adopted and misused label. Wolf-Phillips has

attributed the first use of this term to Alfred Sauvry who, in 1952, referred to the *tiers monde* (Wolf-Phillips 1979). Sauvry was, however, using the term to mean a third force, a political force, in a world where the North Atlantic community represented the first force and the communist bloc of the Soviet Union, eastern Europe and China a second force. It was a distinction based upon ideological commitment rather than conditions of economic attainment. In the latter sense the inclusion of China would have been singularly inappropriate. This origin appears to have been forgotten and later, in the early 1960s, Third World came to mean countries less economically developed than those of the industrialized nations of North America, Europe and the Soviet Union and so the Third World came to incorporate China. The threefold distinction has become even less relevant with the collapse of the former Soviet Union and the Eastern bloc alliance. Several east European countries have now revealed themselves on socio-economic measures as Third World nations. The situation has been further complicated by the current usage of the terms North and South. To anyone with any degree of geographical fastidiousness they are terms which can only be regretted when India and China are allocated to the South and Australia and New Zealand to the North. The terms appear as the title of the first report of the Independent Commission on International Development Issues chaired by Willy Brandt (Brandt 1980). Other terms such as undeveloped, under developed, least developed, less developed, have all appeared and been used by bodies such as the United Nations. Underlying all these labels is a perceived distinction of differences in poverty and wealth, and a belief that there is a process called development during which the condition of poverty is replaced by one of comparative affluence. This book is concerned with examining the nature of this transformation and the attempts made to promote it. It is appropriate, therefore, that it addresses the issues of the Developing World.

Where is this Developing World? The American political scientist Bruce Russett has likened the search for an all-purpose region to that of the alchemist trying to find a universal solvent (Russett 1967). Is there indeed an all-purpose region which encapsulates the Developing World? In 1961 an *Atlas of Economic Development* edited by Norton Ginsburg was published (Ginsburg 1961). Based on an extensive assemblage of the data available in the mid-1950s, it contains a large number of world maps which depict many of the dimensions of economic activities and the characteristics of populations. Each in turn, whether it be a depiction of population density, infant mortality, proportion of the labour force in manufacturing, the use of inanimate energy, per capita food intake, primary school enrolment, per capita income or the consumption of steel, sets out a world pattern. Conspicuous and not unexpected, is the continual emergence of a congruent area of poverty, ill-health, social deprivation and low productivity. It is the area intuitively recognized as the Third World or the Developing World of this book. But not all the maps are coincident. Countries rank higher on some evaluations than others. It becomes clear that there is no sharp, defining edge to the Developing World.

Brian Berry has attempted to bring together these various measures in an integrated analysis (Berry 1960, 1961). Using an early form of direct factor analysis, Berry examines forty-three indices of economic development for ninety-five countries for which comparable data were available. He was able to show that underlying the forty-three indices were four basic patterns which could as effectively describe the similarities and differences as the forty-three. Two basic patterns, the technological and the demographic as he termed them, proved to be particularly efficient in resolving the many indices and he produced a graph based upon the 'second values' of these two patterns. Berry's analysis spanned the whole range of economic levels from nations such as the USA and West Germany, commonly thought of as examples of advanced industrial economies, to nations such as Afghanistan and Ethiopia recognized as poor and disadvantaged. This analysis revealed that there is no natural division of the countries of the world into distinct socio-economic strata. Instead there is a continuum of levels of development from the very poorest and underdeveloped countries to the richest and technologically most sophisticated. There is a continuous transition between levels of development. The Developing World is thus as much a concept as a place. It has a core but no boundaries. It has a beginning but no end. Wherever one draws a boundary it must be arbitrary and it must enclose a range of internal differences.

Table 1 The least developed countries 1993

Afghanistan	Haiti
Bangladesh	Lao PR
Benin	Malawi
Bhutan	Maldives
Burkino Fasso	Mali
Cambodia	Mauritania
Cape Verde	Mozambique
Central African Republic	Nepal
Chad	Niger
Comoros	Rwanda
Equatorial Guinea	Samoa
Ethiopia	Sierra Leone
Gambia	Somalia
Guinea	Tanzania
Guinea-Bissau	Uganda
Guyana	Yemen

Source: United Nations Organisation and World Bank

It is possible to identify a group of nations which occupy the lowest points in this chain of development. In 1971 the General Assembly of the United Nations drew up a list of twenty-four countries which presented the most difficult conditions. They were identified on the basis of three criteria: GDP per capita, share of manufacturing in GDP and the literacy rate. To this list were added other poor countries bringing the total to thirty-two by 1991. They have been described as LDCs, the least developed countries, and are listed alphabetically in Table 1. They represent the core of underdevelopment, the nations at the bottom of the ladder, standing in poverty.

While Berry's analysis demonstrated the international continuum of development it also revealed that the geographical distribution is, to a large extent, discretely aggregated. The nations lowest on his technological–demographic scale were largely, though not exclusively, tropical. The countries of sub-Saharan Africa congregated in the lowest levels of the scale, followed in turn at higher levels by those of North Africa, Asia, Central America, South America, the Soviet bloc and the nations of north-western Europe and North America. There was relatively little overlap between these continental groups of nations though in total they embraced the whole development spectrum without break. It is a distribution suggestive of a ladder of development up which nations can climb, though with some further to climb than others and some encountering particular difficulties. The Developing World is clearly not one single unit in terms of the difficulties it presents nor in the stages of development through which nations will have to pass. There are worlds within worlds.

Per capita GNP is a useful indicator of economic development though it is essential to have additional information to aid in the assessment of its significance. It is mapped for the nations of the world in Fig. 1. Three other indices have been selected and mapped to demonstrate the issues which many developing countries face. One is the proportion of the work-force engaged in agriculture (Fig. 2). The second is the adequacy of food supplies as measured by the daily calorific intake of food as a percentage of the basic physiological requirement (Fig. 3). The third map depicts infant mortality rates (Fig. 4). These maps reveal large sections of the world whose populations are very largely agricultural and yet receive an inadequate supply of food, areas in which a combination of conditions still bring unacceptable risks of death to the very young. They reveal a world in which the combination of natural resources and the efforts of its population still fail to provide the basic means of subsistence. Yet the Developing World extends beyond these poorest areas and that it does so gives reason for hope.

The transformation of nations from a state of poverty to one of improved well-being, from a condition of adequate but self-limiting subsistence to one of continuing and self-generating growth is the nature and purpose of development. It must first encompass the basic physiological needs but must also embrace those other dimensions which make life something to be savoured rather than endured. However, the precise nature of what constitutes development and how it might best be achieved has been, and is, much debated (Brown 1988). The views as to what accounts for underdevelopment determine the solutions offered to rectify it. Todaro lists three groups of development objectives: the provision of basic requirements to sustain life; self-esteem developed by the provision of education, the furthering of cultural development and the maintenance of human rights; freedom from servitude ensured by free

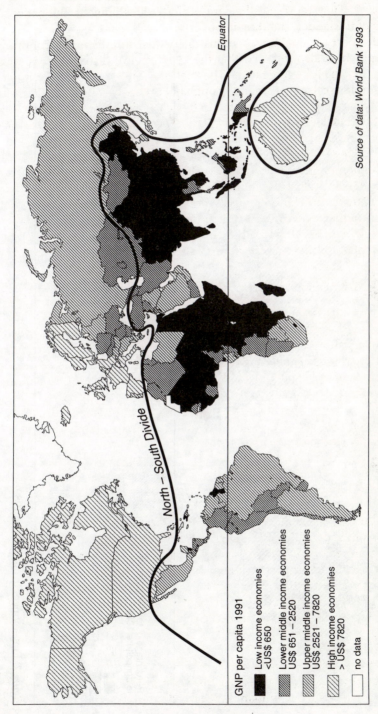

Figure 1 The world pattern of Gross National Product per capita 1991

Equator

Source of data: World Bank 1993

North – South Divide

GNP per capita 1991

■ Low income economies
<US$ 650

Lower middle income economies
US$ 651 – 2520

Upper middle income economies
US$ 2521 – 7820

High income economies
> US$ 7820

□ no data

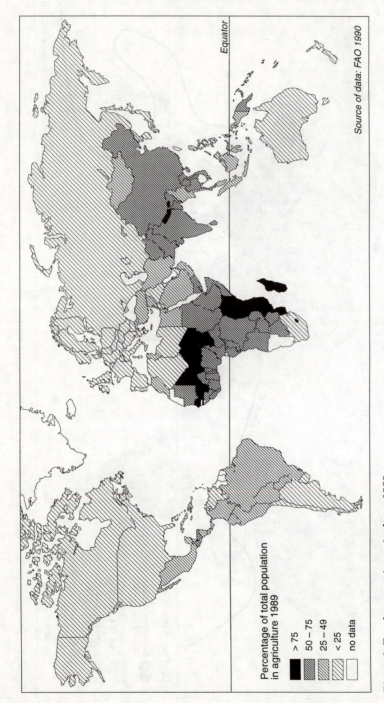

Source of data: FAO 1990

Equator

Percentage of total population
in agriculture 1989

> 75
50 – 75
25 – 49
< 25
no data

Figure 2 Employment in agriculture 1989

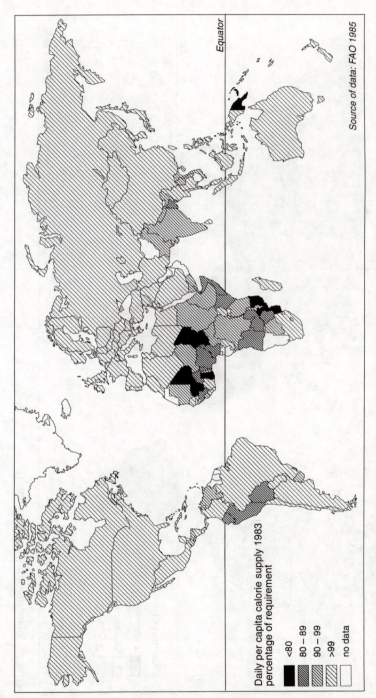

Figure 3 Daily per capita calorie supply 1983

Daily per capita calorie supply 1983
percentage of requirement

■ <80
▨ 80 — 89
▧ 90 — 99
▨ >99
□ no data

Equator

Source of data: FAO 1985

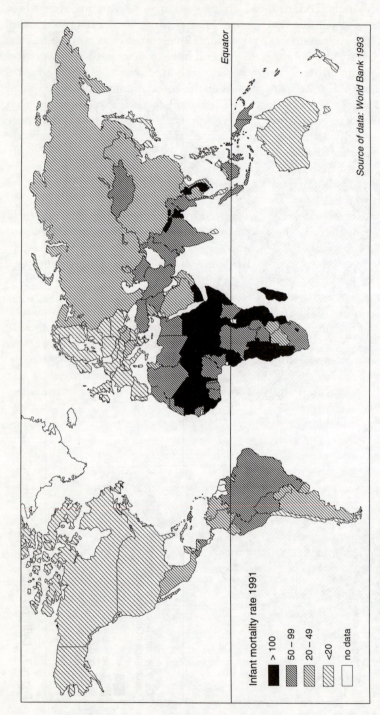

Source of data: World Bank 1993

Equator

Infant mortality rate 1991

■ > 100
50 – 99
20 – 49
<20
no data

Figure 4 Infant mortality per thousand 1991

choice in economic and social issues. Todaro applies these measures to nations as well as individuals (Todaro 1985). There is a danger of equating the goals of development with Utopia and since Utopia does not exist, of raising development to an unobtainable abstraction. Development has been seen as the conquest of poverty and solutions to this end have been promoted only to find that poverty is a manifestation of a wider social and economic ill. Some analysts take the problem of development to the global level, a problem only to be solved by a reorganization of the world order (Harris and Harris 1979). What have been called the orthodox paradigms of development see a changing and more productive economy as the basis of the process of development in which one stage of development passes into another and higher level of development. Rostov's model discussed in Chapter 8 is a classic and influential example of this. Wilber and Jameson refer to such models as 'the parables of progress' (Wilber and Jameson 1988).

The concept of development when viewed from inside the Developing World has often been neglected. The poor peasant of Peru or the slum-dweller of Calcutta would no doubt, if asked, conceive of development as the amelioration of his dire poverty. Others in the Developing World often regard development as the intrusion into their societies of European and North American values imposed upon their own cultures; to them development is a mixed blessing. The Nigerian geographer Mabogunje has stressed the need for developing nations to be free of the dominance of the rich industrial nations (Mabogunje 1980b). However it is engendered, development must allow both peoples and nations the self-respect that comes from participation in the world economy as equals. Any 'web of interconnections' in a post-colonial world must develop not between exploiter and exploited but between partners with a reciprocal respect for the values of each other.

The awakened world of the second half of the twentieth century has become aware of these issues. The many new nations have tried to grapple with them. The practice has proved much more difficult than the preaching. It is these issues which this book attempts to address. In its first part it examines what might be called the dimensions of development and discusses the evolution of ideas and theories as to how development might best be managed. In the second part it examines the translation of ideas into practice in a series of case studies spanning the continents and ranging from the smallest to the largest of nations. Here the theories of development have come face to face with the difficulties and diversities of reality. It has been a salutary experience.

2 The dimension of people: disease and mortality

'We who live in temperate lands find it difficult to realize how baneful
Nature can be to man or to understand that in unreclaimed regions water
may swarm with dangerous germs, myriads of blood-sucking insects may
inject deadly microbes into the human body, and the very soil may be
harmful to the touch'

Pierre Gourou
The Tropical World 1953

Population is variously and contemporaneously regarded in a number of ways.
It is viewed as a resource, a strength upon which political power can be based or
as a factor of production. It is considered as a burden, something to be fed,
clothed, educated, employed and to be characterized by particularly
burdensome sections, the too young and the too old. It is also seen as that which
creates, shapes, nurtures, and indeed is, the very civilization which our varied
societies demonstrate. If social and economic development is about anything it
is about people, and the issues of population underlie all thinking on social and
economic development and come to the fore in the execution of any
development plan. The relationships between the characteristics of population
and the process of development are complex and subtle. It is this relationship
and these characteristics which will be examined. Since they are not peculiar to
the Developing World, it is possible to analyse the population dynamics of the
present industrialized, highly productive, urbanized and wealthy nations from
their pre-industrial period to the present in an attempt to identify causes and
consequences which might illuminate the demographic status of the less
developed countries of today.

The first eighty years of the eighteenth century were characterized in England
and Wales by a very slow population growth at an average rate of 0.36 per cent
per annum. In the last twenty years of the century expansion accelerated to over
1 per cent per annum. The increase in population was discernible to people of
the time who offered explanations. Habakkuk argues that increases in the birth
rate had an important part to play and that the contemporary opinion was that
the population was increasing because couples were having more children
largely because of greater economic opportunities or because of the increased
economic security afforded by the poor relief scheme known as the
Speenhamland system (Habakkuk 1971). The contemporary who applied an

analytical mind to the situation was, of course, Thomas Malthus whose *Essay on the Principle of Population* published in 1798 not only attempted to explain the English situation but put forward a model to illustrate the dynamics of populations which has attracted comment ever since. As his views have a bearing upon the relationship between population and economic activity and as they have been applied by some to the population situation in developing countries today, they merit consideration. He wrote thus:

> Impelled to the increase of his species by an equally powerful instinct, reason interrupts his career and asks him whether he may not bring beings into the world for whom he cannot provide the means of support. If he attend this natural suggestion, the restriction frequently produces vice. If he hears it not, the human race will be constantly endeavouring to increase beyond the subsistence. But as, by that law of our nature which makes food necessary to the life of man, population can never actually increase beyond the lowest nourishment capable of supporting it, a strong check on population, from the difficulty of acquiring food, must be constantly in operation. This difficulty must fall somewhere, and must necessarily be severely felt in one or other or the various forms of misery, or the fear of misery, by a large proportion of mankind.
>
> That population has this constant tendency to increase beyond the means of subsistence, and that it is kept to its necessary level by these causes, will sufficiently appear from a review of the different states of society in which man has existed
>
> (Malthus 1798).

Malthus then argues that in places where there are no food constraints on population increase, he examines eighteenth-century North America, a population could double itself in twenty-five years and possibly in fifteen and could continue to increase in a geometric progression pointing out that 'A thousand millions are just as easily doubled every twenty-five years by the population as a thousand'. He continues:

> But the food to support the increase from the greater number will by no means be obtained with the same facility. Man is necessarily confined in room. When acre has been added to acre until all the fertile land is occupied, the yearly increase of food must depend upon the melioration of the land already in possession. This is a fund, which, from the nature of all sorts, instead of increasing, must be gradually diminishing, . . . It may be fairly pronounced, therefore, that considering the present average state of the earth, the means of subsistence under circumstances the most favourable to human industry, could not be made to increase faster than in an arithmetic ratio.

Malthus brings his two ratios together and draws his alternative conclusions on the condition of man. He was thus arguing that since food subsistence levels could never be increased quickly enough to match an unchecked population increase, the limits of subsistence production would keep down the population either in a direct way, by causing malnutrition and starvation leading to an increase in death rates and infant mortality rates, or in an indirect way by man, the thinking animal, restricting his fertility rates to levels which could be supported. Malthus considered later marriage and sexual abstinence as appropriate means by which this could be achieved. He argued that attempts to increase food production to raise standards of living and health were not an answer since the population would always continue to increase until it pressed against the levels of poverty which would check it by death. He saw the population of England increasing around him and postulated a contained system in which the population would be reduced and held in check at subsistence level.

Poverty is characteristic of many developing countries and is a major index of their condition. Malnutrition is common and starvation by no means unknown. Are they in the demographic strait-jacket of Malthusian containment? Are there other factors in the population equation? The population of England and Wales continued to expand unchecked for over a century after Malthus published his essay, never falling below an annual average growth rate of one per cent from 1781 to 1911 and reaching 1.8 per cent between 1811 and 1821. Significantly, that same century or so was characterized by an unprecedented increase in production and a major transformation of the British economy. This evidence is one refutation of the Malthus thesis. A second comes from an earlier, pre-industrial period of English history, the period from the twelfth to the sixteenth century. Here the evidence so elegantly presented by John Hatcher not so much rejects the Malthusian cycle as modifies it and does so in a way significant for the understanding of the demographic situation in the Developing World (Hatcher 1977).

Hatcher believes that levels of mortality in medieval England were not related simply to the success or failure of harvest even though crises in limited localities were frequent and malnutrition did influence death rates. On a national scale he believes that it was the occurrence of the epidemic and infectious diseases such as the bubonic plagues which swept the country and were unrelated to the state of the economy and levels of subsistence. He agrees that the Malthusian argument holds true in so much as better living standards in pre-industrial England did favour population growth while periods when they were low inhibited population growth, but claims that this is only part of a more complex picture. Infectious and epidemic diseases quite unrelated to economic circumstances superimposed their death rates on any Malthusian cycle. Mortality rates rather than fertility determined the population dynamics of pre-industrial England.

Three seemingly major considerations in the discussion of the sinuosities of demographic change thus appear: first the Malthusian element whereby population levels are contained by the finite and depletable resource base which sustains them, a relationship suggestive of concepts of over-population and under-population; second the question of the changing use, and hence significance, of the resource base which the evidence of nineteenth-century Britain suggests and which allowed a continually increasing population to be supported without Malthusian checks; third the role of diseases, which Hatcher demonstrated. This role will be examined first, distinguishing between those diseases which appear related to living conditions, and hence to economic development, and those which are not. Learmonth's scholarly and wide-ranging work *Disease Ecology* with its rich bibliography is of particular value in this context (Learmonth 1988).

OF FLIES, FLEAS, LICE AND MEN: THE ROLE OF DISEASE

A wide range of zootic diseases are endemic in the Developing World; some are peculiar to it and many are associated with its tropical climates, while others it shares with the rest of mankind. Pierre Gourou writing in 1953 states 'In physical and mental activity and in the reproduction of his kind, man is restricted in the tropics by serious maladies whose existence is entirely due to hot, damp climate' (Gourou 1953).

It is not, of course, easy to assemble direct data on the incidence of diseases for many countries in the Developing World. Indirect evidence such as the rate of infant mortality give some indication though data of comprehensive coverage are of recent origin. In 1960 infant mortality rates averaged 165 per thousand in low-income countries and 145 in the lower middle-income group compared with 30 in the industrialized nations, while corresponding life expectancies at birth were 41, 46 and 70 years in the three groups (see Fig. 4 for 1991). Though the incidence of disease and illness is high in the Developing World it was undoubtedly much higher in the recent past.

European awareness and comment upon the unhealthiness of the tropical world was consequent upon mercantile and colonial contact with its peoples. Ellen Thorp in her account of the history of Nigeria writes of Lagos in the 1850s 'But nightly and daily from the interior of the island came an enemy far more deadly than the leopard, because it was unrecognised – the mosquito, which, breeding in the surrounding swamps, brought malaria and yellow fever' (Thorp 1956). Those who went down with malaria and survived their 'conditioning fever' were also inevitably stricken with dysentery due, though they did not know it, to polluted water.

The scene in Lagos was paralleled in the Niger delta which was the great focus in trade first in slaves and later in palm oil. The 'oil rivers', as they became known, took a great toll of the European seamen and traders. Of all the river ports, Bonny had the worst reputation with its endemic malaria, yellow fever and dysentery. Trading companies from the 1860s until almost the end of the century moored old hulks of sailing vessels, no longer safe for sea passages, in the delta for use as trading depots or 'factories'. In the foul water of their bilges mosquitoes found ideal breeding grounds. The young clerks who came out from Liverpool and London to handle the trade in palm oil from the interior in exchange for flint-lock guns, Lancashire cottons and other manufactures, were soon stricken with malaria or yellow fever and many died before their 'tour' was up. Ellen Thorp quotes a job advertisement of the period: 'Wanted, young man, eighteen to twenty-five, as book-keeper in a West African factory. A few hours work a day, in pleasant surroundings, unlimited shooting and fishing, in a fine tropic scenery, with a boat at his disposal. Free quarters, salary to commence at £70, with chance of rapid promotion.' She comments that successful applicants were soon to discover that the final statement was true 'for his chances of promotion could be as rapid as those in the Navy in wartime. If he himself were not dead of malaria or drink at the end of a few months, a good many of his companions were sure to be, and the young assistant might easily see his Agent buried and find himself in charge of the factory . . .' (Thorp 1956). This notorious coast, which gave rise to the jingle 'The Bight of Benin, the Bight of Benin, where few come out though many go in', is but one example of the experience by Europeans of the disease-inflicted countries of the tropics. The area became known as the 'White man's grave'. What was not said was that it was also the grave of black men. The causes of these many diseases were unknown and in consequence neither prophylactic measures nor cures could be used to counteract them.

It is commonplace to refer to 'tropical diseases' yet while some thus classified are indigenous and specific to the climatic tropics, many are diseases which were once widespread and have been eliminated or strictly controlled in the temperate and, significantly, the affluent and technically sophisticated world. Many 'tropical diseases' are essentially diseases now restricted in their endemicity to the Developing World and are as much diseases of levels of development as of climate. It is for this reason that it is useful to group the 'tropical diseases' into those which are related to the physical environment, those to the social environment, and those which in their ubiquity appear unrelated to either, in order to explore the relationship between disease, population dynamics and development.

A range of widely occurring diseases is transmitted by insects. Since these insects are controlled in their range by their environmental requirements so too are the diseases they carry. Yellow fever, a viral disease with a high mortality rate, is transmitted by the mosquito *Aedes aegypti* whose range is confined to areas where temperatures do not drop below 15°C to 20°C. It is a disease endemic in tropical central and south America and Africa. The mosquito is found in other tropical areas but not the pathogen. Dengue fever, a less serious disease though it can cause death, is also carried by *Aedes aegypti* but the virus is only transmitted when the temperature exceeds 20°C (Wisseman and Sweet 1961). The vectors of filariasis, which obstruct the lymphatics and can produce elephantiasis, are another group of mosquitoes, one of the chief in Africa being *Anopheles gambias* (Kessel 1961). The disease can be crippling, cause blindness and, as in most parasitic diseases, reduces both energy and resistance to infections. Flies other than mosquitoes transmit filarial diseases in their bite. The particularly distressing River Blindness, onchocerciasis, is a fly-borne disease carried by various species of *Simuliidae*. The fly's habitat is along fast-flowing well-aerated streams. The disease occurs in the near-equatorial areas of Central and South America but it is most widespread along the rivers of West and Central Africa and along the Nile. Some 20 million are estimated to be affected with this fly-borne blindness (Walsh 1985, WHO 1987). Outside the tropics, low temperatures prevent the evolution of the necessary life cycle of filaria and fly. Leishmaniasis, a protozoan infection transmitted in sand-fly bites and known variously as Kala-azar, Delhi boil or Oriental sore, is again widespread in the tropics and is debilitating rather than killing. In Africa, the several species of tsetse fly *(Glossina* spp.) carry the disease trypanosomiasis which debilitates and eventually kills humans. Importantly, species of the trypanosomes also infect cattle and have significantly determined the distribution of cattle in Africa. In thus influencing agricultural productivity and the availability of animal protein, Nagana, the trypanosomiasis of cattle, has made a major impact upon health and mortality in the continent. At high altitudes and desert conditions in tropical Africa the several species of tsetse fly cannot exist, but in the rain forests and Guinea savanna zone it is widespread, while in the drier savanna areas its distribution is riverine (Matzke 1983, Molyneux and Ashford 1983). A form of trypanosomiasis transmitted by the bite of bed bugs and known as Chagas' disease is found in scattered localities throughout Central and South America where the domestic rather than physical environment is its habitat control.

By far the most important insect-borne disease of the Developing World and the tropics is malaria transmitted in the bite of one of the many species of Anopheles mosquito. Their habitats are not confined to the tropics and some Anopheles species can and do exist in temperate areas. They are, however, sensitive to low temperatures, and cold winter seasons both restrict their occurrence and facilitate their elimination. The mosquito can, of course, transmit the disease only if it has access to the causal plasmodia in the blood of the infected human beings. Where the two exist together the disease can flourish and be spread. The disease varies according to the plasmodium involved but all can cause death and fatality rates among infected children are particularly high. In the mid-1960s the World Health Organisation estimated that some 100 million persons suffered from malaria, of whom one million would die each year. It has been suggested that endemic and epidemic malaria has played a major role in the population dynamics of the nations of the world, notably in the sub-tropical and tropical areas, over many centuries (Learmonth 1988). These, then, are the major diseases of the Developing World which are transmitted by insects and are associated with their habitats.

A second group of diseases is more closely related to the habits of man, to eating, drinking, excreting, of how man lives and where he works, of his personal hygiene and sanitary provision. Since these are features of cultures, life styles and standards of living, they are very much related to levels of development and the development process, and characteristics of man, the social animal. The worms which commonly infect and are parasitic on man in many parts of the Developing World are such manifestation of insanitary conditions. Open latrines, inadequate sewage facilities, polluted water and too little water, are all associated with the infestation of hook-worms, round, whip and Guinea worms. All these infestations, and they are commonly multiple, are debilitating and drain energy, causing anaemia and generally weakening large percentages of the population. The killing bacterial diseases of cholera and typhoid are transmitted through water, milk and food, with flies assisting in insanitary and crowded conditions of living and eating. The dysenteries, both bacillary and amoebic, occupy a similar niche, while infective hepatitis is essentially transferred by the faecal–oral route.

These diseases are widespread in the Developing World. Cholera is believed to have been confined to the Indian subcontinent for a long period of time but in a series of four pandemics in the nineteenth century spread throughout wide areas of the world. In the early twentieth century the imposition of quarantine measures has prevented the intrusion of pandemic cholera into the western hemisphere and Europe. Today cholera is particularly associated with India, Bangladesh and Celebes in Indonesia where it can be considered endemic. It is not an environmentally specific disease but as Jacque May puts it 'it is a consequence of unwashed hands' (May 1958). The typhoid and paratyphoid salmonelloses diseases, classic examples of water-borne diseases, while occurring world-wide are very much associated with inadequacies in clean water supplies, insanitary living conditions and unhygienic food handling. As these situations are more prevalent in poor countries they are diseases of both town and country in the Developing World. Bilharzia, or schistosomiasis, is at once both a disease of insanitary conditions and of sub-tropical and tropical areas. The pathogen, the schistosome, requires both man and freshwater snail as alternate hosts at stages in its complex life cycle. Man is infected by drinking, washing in or working in, water containing the cercariae stage of the life cycle, which he completes by voiding his wastes into streams, ponds or irrigation ditches. This extremely debilitating disease is found throughout the whole of Africa, in the West Indies, Venezuela, Guyana, Surinam and French Guiana and wide areas of eastern Brazil, in irrigated areas in the Middle East, throughout the Indian sub-continent, Burma, most of south-east Asia and in parts of Indonesia and the Philippines. As the still water of irrigated fields provides ideal conditions, bilharzia is a disease of tropical, irrigated lands and where economic development has extended irrigation, so too has the disease spread (Kloos and Thompson 1979, Cairncross and Feacham 1983, Molyneux and Ashford 1983, Weil and Kvale 1985).

The third group of diseases are contagious diseases, commonly bacterial and viral, transmitted from person to person without necessarily the intermediary of insects, food or drink. They include leprosy (once not uncommon in Europe), yaws (the spirochaete infection similar to syphillis though not venereal and associated with tropical regions experiencing average annual temperatures exceeding 25°C), smallpox, influenza, measles, cerebro-spinal meningitis and tuberculosis. All are diseases which can kill, quickly or slowly, and most are not environmentally specific. Smallpox, until eliminated in the 1970s, reaped a heavy and continuous harvest and influenza has killed millions in its pandemics. Measles, rarely a killing disease today in the industrialized world, is still characterized by high mortality in the tropics and when introduced into communities which had previously not known it, death rates and demographic consequences similar to those of the Black Death have been experienced. Tuberculosis most certainly is not a disease which has been confined to the Developing World but it is one very closely associated with poverty and with resulting poor nutrition and crowded housing conditions. It is in consequence very much related to levels of economic development and has been widespread in many developing countries.

All the diseases cited are subject, in varying degrees, to amelioration by medical treatment. It is frequently stated that the medicine of the scientific industrialized world has been responsible for the dramatic upsurge in the population of developing countries because medical science has reduced mortality levels as it did in Europe a century or so ago. This will be examined in terms of the nature of the process and its timing and with regard to the three groups of diseases which have been described.

DISEASE AND POPULATION DYNAMICS IN BRITAIN

The analogy with the demographic situation in Europe merits discussion. It will be recalled that the population of England and Wales began to rise towards the end cf the eighteenth century and continued the upward trend throughout the nineteenth. Though the data for the eighteenth century are unsatisfactory, some historical demographers have attributed the increase beginning then to an increase in birth rates and a fall in death rates, with the latter as the most influential and attributable to improvements in medicine, social hygiene and hence health. Such developments were therefore not a direct consequence of the great structural change in the British economy which was taking place. Others claim that it was precisely these economic changes which encouraged larger families because of the greater employment opportunities they offered. It is difficult to see, however, how already high birth rates could be raised much further since there are physiological limits to reproduction. What can yield a big and rapid increase in population in circumstances of high birth rates is a fall in death rates. Holding the view that the latter is the more convincing explanation, McKeown and Brown have examined the role of medicine in the population dynamics of the eighteenth and early nineteenth century in England (McKeown and Brown 1965). They consider that the establishment of medical schools in the eighteenth century and the greater understanding of the body's structure and functions which resulted, desirable though it was, had little direct impact upon mortality rates nor did developments in surgery improve life expectancy. The nature and causes of diseases remained largely unknown. The setting up of lying-in (maternity) hospitals in Britain and Europe increased rather than reduced mortality of both the mother and the newborn by enhancing the risk of cross-infection of puerperal fever in circumstances where the role of hygiene and cleanliness was not appreciated. Likewise McKeown and Brown consider the role of General Hospitals to have been detrimental to health certainly up until the mid-nineteenth century. They became places where diseases could be contracted since the nature of disease transmission was not fully understood and most medical drugs of the day were ineffective. With few exceptions, such as the pioneering work of Jenner in developing an effective vaccination against smallpox, the development of medical science up until the middle of the nineteenth century appears to have had little effect upon mortality rates yet these rates did fall with all evidence pointing to a steady decline between 1775 and 1850. Crude death rates stood at around 30 per 1000 in 1800, were 23 in 1850 and remained at that rate until 1870 before falling steadily to 13 in the 1920s. If deaths from infectious diseases were not markedly reduced by medical treatment, what was the cause? Mortality could be reduced by changes in the virulence of the disease or by the human population becoming more resistant. The virulence of scarlet fever and measles in Europe has diminished but in the twentieth rather than the nineteenth century. A better fed population could possess a higher resistance and so the economic development of the eighteenth and nineteenth centuries could, and probably did, have a role to play in disease diminution and in lowering mortality. Finally and importantly, improvements can serve to inhibit the transmission of diseases. These factors vary, however, in their significance from one disease to another.

Habakkuk considers that the falling death rates were due to improvements in environmental conditions (Habbakuk 1965, 1971). Did living conditions improve? Did the increasing proportion of the population living in the new industrial towns run a lesser risk of infectious diseases than the largely rural population of earlier periods? The towns of the nineteenth century left much to be desired in health terms but it could have been that room space was greater, and sewage disposal improved and separated from water supplies; and that as these utilities became more efficient the diseases of insanitation – the dysenteries, typhoid, hepatitis and cholera – declined and the incidence of tuberculosis, the great killer of nineteenth-century cities, diminished as housing bye-laws were enforced and, as in the latter part of the nineteenth century, isolation hospitals removing infectious cases from the community came into being. The evidence is incomplete. What is certain is that in England and Wales the vulnerability to harvest failure disappeared by the nineteenth century as subsistence agriculture was replaced by more productive commercial forms and as the dependence of local demand on local production was reduced. Where it had not, as in Ireland, crop failure could produce death, disaster and emigration. Better nutrition, a rise in living standards, a greater appreciation of the need for cleanliness and hygiene translated into practice at family and public level and manifest in completed town sewerage and water-supply schemes and building regulations, all reduced the significance of the diseases of the environment and all were improvements characteristic of the whole of the nineteenth century. Medical science improved but the impact of prophylactic inoculations and therapeutic drugs was yet to come and in many diseases post-dated 1945.

DISEASE CONQUEST AND FALLING DEATH RATES IN THE DEVELOPING COUNTRIES

Has this pattern of declining mortality initiated by improvements in living conditions and in public health facilities been replicated in the Developing World? In some countries it has been achieved but they are few. In most, public utilities of consequence for health, namely pure water supplies and adequate sanitation together with housing provision, are far from satisfactory. In 1897 the Annual Report for Lagos stated:

> Discussion is still going on as to a feasible scheme for the sanitary reform of Lagos town in connection with water supply and drainage that can be carried out reasonably within the means of the Colony. It will be easily comprehensible, that, on a malarial island a mile or two long by half a mile broad, with a population of 50,000 souls living on it and a much larger number of bodies dead and buried in it for many years past, disturbance of the soil is to be avoided by every possible means. No scheme has yet been approved
>
> (Thorp 1956).

Today, Lagos with several million inhabitants still has no comprehensive sewerage system, a feature common in both major and minor cities throughout the Developing World. Adequate water supply systems of pure water are likewise still rare and the diseases associated with these situations are still prevalent. The English pattern of a steady reduction in mortality from food, faeces and water-borne diseases as standards of living improved, does not characterize the rapid decline in mortality rates which has taken place in all developing countries since the Second World War. This decline, and it has been as dramatic as it has been swift, is the result of the application of modern scientific medicine with not only its understanding of the nature of diseases and their transmission but also armed with a weaponry of drugs and insecticides to combat them. It is a development which has been felt at one and the same time in both the affluent industrialized world and the poor developing countries.

The application of these remedies has varied in cost, ease and effectiveness across the disease spectrum. The most effective are those which immunize the individual. If the protection is secure, economically feasible and long-term, it obviates the need for other measures which may be more difficult and costly to implement. However, the protection offered by modern vaccines varies considerably. Against the serious diseases associated with insanitary conditions, cholera and typhoid vaccination gives only short-term and, in cholera's case, uncertain protection. Effective protection is afforded against diphtheria, measles, whooping cough, poliomyelitis and against tuberculosis, the disease long-associated with overcrowded and poor living conditions. Smallpox defeated by the world's oldest vaccine has been totally eliminated. Notably absent from these conquests are trypanosomiasis, bilharzia, leprosy, filariasis, and most importantly malaria. All, except leprosy, involve an insect vector or an alternative host and all have proved very difficult to eradicate. Yellow fever is the only major insect-borne disease to be adequately counteracted by inoculation, an immunization which was developed at the beginning of the twentieth century. Bilharzia is in many respects a disease of rural insanitation. To control it requires either the treatment by drugs of the infected population, which as yet has not been successful, or the elimination of the snail, or the provision of latrines and uninfected domestic water. These provisions are complex, expensive and difficult to administer. Bilharzia remains a major problem. Trypanosomiasis has presented problems. The tsetse fly can be killed by insecticides or its habitat destroyed. In densely-settled, cultivated areas it cannot live but such areas are few in Africa and the disease, particularly in the form affecting cattle, is unsubdued. Malaria, the killer of many throughout the Developing World, presents the greatest challenge. During the Second World War drugs were developed which by killing the pathogen in the bloodstream, if taken regularly, can protect the population at risk. The anopheles mosquito can be killed by insecticides of which DDT was the first effectively and extensively used. The breeding environment of the many species of malarial mosquito can be destroyed though in practice this is a near impossible task. All these approaches are commonly used, with insecticide spraying as the most widely applicable and cost-effective. The World Health Organisation has mounted a series of malarial eradication programmes which have done much to reduce the incidence of malaria but the rate of success has been varied for several reasons. First, not all infected areas, such as densely forested country, are amenable to spraying and second, mosquito species are becoming resistant to insecticides. Third, resurgence has occurred, where the rigour of the eradication measures has slackened or lapsed. New drugs are required to be developed as the malarial plasmodia themselves develop resistance. Finally and importantly, the movements of both man and mosquito greatly increase the problems presented in eradication. Infected persons, host to the plasmodium, move into cleared areas where the disease-free mosquito can take up the pathogen and re-infect the cleared population. Mosquitoes themselves know no political boundaries and an eradication programme in one country may be less than successful because its neighbour has no such programme. None the less the incidence of malaria has been dramatically reduced and this has been reflected in falling death rates (Prothero 1965; Learmonth 1957, 1978, 1988).

In Sri Lanka, because it was an island where population movements could be controlled, because its dimensions allowed a comprehensive cover and because the vectors' habitats were easily identified and treatable, the eradication programme using DDT sprays, a new insecticide against which no resistance had yet developed, was highly effective. Death rates showed a marked decline after 1945 but the precise role of malarial eradication in the decline has been much investigated and deeply debated. In a sophisticated analysis, R. H. Gray has shown that malarial control in Ceylon was responsible for no less than 23 per cent of the post-war decline in mortality. As Gray indicates, the demographic consequences of malarial control in other countries such as Mauritius, Mexico, Venezuela and Guyana have been similar to those in Sri Lanka (Gray 1974). Malaria is, however, far from defeated and after the earlier successes is on the increase in many areas. Some estimates put the number of sufferers at 200 million. In India, for example, after a very successful eradication campaign, resurgence has occurred widely; the resistance of the mosquito to DDT and of the plasmodia to chloroquin, together with the disruptions caused by the Indo-Pakistan war of 1965, played a major part in the effectiveness of the campaign (Learmonth 1988).

Disease eradication by the application of modern scientific techniques of immunization, curative drugs and insect control has markedly reduced mortality levels in developing countries. The impact has been immediate and dramatic. It has not been the product of the economic and social developments which seemingly brought about the first and much more gradual falls in mortality characteristic of the demographic history of the industrialized nations. That stage is yet to come and with it, hopefully, the conquest of the diseases of insanitation and poverty. This reversal of the sequence has profound consequences for the Developing World's nations. The population of Ceylon in 1946 was 6.65 million. By 1953 it was 8.09 million and in 1963 had reached 10.58 million; an increase of 59 per cent in fourteen years. Sri Lanka's population in 1990 was 17 million. The increase is far from exceptional. It brings to the fore again the issues raised by the Malthusian model of population growth and population containment.

Malaria has been joined by another disease of profound significance for both the demographic evolution and the social and economic development of the Developing World. The disease is AIDS (Acquired Immune Deficiency Syndrome), a disease of the blood first identified in the USA in 1981 and transmitted, in the main, by sexual activity. Its occurrence is believed to be world-wide though the extent is not fully known because of the nature of the disease which can remain dormant for several years. The World Health Organisation estimates that in 1990 the world total of AIDS sufferers was between 8 and 10 million adults with some 5.5 million in Africa. The proportion of AIDS victims occurring in developing countries is reckoned to reach 75 per cent by the end of the century (PANOS Dossier 1986, Population Reports 1986, Learmonth 1988, Bongaarts and Way 1989, Larson 1990, World Bank 1991). The disease is eventually fatal and there is, as yet, no cure. As a disease of the sexually active it infects the age groups which make up the bulk of the labour force. The prospects both for the families infected and national economies are grave. It is too early to assess accurately the magnitude of the situation or to see if changes in sexual behaviour or the development of curative medicine will curb this modern pestilence but it is a grim example of disease overriding the Malthusian equations in the manner identified by Hatcher (Hatcher 1977).

3 The dimension of people: birth rates and population dynamics

'Impelled to the increase of his species by an equally powerful instinct, reason interrupts his career and asks him whether he may not bring beings into the world for whom he cannot provide the means of support'

Thomas Malthus
Essay on the Principle of Population 1798

Economic, social and medical factors have clearly influenced trends in death rates in developing countries; the comparisons and contrasts with the British experience help to illuminate these interactions. This is but one side of the equation; the other concerns birth rates. Changing birth rates will now be considered, again drawing parallels and contrasts with developments which have taken place in Britain.

In Britain birth rates remained high throughout the first eight decades of the nineteenth century and were over 35 per 1000 in 1880. Thereafter they began a steady decline, without interruption until 1940, a decline taking sixty years from 35 to reach 14. Overall national productivity was up and certainly from the 1880s living conditions were improving, yet the population did not respond in a Malthusian way and birth rates went down. This decline in fertility was characteristic of all the countries now characterized as developed and largely industrialized. What has made the decline difficult to interpret is that it does not correspond with the same stages of economic development in this group of European countries. In Britain industrialization and associated urbanization was characterized by high birth rates which did not begin to fall until 1880. In countries which industrialized after Britain, the same general fall in fertility took place from the late 1870s down to the 1940s. There were some variations, of course, but the trend was essentially the same as in Britain at the same time. This meant that in some cases it took place *during* their process of economic transformation which had taken place in Britain earlier. Falls in fertility took place in the Scandinavian countries, in Belgium and the Netherlands and also in France where birth rates had throughout the nineteenth century been lower than in England. All differed from England in natural endowments and in the chronology of their economic development.

Death is rarely a matter of choice. Whether or not to produce children can be. It is certainly incorrect to say of European populations that their physical ability to reproduce, their fecundity, diminished. Mitchison writes that changes in age distribution, resulting from higher life expectancy and from emigration, did not reduce the proportion of women in child-bearing age groups by more than 2–3 per cent between 1861 and 1901. Further she claims that although a rise in the age of marriage did take place it was not enough to affect fertility in a major way but contributed to that slow fall in fertility which did take place in England and Wales from the late 1870s on (Mitchison 1977). Clearly married couples were choosing to limit the size of their families and this was characteristic of all west European countries. The questions of how and why fertility rates fall, therefore, revolves around why potential parents choose to have fewer children. Habakkuk has attempted an explanation and considers a number of interacting factors that can be related to social and economic developments which may have influenced parents (Habakkuk 1971). Industrialization had meant more material wealth for the manual workers but with even more available for the middle and upper classes. People, he suggests, were less tied by customs and traditions since they lived in towns and had broken with their rural past. In consequence, their decisions were more deliberate and they considered the issue of the relationship between the size of their families and their standard of living. As the economic structure evolved, so the output of manufacturing industry broadened to include consumer goods and as they became available to a wider section of the population, they became first desirable and then, with the income elasticity of demand, necessities, and a call upon resources thus competing with cost of children. Also the acquisition of goods, the retention of money for investment, and the increasing opportunity for house purchase meant that those who saw these opportunities realized that their attainment was easier with fewer children. This development would, of course, be more likely to manifest itself in the middle classes but as Habakkuk argues, the movement of people, the reduction of parochialism (both associated with urbanization), the spread of education and the growth of newspaper circulation all meant that new ideas, opportunities and knowledge of the norms of living of others became more rapidly and widely diffused.

Tranter, in his work on the population of England and Wales since the Industrial Revolution, is of the view that the enormous increase in opportunities for the employment of children positively encouraged both early marriages and the formation of larger families in the early phases of the Industrial Revolution. As the economy further matured, changes in society led to circumstances which tended to reduce birth rates. Tranter refers to declining infant and child mortality rates, the way in which the state assumed greater responsibility for the elderly, and to the change in the status of women. He stresses that the most important influential factor was the growing desire for material possessions and social improvement (Tranter 1973). The temporal matching of economic events and demographic changes is, however, far from exact in Britain and when applied to other countries further mismatches occur. The relationship is exceedingly subtle and yet its analysis is important if the European experience is to further an understanding of the demographic changes taking place in the Developing World.

Causal factors, it may be hypothesized, fell into a number of interrelated groups. First, those related to mortality. Throughout Europe significant falls in mortality preceded falls in fertility. The simple explanation refers to 'insurance births'. Since many would die, wives produced a large number of children to ensure sufficient survived to maintain the family and its labour force. When mortality rates fell, for the reasons discussed, the need for this 'insurance' margin was removed and fewer children were born because births were controlled. In reality the role of the mortality factor was more complex. Crude mortality rates conceal variations in age-specific rates. Mortality in Britain declined first, and sharply, in the 5–15 age group, then in the 15–24 in the late 1870s with the death rate for older persons not falling until the 1880s and 1890s. Infant mortality rates, which were very high in the mid-nineteenth century (148 in 1860), fell hardly at all and were 146 in 1900 (Greenwood 1936). This meant that couples perceived mortality as remaining high and did not limit their families though a greater proportion of those who survived infancy lived. The drop in family size was thus retarded. It has been argued that among the better off, the middle classes, infant mortality did decline earlier so that their response in limiting their families likewise took place earlier. When overall infant mortality rates fell in the twentieth century then family sizes diminished and smaller families became the accepted norm. Age-specific mortality rates are thus more significant than crude rates and their role is an interaction between economic and social factors.

The second group of factors are the economic factors which interlink with, because they are monetary evaluations of, other factors. In economic terms children, if they can be employed, are regarded as a resource by the family. Children can also be regarded as a form of consumption, a commodity which costs money, and so they are part of the demand pattern of the family and they compete with expenditure available for other goods. Families will therefore arrive at a mix of goods, including children, which they consider desirable. As income changes so the mix changes; as other conditions influencing society change, for example the mortality factor discussed, so again does the mix. If educational opportunities change this will feed back into the family decision. The value of more children which the family cannot afford to educate could be set against the improved and more valuable resource of children who after a non-productive period of education would, because of their training, have a greater capability of income-earning. Thus the loss of an immediate resource-gain would be set against a later but larger one. Similar economic ideas could be applied to the role of women. The woman giving birth to and bringing up a large number of children may be forgoing an income. She incurs an opportunity cost which must be taken into account in supporting the alternative of bearing fewer children. It can be argued that if incomes go up the need for investment in large families is lowered.

The third group of factors may be called social or cultural, and likewise they involve both the mortality factors and economic valuations. For example the concept of status may in some societies find expression in the range and quality of material possessions. The middle class in nineteenth-century England might have wished to remain distinctive from the increasingly better off working classes and, to ensure this, substituted goods for children in their consumption pattern. The professional and middle class of Victorian England could also have been motivated by concern for their children, by the wish for them to benefit from the enlightenment of education or to have the opportunity for more fulfilling or remunerative occupations after, at considerable expense, they had been educated. Fewer children would make these goals more reachable. As a greater proportion of the population saw the possibility of such goals, the norm of the smaller family would characterize most strata of society. Changing attitudes to women and their role in society could have considerable repercussions upon birth rates and family size. The educated woman was more likely to wish to play roles in addition to those of wife and mother, to identify the opportunities for and needs of her children in terms of a fuller, healthier and educated life and to play an active part in society herself. All would be reasons for viewing the smaller family as desirable.

It may be wondered that there has been no mention of the development of contraceptive techniques. All the evidence shows that forms of family limitation had been practised over many centuries in many societies, and techniques of contraception have improved markedly in their effectiveness particularly in the twentieth century. The central issue is, however, of the conditions which lead to the practice of family limitation, to the use of the available technology, and it is these factors which have been discussed.

The multiple nature of these interlocking factors all influencing the trend towards fewer births meant that they could, and probably did, vary in their strength from country to country in Europe and that this accounts for the variations in the timing of the decline in fertility and its coincidence with economic and social developments. Clearly these groups of factors can apply to the situation in the Developing World and equally, the nature of their significance and their timing will vary throughout its many countries and cultures. In some cultures status may not be conferred by material possessions but by size of families. In others the impact of increasing numbers of children, as mortality rates fall, may bear not only on the nuclear family but upon the delicately balanced social systems of the extended family and wider society established to care for the welfare of dependants and unable to cope with a big increase in the younger dependent section of society. Long-established systems may begin to crumble under the pressure of the enlarging younger population. For all these reasons it is possible to argue that a similar fall in birth rates to that which took place in Europe may be characteristic of all the countries of the world. It can also be argued that the timing and character of the demographic change may well be as different as are conditions.

THE DEMOGRAPHIC TRANSITION

The population dynamics characteristic of Europe from the eighteenth to the twentieth century have been encapsulated in the well-known model, the demographic transition. It is shown in Fig. 5. A period of high birth rates and high death rates, essentially controlling the size of population, is followed by one in which a fall in mortality is experienced and, as birth rates remain high, the population increases, essentially the situation during most of the nineteenth century in Britain. There then follows a phase in which birth rates fall, death rates approach a very low level but the population continues to increase as the youth of its past become parents. Finally with both birth rates and death rates very low the population settles at a stationary level with death rates beginning to rise a little because the population is older. This transition has confounded Malthus; it is necessary to examine why.

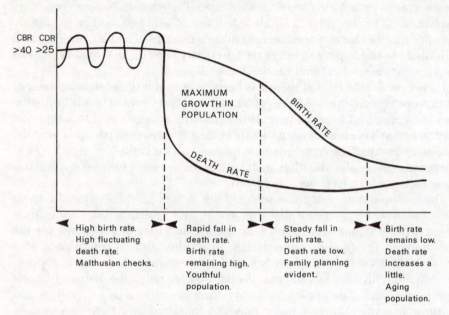

Figure 5 The demographic transition model

The Malthusian theorem did not work out in the country of its origin. Underlying his ideas on limits to the rate of production was the idea of diminishing returns. Applying a variable factor of production, it could be labour or manure or some other item, to a fixed factor of production, namely land, would increase output by a certain amount for each unit of input. At first each unit of input would produce a big increase in yield but this amount of the increase would progressively decline as more units of input were applied, until a stage was reached when one additional unit of input produced merely one unit of output. Beyond that point the curve tapers off and the amount of return on output becomes increasingly smaller than the unit input until eventually additional inputs actually produce a decline in output. The marginal product is the additional amount produced by each additional input, and its curve shows precisely this diminishing effect until it passes below zero and becomes negative. Malthus saw this happening and postulated that the diminishing return would produce famine and malnutrition, death and a diminution in population to a point where production could sustain it. The population would then increase again and the whole process repeated so that the population was kept in check. This concept could, of course, be applied to other finite resources in addition to land. In the simplest terms two things happened which modified the Malthusian model. First the diminishing return effect did not manifest itself because that great assemblage of changes in technique, organization and economic restructuring called the agrarian and industrial revolutions, enormously increased the ability to produce per unit of population. The curve of production was raised above that of population increase so that the diminishing return curve did not bisect that of population and reduce its per capita benefits from production. Despite the fact that the population increased, not because of increased birth rates but because of reduced mortality, itself consequent upon economic changes and the reorganization of society, the scale and nature of economic development continued to stave off the critical conflict between size of population and resources to support it which Malthus had envisaged. Diminishing returns were continuously postponed by economic development, a feature of considerable significance for the Developing World. Secondly, the upsurge in population was contained not because it was cut back by famine and poverty but because, for the reasons discussed, couples chose to have smaller families. They made the choice under conditions of greater material prosperity and security; again a pointer for Developing Countries. Resources had been used more effectively and the nature of society had changed. Social and economic development had produced a demographic transition. The questions remain. Is the demographic transition of the kind experienced in Europe, North America and indeed Japan, the key to the understanding of the demographic situation in the Developing World? Is it an inevitable sequence of events? Will it, in the varied conditions of the many developing countries, take the same form as in Europe? Is a demographic transition of such a kind desirable? Has it a significance for the development process?

RESOURCES AND PEOPLE AND THE QUESTION OF OVERPOPULATION

In all countries of the Developing World populations are expanding and in many have been expanding rapidly since the Second World War. The average annual growth rate in the decade 1960–70 in the low-income economies was 2.3 per cent and in the lower middle-income 2.6 per cent, figures much higher than the maximum rate of 1.8 per cent ever reached by England and Wales. No fewer than twenty-six of the eighty-three countries in these World Bank classifications averaged increases of 3 per cent and over during the period 1980–91 (Fig. 6). Significantly eighteen of the twenty-six were in Africa. Their death rates have been quickly and markedly reduced by prophylactic medicine and the conquest of many major, infectious, zootic diseases; a conquest unrelated to the stage of

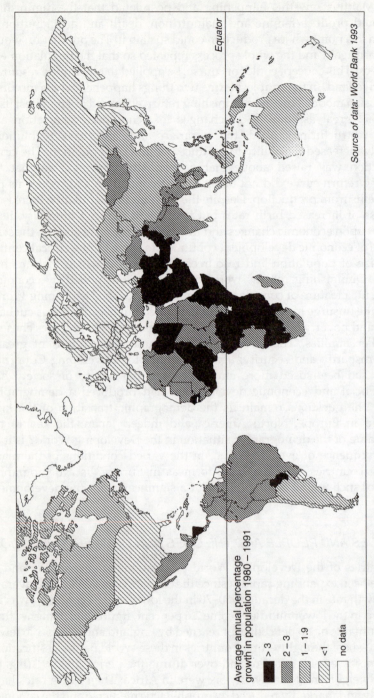

Source of data: World Bank 1993

Average annual percentage
growth in population 1980 – 1991

> 3
2 – 3
1 – 1.9
<1
no data

Figure 6 Average annual percentage increases in population 1980–1991

their economic development and one mainly of diseases unrelated to living conditions. Their birth rates, in contrast, have remained high and, if the European experience is any guide, this is a feature very much related to the socio-economic development process. It could be said that many developing countries are in the second phase of the demographic transition. The question is whether all, or some, will move into the stage where both deaths and birth rates are low and, if so, how quickly. This is a matter of some concern. The European transition is not inevitable. The conditions that pertained from the eighteenth to the twentieth century in Europe, which saw population expansion associated with, and in large measure contemporaneous with, economic and social development as resources were assembled and utilized in a more productive way to sustain expansion, may not emerge in the countries of the Developing World. The population expansion there has largely preceded massive economic development because it was the product of non-associated external medical innovations. If, on the one hand, production can be raised to encompass the increased numbers and if, on the other, birth rates can be controlled to retard the rate of increase, a controlled transition can be brought about. If these goals are not achieved, the spectre of Malthus will haunt the villages of the Developing World and stalk the streets of its growing cities, coldly calculating its ratios. It is to be fervently hoped that AIDS will not be the means by which population increases will be reduced, truncating as it does both families and societies.

Many of the subsequent chapters in this book will be concerned with the production side of the equation, with the lifting of the production curve of diminishing returns so that it encompasses the population dimension. Indeed underlying the ideas of Malthus, the demographic transition, concepts of over and under population, is the relationship between resources or supply and population or need.

Views on changing populations have varied in time and place. An increasing population, characteristic of most developing countries, could be regarded as an increasing asset, and source of power. Mussolini, the fascist dictator of Italy, is reported to have said in the 1920s that the population of Italy must grow if she was to be more powerful, claiming 'with a falling population one does not create an empire but one becomes a colony'. A declining population can be regarded as a disaster. E. M. Hubback writing in 1947 states 'A Britain with a much smaller population would no longer be an important influence in world affairs either in peace or war. She would lose her status and power' (Hubback 1947). In the past at least, many governments in developing countries have held not dissimilar views. Some have said that the governments, economists and others in the industrialized countries who urge upon developing countries the pressing need to reduce birth rates and population growth are simply expressing the wish of the industrialized countries to remain dominant. They claim that larger populations are essential to stimulate economic development and state that industrialization in Europe was accompanied by the large growth in population which made it possible. Others in developing countries, expressing views like Mussolini and Hubback, state that young men are needed in greater numbers to sustain their status and power, and they are commonly referring to potential conflicts with each other. Similar concepts of the relationship between power and population size emerge in multi-racial countries, or countries with several religious faiths, each race or faith fearful that if its numbers decline it will become underprivileged, disenfranchised or persecuted. Elements of truth exist in many of these views; fears of loss of ethnic, religious or indeed national identity are very real, if incomplete in some cases and unrealistic in others. They have certainly influenced the demographic policies of many countries. But to ignore the fast rising numbers, which have preceded the laying down of foundations for a sustained economic growth, is to become burdened with poverty rather than power. What has emerged, not only in the industrialized nations but also in the great majority of countries in the Developing World, is the need to reduce fertility rates and to do this quickly in order to make the size of the development task more manageable. This size has two demographic dimensions: the rate of population increase and the volume in absolute numbers. In some countries the numbers might be small but the speed of increase is greater than the rate of possible resource development. In others the rate of increase might be less but the population base is so large that the annual yield is vast. The population of tiny Fiji rises each year by an amount equal to the size of its second largest town. In the Fijian dimension it is significant. The population of India between 1989 and 1990 rose by some 16.9 million, i.e. more than the size of Australia's population. The crucial relationship is the match between rates of population increase and rates of social and economic development.

4 The dimension of people: the demographic transition in the late twentieth century

'Fertility is an area of human behaviour where individual tastes, religion, culture and social norms all play a major role. Yet evidence from large groups of people suggests that differences in fertility can be largely explained by differences in their social and economic environment.'

The World Bank
World Development Report 1980

It is possible to make a general statement as to the population dynamics of the developing countries in aggregate and to compare this with the situation in the industrialized economies, to compare the demographic transition. But the Developing World is vast; it embraces most of mankind and its nations. It is, therefore, important to examine its several components and to reveal any differences in both the form and speed in the demographic transitions which are taking place.

In its Development Report for 1980, the World Bank traced the progress of crude death and birth rates from the eighteenth century into the twenty-first for two aggregates of nations: the industrialized countries, including those of eastern Europe and the Soviet Union, and the developing countries. Projections beyond 1980 were based upon reasoned generalized assumptions of the performance of the population dynamic for each country. The graphical depiction of the situation is reproduced in Fig. 7. It indicates that a demographic transition, similar in its trends but displaced in time, will encompass the Developing World with net reproduction rates reaching unity between AD 2005 and AD 2045 and with stationary population being achieved between AD 2065 and AD 2175 (World Bank 1980a). The increase in population is at its maximum during the second half of the twentieth century, when high, though falling, birth rates are associated with low, and still falling, death rates. Since China and India together accounted for 73.6 per cent of the developing countries' population in 1978, their demographic performance dominates the generalized transition shown in Fig. 7. The situation in other countries, areas or continents could differ appreciably and it is the situation within individual countries which bears upon their economic problems and possibilities. These variations will now be examined.

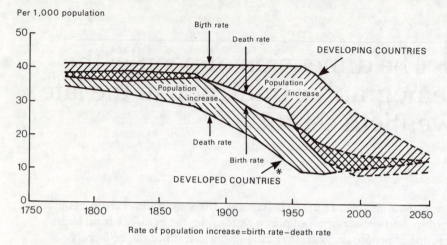

Per 1,000 population

Rate of population increase = birth rate − death rate

Crude birth and death rates. The projected increases in death rates after about 1980 reflect the rising proportion of older people in the population.
* Include industrialized countries, the USSR and Eastern Europe.

Source: World Bank 1980

Figure 7 The projected demographic transition AD 1750–2050

A TRAVAIL OF TRANSITIONS

The European model of the demographic transition contained the following sequence of cause and effect:

1. Socio-economic changes dominated mortality rates up until the First World War.
2. Socio-economic changes influenced birth rates.
3. Resulting declines, first in mortality and subsequently in fertility, produced the demographic transition with modern medical science playing its part in only the final stages.

The presumed demographic transition model for the Developing World contains the following elements:

1. Prior to the Second World War disease-ridden societies displayed the medieval European balance of high birth rates checked by high death rates.
2. Medical science has dominated mortality rates since the Second World War and the rate of their decline has been much greater than in the European model because of this.
3. Socio-economic changes will determine any further falls in death rates.
4. Socio-economic development will influence birth rates and induce their decline.
5. The demographic transition will be brought about by the interaction of (2) and (4) and, in its final stages, will be characterized by the operation of factor (3).

The European model of demographic transition is based upon events which have already taken place in the affluent industrialized nations of the world. The Developing World model presumes future developments. Yet the presumption that continuing medical advances will lower death rates has been challenged by the failure to contain malaria and by the emergence of AIDS. On fertility trends it is also a model based on hypothesis. Its nature, or natures since many alternatives are possible, will depend upon the ways in which the variety of social and economic factors come together in the many countries of the Developing World and influence birth rates. The European model could be very much culture-related and the multitude of cultures in the model give birth to a multitude of transitions. In this cultural matrix there are many variables. Religion could play a part, the place of women in society could be influential, the adoption and rigour of family planning programmes could vary, and the degree of urbanization might be influential. Culture in the widest sense, the civilizations of the world with their distinctive mores, could condition the way in which family size and birth rates are viewed. Even the nature of governments could play a part. The power-hungry totalitarian state might well take the Mussolini view of population. The state which put social provision and equity to the fore could strongly influence attitudes to family planning programmes and their provision. These are speculations but they are reasonable. Some are verifiable, others are not.

While in the immediate post-war period and into the 1950s there was much uncertainty about the demographic situation in the Developing World, since 1960 the wider availability of more reliable data is making the position clearer. A demographic transition in its broad outline similar to that of the European model is taking place but there are variations in both the speed of the transition and its dimensions. This situation will now be examined. The graphical device used to analyse the data is a development of that devised by Jakubowski (1977). It allows both temporal and locational comparisons of individual countries and groups of countries. Countries in the pre-transitional, or early transitional stage, are characterized by high death rates and birth rates and will be located in the top left-hand corner of the graphs. Countries such as those of Europe which have passed through the transition to a period of very low death rates and birth rates appear towards the bottom right-hand corner. As their populations age, a function of their vital statistics, locations shift a little to the left into slightly higher death rates. The crescentic sweep of the transition curve can be detected in Fig. 8.

The first and coarsest analytical sieve is that of continental location. In 1965 the countries of Africa, with but the two exceptions shown on Fig. 8A, and most Asian countries were bounded by dimensions of crude birth rates exceeding 40 per thousand and crude death rates in excess of 15. The birth rates were those of England a century earlier and the death rates of England in the nineteenth century, in most cases. Central and South America was displaced to the right, that is by lower death rates, but most countries still demonstrated birth rates in excess of 38. There were some notable exceptions and the Latin American

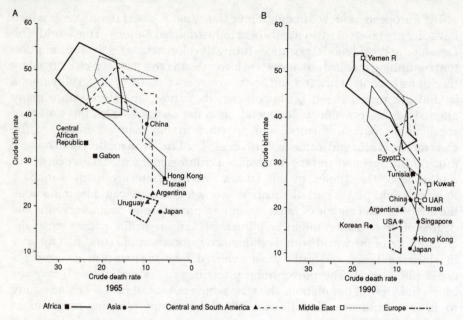

Figure 8 Crude birth rates and death rates by continents: (A) 1965; (B) 1990

countries of Argentina and Uruguay revealed population characteristics similar to those of the USA and the USSR. The European nations were characteristically bunched together in a zone of low death and birth rates and were joined by Japan. The countries of the Middle East all showed very high birth rates but death rates ranged from 27 to low European levels.

Twenty-five years later the situation had changed markedly (Fig. 8B). European birth rates had fallen still further and in all other continental areas death rates had declined. The compactness of the continental groupings had also been much reduced. Africa, and the Middle East each covered a much wider range of birth rates; Asia did so to a much greater extent; Central and South America as a whole had reduced birth rates and the range of death rates contracted considerably with the weighted average reaching the low figure of 7 per thousand. What is discernible in all the continental areas is the characteristic crescent shape of the demographic transition with the crescent flattening as the range of death rates contract and lengthening as individual countries within the continents reduce their birth rates much more than others. The low birth-rate countries reach down to the European situation and the transition is much in evidence. It is necessary next to examine the reasons for the range in values within the continents.

In Fig. 9A, the weighted means for crude birth and death rates have been plotted for each of the economic groups of the World Banks classification. India and China, both in the low-income group, are shown separately. The transitional crescent is evident with the economic groups ranged along it with the low- income countries

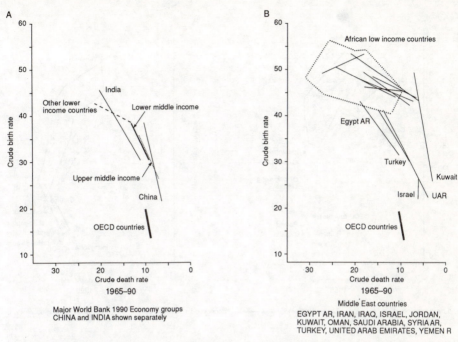

Figure 9 The demographic transition: (A) major World Bank economy groups: (B) Middle East countries

at its beginning, then the lower middle- and upper middle-income groups with still some way to go to achieve the European situation which is evident in both the capitalist industrial economies and those of former communist Eastern Europe. There is therefore a general demographic sorting according to GNP per capita income but there are exceptions. Among the high-income oil exporting countries, Saudi Arabia with an income as high as or higher than European countries shows, unlike Kuwait and the United Arab Emirates, a demographic path similar to that of the low-income countries. Secondly, India and China, though low-income, have distinctive paths. India is closer to the lower middle-income pattern but is moving rapidly towards the upper middle-income mean. China is truly exceptional. The world's most populous country has achieved the demographic transition in a brief span of some two decades. Figs 10A and 10B show the situation more fully by demarcating the range of values of individual countries within each income group.

In 1965 there were a number of anomalies in this income classification. Among low-income countries Sri Lanka was distinguished by very low birth and death rates for such a poor country. Gabon exhibits a higher death rate than most upper middle-income countries while the East European states of Bulgaria, Poland and Romania show Western European demographic features rather than those of the lower middle-income group to which they belong. The 1965 data referring to the anomalous Central African Republic are suspect since but a few years before and after 1965 the data put the country in the low-income

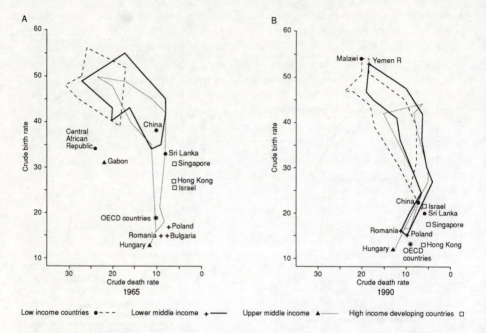

Figure 10 Crude birth rate and death rate dimensions by economy groups: (A) 1965; (B) 1990

group of which the country is a member. Countries which are in the high-income group but classified by the United Nations and other authorities as developing (namely Singapore, Hong Kong and Israel) all feature very low birth rates though their death rates remain higher than in Europe. The position of the world's largest country, China, deserves special comment. As a low-income country, China was not differentiated from other low-income countries in 1960 but five years later was well along the transitional crescent. During the next twenty-five years the income groups began to pull apart with the higher income countries, for the most part, proceeding more rapidly through the demographic transition. By 1990, the position shown in Fig. 10B had been reached in which all developing countries showed death rates much lower than twenty-five years earlier. Birth rates for all income groups, save that of the world's richest countries, were stretched over a wide span as within each group some countries more successfully reduced birth rates. Already a degree of bunching appears as countries in all income groups reach down to European levels. Such an assembly would be the logical outcome of the demographic transition. The situation in 1990 raises the question as to why some countries are quicker in their progress through the demographic transition than others. Clearly the Developing World is not homogenous in its demographic features.

The analysis can be refined further by plotting the demographic shift for each country within each continental income-group (Figs 11–13). The low-income economies of Africa which account for 27 of the 40 nations in the continent (of over 1 million population) are markedly similar in their demographic characteristics (Fig. 11A). All show continuing high birth rates with no marked decline and some increases. Death rates have fallen but in fifteen countries the rate still exceeds 15 deaths per thousand and in no African low-income country does it fall below the income group's average of 10. In Asia the position is different. Only three countries remain in the African situation, the others have moved markedly through the transitional phases, notably the two most populous countries, India with a birth rate much reduced though still high by European standards, and China. Of all Asia's low-income nations Sri Lanka most closely approaches the European pattern. In the next higher income group African countries show similar features to their low-income continental neighbours though the steepening of the transition lines indicates in some cases a more rapid transition (Fig. 11C). Birth rates nonetheless are still high with only Tunisia's below 30. Africa is a continent still in the early phases of the demographic transition. The Asian income equivalents are few in number and with one exception are much further advanced in their demographic transition (Fig 11C). In the countries of Central and South America and the Caribbean the majority of lower middle-income nations are approaching the terminal phase of the transition (Fig. 12A). The upper middle-income countries of Asia and South and Central America are shown in Fig. 12B and display very similar characteristics to their lower-income neighbours. Uruguay's position is indistinguishable from that of the OECD countries. South Korea's much reduced birth rate has been accompanied by a rising death rate.

The Middle East, largely characterized by the Arab and Islamic countries, presents a mixed situation. Undifferentiated by income group these countries are plotted in Fig 9B. In seven of the twelve countries high births, at levels comparable with those in Africa, have persisted. Death rates have been reduced to low levels except in the Republic of Yemen. Kuwait and the United Arab Emirates together with Israel are approaching the European demographic situation.

The changing demographic situation in the Developing World between 1965 and 1990 is best summarized in Figs 13A and 13B. In 1965 all these countries demonstrated high birth rates with the exception of the countries separately plotted. The low-income groups for all continents were characterized by high death rates. Better off middle-income countries exhibited lower death rates. The whole cluster was markedly differentiated from the world's wealthy and largely European OECD countries though a broken chain of individual nations – China, South Korea, Sri Lanka, Trinidad, Argentina and Uruguay – formed a continuum to the industrial market economies. By 1990 the dynamic demographic situation of the previous quarter century was evident. The same order of sequence was present but the countries of the Developed World from the least changed African low-income countries to the upper middle-income economies of South America stretched completely across the demographic spectrum as some countries pulled away from others in the extent to which they

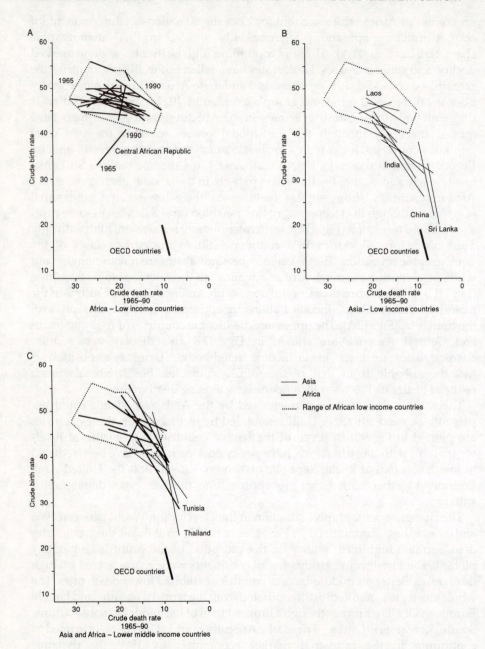

Figure 11 The demographic transition 1965–90: (A) African low-income countries: (B) Asian low-income countries; (C) Asian and African lower middle-income countries

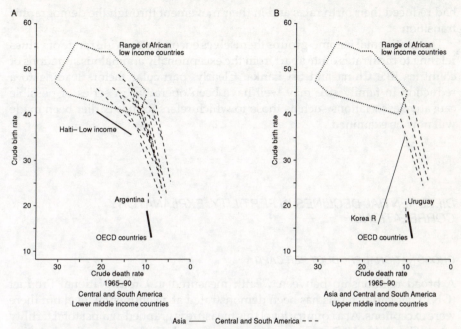

Figure 12 The demographic transition 1965–90: (A) Central and South American lower middle-income countries; (B) Asian and Central and South American upper middle-income countries

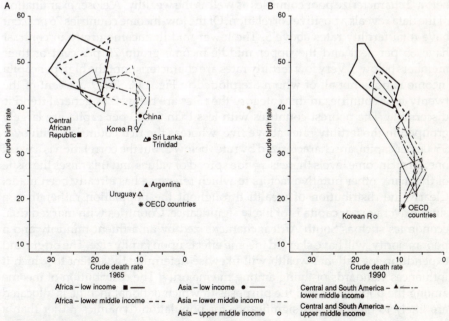

Figure 13 Crude birth rate and death rate dimensions by continental income groups: (A) 1965; (B) 1990

had reduced their birth rates and in their movement through the demographic transition.

The continental income-groups themselves encompass a wide range of values relating to birth rates quite apart from the exceptionally anomalous situations of countries like China and Sri Lanka. Clearly, particular factors leading to a reduction in family size may well have been operative. A series of possible causative associations such as those to which reference has earlier been made will now be examined.

DIFFERENTIAL DECLINES IN FERTILITY: EXPLANATORY CORRELATIONS

GROSS NATIONAL PRODUCT PER CAPITA

A broad relationship between wealth measured in Gross National Product (GNP) per capita terms has been demonstrated at a continental level but there were exceptions. At a country level this measure is graphed against total fertility rates (Fig. 14A). At first glance there appears to be little relationship between GNP per capita and fertility rates. The low-income countries exhibit a range from 8.3 in the case of Ruanda to a rate of 2.4 in Sri Lanka. Total fertility rates of below 2 characterize poor countries as well as the wealthy. A closer examination of the data reveals a positive correlation. Of the low-income countries 76 per cent have total fertility rates above 5. The lower middle-income group in contrast have 65 per cent and the upper middle-income group 70 per cent of their members below 5. Very low fertility rates are characteristic of the high per capita income nations not all of whom are plotted in Fig. 14A. In 79 per cent of the twenty-four countries in this category the rates are below 2. A general trend is discernible. The poorest countries with less than $1000 per capita are largely grouped in the fertility rates above five, whereas the rich countries, with over $5000 per capita, are characterized by rates below 3. But the correlation is a weak one. At all income levels there is a wide spread of values and this raises the issue of the many other putative factors to which reference has already been made. Clearly the distribution of wealth throughout the population rather than a simple average per capita is of likely significance. Countries with marked dual economies such as South Africa, characterized by an affluent minority and a poor majority, will have skewed income effects upon family size. The origin and expenditure of national wealth will likewise determine the extent to which it influences standards of living among the majority. The distribution of income among the population and the purposes to which national income is allocated are likely to have greater significance for population dynamics rather than a simple average.

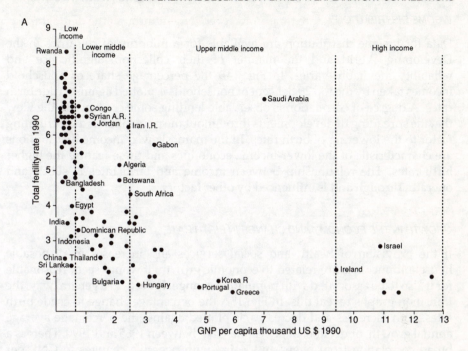

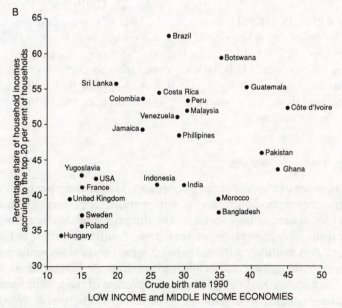

Figure 14 (A) Average per capita GNP and fertility rates 1990; (B) household income distribution and crude birth rates 1990

INCOME DISTRIBUTION

Data for income distribution are available for a minority of countries in the Developing World and the manner of their collection, significance and reliability inevitably varies. In Fig. 14B the percentage share of household incomes taken by the top 20 per cent of households is plotted against crude birth rates. A degree of correlation is observable lending some credence to the view that the spreading of developments throughout the population is a contributing factor to the lowering of birth rates. In the main, skewed income distributions are characteristic of the lower-income economies and these feature the higher birth rates. The relationship between income and birth rates is subtle and operates through and is influenced by other factors.

GROWTH IN THE ECONOMY AND CHANGE IN BIRTH RATE

If the provision of health and social services and increases in disposable household incomes are related to economic growth, it would seem reasonable for it also to be associated with population change. Indeed, in a general way, the transition graphs reveal this. In Fig. 15A the percentage change in crude birth rates, both increases and decreases, is plotted against the percentage average annual growth, or decline, in per capita GNP between 1965 and 1990. There is a broad correlation from Niger and Zaire, with negative changes in GNP per capita and increases in the crude birth rate, to China, Hong Kong, Singapore and South Korea at the other extreme with very high increases in GNP per capita and dramatic reductions in birth rates. However, in the tiers of lowest declines in birth rate and in the countries where birth rates have increased there is little correlation, and per capita GNP values range from declines of over 2 per cent to increases of almost 5 per cent. This lack of correlation is almost entirely associated with sub-Saharan countries. Again it would appear that birth rate trends are wealth related though the relationship is too complex and indirect to yield a strong, simple correlation.

AGRICULTURAL SOCIETIES AND URBANIZATION

The agricultural countries of the world are poorer and, for the most part, have diffused rather than concentrated populations, which are, in consequence, more difficult to serve in terms of education, social and health provision. In discussing the British demographic transition reference was made, it will be recalled, to the view that agricultural communities adhered to old customs and values while the new town dwellers, in leaving their old societies, also abandoned their old ways including attitudes to family size. In Fig. 15B, the proportion of the labour force engaged in agriculture is compared with crude birth rates. There is a strong correlation between the two indices from countries characterized by over 60 per cent of their population engaged in agriculture and birth rates of over 40 per thousand and the largely industrialized countries with less than 20 per cent

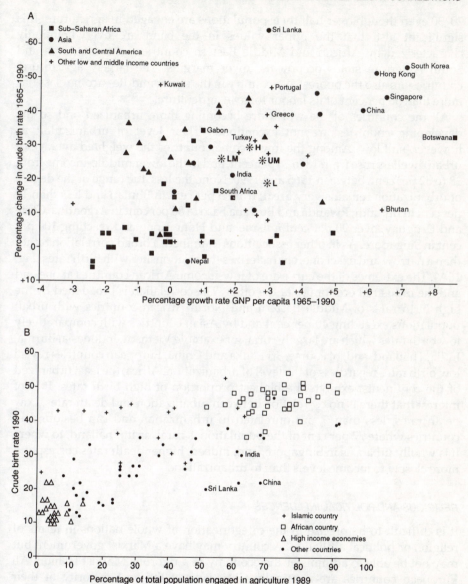

Figure 15   (A) Changes in crude birth rates and changes in GNP per capita 1965–90; (B) crude birth rates 1990 and population working in agriculture 1989

involved in agriculture and birth rates below 20. These latter countries are also the wealthiest. Islamic countries are distinguished by their high birth rates; all but three of the seventeen plotted have birth rates exceeding 30 although agricultural employment ranges from less than 10 to over 60 per cent. China, Sri Lanka and Thailand are also exceptional in falling within the birth-rate band of

45

20–30 even though over half their populations are engaged in agriculture. It is significant also that the upper values in the band are overwhelmingly characterized by African and Middle Eastern countries whereas Asian and South American states occupy its lower margins. The largely agricultural countries are also the poorest. No country in the upper middle-income class has more than 50 per cent of its labour force in agriculture.

All the countries of the world are becoming more urbanized and so the developing countries are not exceptional. Their level of urbanization is, however, still low. Among the low-income countries the weighted average of urban dwellers rose from 18 to 38 per cent, and in the lower middle-income from 38 to 52 per cent, between 1965 and 1990. None the less, the range of the degree of urbanization remains very great; from 5 per cent in Bhutan and less than 10 per cent in Burandi, Rwanda and Burkina Faso to 86 per cent in Argentina, Chile and Uruguay, over 90 per cent in Israel and Hong Kong and reaching 100 per cent in Singapore. As in other associations examined, a broad correlation rather than a narrow and exact one, characterizes the relationship with birth rates (Fig. 16A). The extremes of the curve are of low-income African countries at one end and the industrial economies at the other. The correlation is broadened by the high birth rates of Middle Eastern and North African countries with urban populations exceeding 40 per cent and by Asian countries with comparatively low birth rates which are largely rural, for example Vietnam, Indonesia, Burma, India, Thailand, and, of course, Sri Lanka and China. European countries reveal low birth rates regardless of the level of urbanization. Africa, the least urbanized of the continents, exhibits the highest proportion of high birth rates. It is of interest that there is no correlation between urbanization and death rates. Low death rates, less than 9, are unrelated to urbanization and can be found in countries where 77 per cent of the population is rural, as in Thailand, to where it is wholly urban, as in Singapore, and, indeed, higher death rates correspond more closely to income levels than to urbanization.

RELIGIOUS AND POLITICAL INFLUENCES

It is difficult to be precise in the categorization of whole nations in terms of religion or political affiliation. A country may have a Marxist government but may not be able to administer the economy in a rigorous Marxist fashion. All European countries are nominally Christian yet the vast majority of their populations do not attend church and many do not believe in the Divinity though most adhere to Christian moral values. In both examples it is the practice rather than the posture which is significant. Countries in which Buddhism is the main religion tend to have birth rates below the average of their income category and the Hindu countries likewise. Both are essentially Asian religions but there are marked exceptions to this tendency with birth rates of 40 per 1000 in Nepal, 43 in Lao, and 39 in Bhutan contrasting with 27 in Thailand and 20 in Sri Lanka. Islamic countries where the majority of the population are devout believers are, as a group, characterized by above average birth rates for their income groups.

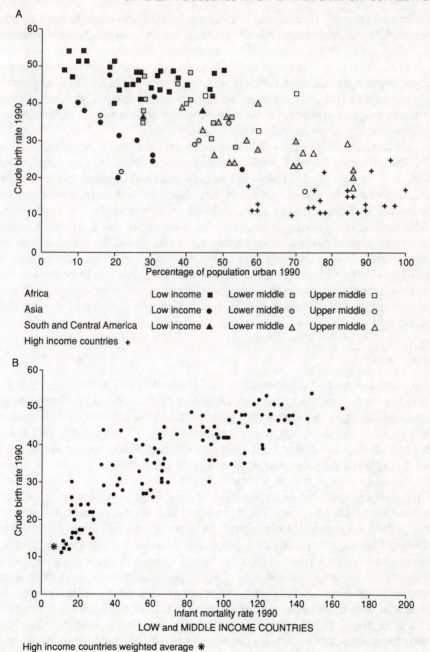

Figure 16 (A) Crude birth rates and urban population 1990; (B) crude birth rates and infant mortality rates 1990

There are exceptions. Tunisia and Turkey are both below the average for their lower middle-income group. Islamic nations are the world's fastest growing population group but it is too facile to attribute this characteristic solely to religion. As Weeks has demonstrated in his work on the demography of Islamic nations, Islam is more than a religion, it is also a way of life and thus embraces all aspects of society, including the relationships within the family (Weeks 1988). It is these cultural attitudes as well as religious beliefs which influence the demographic trends discernible in the Islamic world.

In political terms it is not possible to cite easy correlations. It is true that China and Cuba with their Marxist–socialist systems both exhibit very low birth rates within their income groups but so too do Sri Lanka and Jamaica. In the Middle East the socialist Peoples Democratic Republic of the Yemen and the non-socialist Yemen Republic have both exhibited the characteristic high birth rates of Middle Eastern Islamic countries. The communist state of the Korean Democratic Republic has a much higher birth rate than the capitalist Korean Republic while in Europe birth rates in Western non-communist countries are essentially similar to those in the former Soviet bloc. What is important is the extent to which governments, whatever their political or religious label, incorporate into their views of, and plans for, their societies a deliberate social provision.

SOCIAL PROVISION

Social provision may be expressed in terms of education, of medical services for all and not simply for the urban rich, of maternity care and of family planning advice and facilities, together with a policy related to a more even distribution of income. Among the countries of the Developing World, China, Sri Lanka, Cuba and Jamaica have been noticeably to the fore in this respect and this does much to explain their particular positions in the demographic transition. Three measures to assess this feature will be examined.

First, data are available on the number of persons in a country per physician as a measure of medical provision. It proves to be a very poor indicator of birth rates. In the low-income countries birth rates in excess of 40 are to be found in countries like Pakistan with one physician for every 2900 people and in countries like Niger, Upper Volta, Chad and Ethiopia where there is only one physician for between 38,400 and 78,000 persons. Lower middle-income countries again demonstrate a wide range of values. China, with a birth rate of 22, has a large number of doctors, one for every 1010 persons, but Saudi Arabia with a birth rate of 43 has even more per head of population, one for every 730. Clearly much depends upon who has access to the physicians. Few developing countries have nationwide health services like the Chinese; in many, their physicians have urban practices and treatment is not free.

Second, the impact of medical care, of social provision, of improvement in living conditions and awareness of hygiene can be measured by levels of infant mortality as it was in Britain. Low levels consequent upon a complex of attitudes and facilities could well, as has been argued in the European context, lead to a lowering of birth rates as fewer 'insurance' births become necessary. Figure 16B illustrates a clear relationship existing between infant mortality and birth rates across the whole spectrum of low- and middle-income countries though, for obvious reasons, at very high mortality levels, high birth rates are characteristic regardless of the precise level.

Thirdly, it has been argued that the more educated the population, and in particular the women, the more likely they will be to evaluate the need to plan the size of families. Data to adequately measure comparative levels of education are lacking but utilizing data relating to females, the percentage of the primary school age group which attended school in 1965 has been compared with crude birth rates in 1990 and depicted in Fig. 17A. An association between a more educated female population and a reduction in birth rate is evident. There are anomalies such as Zimbabwe but it is of interest to see that in countries such as Sri Lanka, Jamaica and Costa Rica, all with low birth rates for their respective groups, a large proportion of girls attend primary school. In contrast in many of the Islamic countries, Saudi Arabia, Libya, Sudan, Iran, Pakistan, Bangladesh, less than 45 per cent of girls attended primary school in 1965 and in all these countries birth rates in 1990 were high. It should be stressed that in most Islamic nations in the Middle East and North Africa the population of girls attending primary school is now very high. The 1965 data for China are not available.

FAMILY PLANNING IN THE DEVELOPING COUNTRIES

It was suggested earlier that it was important to identify the factors which motivated couples in Britain to limit their family size rather than to discuss the role of contraception, contraceptives being a means of, rather than a reason for, limiting births. Family planning programmes and the use of contraceptives in developing countries must, however, be considered for two reasons. First, one can see the extent to which contraceptive use is reflected in birth rates, or fertility rates, since this may summarize the interaction of the various possible influential factors which have been examined. Secondly, and more importantly, there is a greater need for governments in developing countries to be involved, directly or indirectly, in population dynamics than European governments were in the nineteenth century. This stems very largely from the demographic development in the Developing World which has been outlined. A dramatic fall in death rates with a subsequent and slower fall in birth rates has, in most developing countries, preceded the establishment of a self-sustaining, flourishing and more productive economy. In Europe population increase accompanied a massive economic growth. In many developing countries, and most notably in the poorest, their present position in the demographic transition is one of maximum population increase. Of the 42 countries in the World Bank's low-income category no fewer than 30 experienced higher average annual increases in population in the period 1980–90 than in the period 1965–80. Their

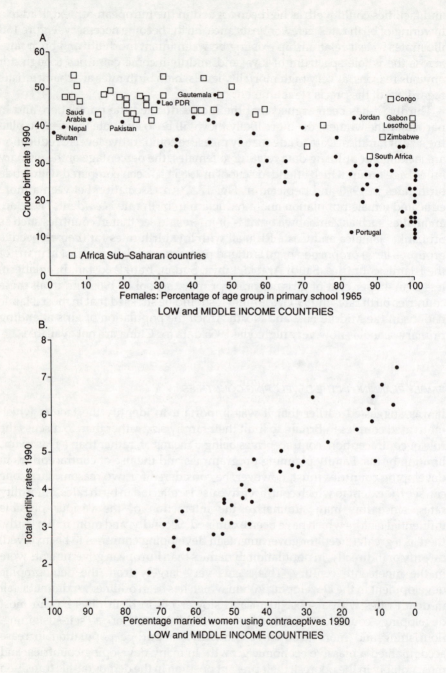

Figure 17 (A) Crude birth rates 1990 and females in primary school 1965; (B) total fertility rates and use of contraceptives 1990

young populations, with a high proportion in reproductive age groups, are maintaining high birth rates. Population size is a dimension in development planning. To raise the diminishing production curve above that of a rising population is difficult enough; to raise it in time to encompass a rapidly rising population presents difficulties which tax the most resourceful of countries. The logical and necessary accompaniment must be control of the population curve by a reduction in accepted norms of family size. This may, and commonly does, involve the education of women in the specific matter of family planning, the provision of facilities and a degree of governmental commitment.

The impact of family planning practice upon fertility rates is clearly shown in Fig. 17B. It is a solution which must be adopted in countries where the speed of population increase, or where the absolute size of annual increases, is inhibiting social and economic development. Two sets of forces are thus at work bringing about the reduction in birth rates in developing countries. One is the improvement in economic, social and environmental conditions which make large families less necessary and small families desirable. The other is the work of private and government agencies to promote the use of contraceptives together with government demographic policies enforced by disincentives such as taxation or housing penalties for large families or made attractive by privileges and facilities offered to small families. The strength of both sets of forces, as has been seen, varies considerably among the many nations of the Developing World. The World Bank estimates that up to 60 per cent of the variation in fertility changes between 1960 and 1977 in developing countries is accounted for by social and economic developments while an additional 15 per cent reflects the range of effectiveness of family planning programmes (World Bank 1980a; see also World Bank 1984).

A demographic transition from high birth and death rates to low birth and death rates has been shown to be taking place in many countries of the Developing World. It is a process which takes time and though some countries, notably China, have progressed very rapidly, others, largely in sub-Saharan Africa, have reduced birth rates very little if at all. The world's two largest nations China and India, notable for their rigorous measures to reduce births, have themselves experienced reversals in the downward trend of their birth rates in the 1980s. In other countries the decline in birth rates has levelled off (Population Today 1988). The populations of the Developing World will continue to grow. It is estimated that the total population of the less developed countries will rise from the 3680 millions of 1985 to 7114 millions by 2025 (United Nations Organisation 1989).

5 Agriculture: the environment and indigenous farming systems

'And he gave it for his opinion, that whoever could make two ears of corn or two blades of grass to grow upon a spot of ground where only one grew before, would deserve better of mankind and do more essential service to his country than the whole race of politicians put together'

Jonathan Swift
Gulliver's Travels 1726

The countries of the Developing World are still essentially agricultural even though the proportion of their populations engaged in agriculture has continued to fall over the past thirty years. By the last decade of the twentieth century, in twenty-two of the eighty-four countries in the World Bank's low- and lower middle-income categories, the proportion of population engaged in agriculture was over 70 per cent. In the two giants of the Developing World, India and China, the proportions were 63 and 68 per cent respectively (Fig. 2). The significance of agriculture as a source of employment declines as per capita GDP increases. Agricultural nations are commonly the poorest. These predominantly agricultural populations are in many cases inadequately fed (Fig. 3). In the Developing World as a whole per capita daily calorific intake was only 65 per cent of that in the industrialized nations in 1970. The situation has since improved with the percentage rising to 72 by the late 1980s. None the less undernourishment remains widespread with some 37 nations undernourished in the mid-1980s (Grigg 1982; FAO 1992). The Far Eastern region accounted for the largest absolute number of undernourished because it is the most populated area; in proportional terms the highest incidence was in Africa. In the mid-eighties nearly one in every three Africans was undernourished and the situation is currently deteriorating. All other regions showed improvements the greatest being in the Middle East where undernourishment was halved between 1971 and 1986. The FAO regards this situation in which malnutrition is a major problem in 20 per cent of developing nations as 'a major indictment of virtually all aspects of development' (FAO 1991). Furthermore many of the poorer countries are producing less food per capita than before. The average food production index was lower in 1988–90 than in 1979–81 in 29 per cent of forty-three low–income countries, the twenty-three African nations contributing notably to this situation. The fact that staple food production is below demand is not necessarily a problem if food imports can be paid for. The ability to pay will depend upon foreign exchange earnings and saving capacity. The self-sufficiency ratio in cereals is the production of cereals divided by the figure which represents production less exports plus imports by volume. A small fall in this ratio can have a very marked effect upon the balance of payments particularly in countries with a large population; for example, in Bangladesh where over a period of twelve years the self-sufficiency ratio fell from 94 to 89, the resulting cost of cereal imports to make up the deficiency absorbed no less than 58 per cent of Bangladesh's export earning which could have been used in development investment (FAO 1982). Where large export earnings such as from oil are possible or where manufacturing exports have been developed, food can be imported with little adverse impact upon the economy. In the case of most developing countries this is not possible.

Two important issues stem from the situations which have been described. First, many developing countries, overwhelmingly agrarian though they may be, are unable to feed their populations securely and adequately. Second, the inadequacies of the agricultural production of food are retarding other economic developments. It is a situation which presents, head on, the question of raising the production curve, with its diminishing returns, above that of expanding populations. This, in turn, requires an assessment of the systems of farming which characterize the Developing World and the consideration of the extent to which they are adequate bases for a more productive development.

Agriculture and the rural economy are thus central to the problems of many developing countries. Food intake is lower in a calorific sense, is deficient in protein and lacks essential vitamins for optimum bodily health certainly among the Developing World's poor and they are many. Often food supply is seasonally varied and in some cases markedly so. Physical health does, as has been seen, reflect societal and environmental circumstances and this includes nutrition. Brain development can be retarded by malnutrition in infancy and its effect can be permanent. A famine can affect a whole generation. Poor nutrition does affect energy output, ability to work and susceptibility to disease. The securing of an adequate, balanced and nutritious diet is essential to the physical and mental well-being of the people of the Developing World and agriculture is clearly central to this situation. It is important in other respects in the processes of economic development but it is basic to the well-being of societies.

The majority of developing countries are not only less industrialized but are also less productive in agriculture than many so-called industrial countries. As their populations grow rapidly, so the pressure on inefficient and insecure food resources grows. The problem is not simply a matter of growing more food but also increasing the efficiency of harvesting, storing, transporting and marketing it and of using it more effectively. It is also a matter of the distribution of income and of the access to food of all sectors of the population, both rural and urban.

The change in attitudes to new methods of production and to new organizational structures is a major one. Agriculture in the developing countries is very much a way of life of whole societies rather than simply a sector of an economy or an area of resource allocation. It embodies within it social structures, values and beliefs, prestige symbols and religious observances – indeed all the strands of the fabric of life. To change within this situation is, perhaps, more difficult than moving out of it. To change within the extended family and the village may well present more of a problem than to migrate to a town to work in a factory or join the urban unemployed. Agriculture deals with the animate as well as the economic, with the life cycles of plants and animals, with the living soil and with climates the vagaries of which are difficult to predict and whose deficiencies are difficult to remedy. It is perhaps small wonder that governments so readily welcome, and invest in the factory, steel works and power station. It is not that these are not important, they clearly are, but that their apparent relative ease of construction and functioning, and their openly declared manifestation of modernity and change may well lead governments to neglect the fundamental significance of agriculture with its multiplicity of production units and its association with a past rather than a future.

Agriculture in all economies and societies has, therefore, three components. One is physical; the role and influence of the physical environment. The second is social, reflecting systems and values at family and village level but also at national level as expressed in political ideologies. The third is economic. This is the dimension which not only calibrates the forces of the other two but also links agriculture to production and consumption activities at local, national and international levels. The farming systems of the Developing World embody these dimensions.

FARMING AND THE PHYSICAL ENVIRONMENT

The Developing World is essentially a world of tropical climates though some of its countries extend into the sub-tropics. The issue has often been raised as to whether the climatic and edaphic environments of the tropics present particular difficulties which not only explain the relatively lower agricultural productivity of developing countries but also restrict further development.

The extreme situations of difficult environments are obvious whether in the polar regions of North America, Europe and Asia or the hot deserts of the tropics. In both cases populations are small and indigenous systems of food production from hunting, grazing or tillage are highly specialized. In countries where such areas are but a part of their territory, as in for example India, Pakistan and Chile, they tend to remain undeveloped. There are, however, developing countries which almost in their entirety are dominated by drought. Burkina Faso, Mali, Chad and Niger are examples of independent countries desperately trying to manage their economies in the most difficult environmental conditions.

Apart from these obvious extremes, the question as to whether environmental differences are important determinants of agricultural and other developments has been much debated by some and ignored by others. Andrew Kamarck, director of the Economic Development Institute of the World Bank, has written:

> The effects of tropical climate are not absolute obstacles to economic
> development, but they do make many of the problems of economic

development in the Tropics sufficiently different from those in the Temperate Zone countries so that an additional hurdle has to be overcome and, consequently, all other relevant factors being equal, the pace of development in tropical countries tends to be slower

(Kamarck 1976).

Pierre Gourou has made equally positive statements as to the role of the physical environment, stressing the poverty of many tropical soils: 'The poverty of tropical soils is often revealed quite unexpectedly by a yield often inferior to that in temperate soils' and 'For climatic reasons tropical soils are poor and tend to become poorer quickly' (Gourou 1953).

Tosi and Voertmann have addressed themselves to this issue but with somewhat differing conclusions. They write:

There is an informed judgement that the basic environmental traits which have historically restricted the productivity of tropical agriculture also seriously impede the prospects of future development. But it is the opinion of the authors that this pessimism reflects a restricted conception of a high-productivity system tied to the crops, techniques, and organisation of the grain and pasture system. It is true that a high proportion of the plant species which form the basis of the high productivity system either will not grow at all in most of the tropical regions or else are very much less productive there. It is true that tropical soils are, for the most part, different in both physical and chemical make-up than the temperate region soils currently associated with high-productivity. It is also true that the dietary habits, food processing and raw material processing systems as well as agricultural machinery, land management techniques, and much else, are biased in favour of annual crops which, due to climate and costs differences have not prospered in the tropics generally. But all this is merely to admit that the land-use system, and indeed much else in the economic social and cultural life of the high income economies have been specialized by an environment structure quite unlike that of the tropics

(Tosi and Voertmann 1964).

In other words, Tosi and Voertmann believe that a different situation may require a different approach from that in temperate areas and that which appears to be a deficiency or an inferiority may be simply a product of judging it by criteria which are not applicable. In their assessment, the humid tropics possess an enormous potential for production in terms of heat, radiant energy and water for plant growth.

A similar theme, but with different conclusions, is developed by Jen-hu Chang. His work is concerned with the humid tropics commonly associated with the rain forests and largely equatorial in location. Referring to the errors made by the disastrous ventures of the Ford Company's rubber plantation in the Amazon basin and the British groundnuts scheme in East Africa, Chang (1968) writes: 'Clearly, the principles of land use that have evolved from mid-latitude experience cannot be applied directly to the tropics; the efficiency of tropical agriculture must be judged by a different yardstick.' The equatorial forest areas he discusses are characterized by low population densities and low agricultural productivity in South America, Central Africa and in parts of the East Indies but by high population densities in other parts of the East Indies where rice, higher in calorie and protein content than other hot wet crops such as cassava though lower in protein than temperate grains, is successfully grown. He writes that 'in general, soils in the humid tropics have a distressingly low fertility'. There are, however, exceptions to this dictum, notably on soils developed on mineral-rich recent lavas such as those of Java and the soils of flood plains and deltas of rivers which drain mineralogically varied areas. Unfortunately, in wide areas of tropical South America and Africa, the source materials for soils are the long-weathered Precambrian rocks on ancient peneplanes. Of climate in the humid tropics Chang states:

'the much attenuated solar radiation, the persistently high night temperatures, the lack of seasonality, and the excessive rainfall, combine in one way or another to reduce the potential photosynthesis and to limit possibilities of diversified agriculture. The attainment of the potential yield prescribed by the thermal and radiation regime of a place is rendered more difficult by such factors as the intense oxidisation and leaching of soils, the lack of nitrogen-fixing legumes, the high cost of fertilizers, the troublesome weed problem, and the prevalence of pests and diseases. All in all, the humid tropics are regions of 'debilitation' in Fleure's sense where human effort has rarely been accorded favourable response

(Chang 1968; the reference is to Fleure 1919).

Chang recognizes that certain systems can succeed in the humid tropics but, unlike Tosi and Voertmann, does not see as yet a satisfactory solution. The systems which have developed successfully beyond shifting cultivation have commonly been tree and perennial crops for export which have little impact on food requirements, neither does Chang regard them as presenting a basis for a stable economy. The one big exception is paddy rice which can give higher yields, which conserves soil fertility and which has for centuries sustained large populations.

We have, therefore, on the one hand the view that it is not the tropical environment which presents great difficulties, it is simply a matter of doing the wrong things, and on the other the widely held view that the environment is difficult and restrictive. Of the humid tropics Chang writes 'the natural choices are fewer and less appealing in the humid tropics than in most other climatic regions on the earth' (Chang 1968). This latter view would appear to be the more realistic. More recently the work of Eden on the Amazonian rain forest indicates the fragility of its agro-eco-systems and the extent to which they are vulnerable to the various forms of development (Eden 1978). It would seem that the 'informed judgement' to which Tosi and Voertmann questioningly refer is not only informed but accurate and realistic.

Much of this discussion has been concerned with the humid tropics. Great parts of the Developing World lie in the less humid, seasonally arid tropics. Here the environment also presents difficulties. The availability of water, not temperature, determines the start and finish of the growing seasons. Lower rainfalls are commonly characterized by greater variations in annual amounts; by, in other words, lower reliability. In the tropics not only is this the case but the soil water regime, in contrast to that in temperate regions, tends to accentuate rather than compensate for, rainfall deficiencies. In Western Europe, the lower temperatures of winter mean that precipitation exceeds evapotranspiration and tops up the soil moisture levels which thus provide a reservoir of moisture to be drawn upon during the warmer growing season with its higher evapotranspiration losses. In the seasonally arid tropics soil moisture is adequate during the growing season but the rains cease before harvest; a season with no rainfall follows, soil moisture is lost and the sowings for the next season take place in soils with no water reserves. In such circumstances, the time and nature of the first onset of the rains is crucial. If, as in the savanna zones of West Africa, the first downpours can be followed in some years by rainless days or weeks before a second fall, the precious seed corn having been sown can be lost. The rain itself, essential though it is, can cause problems. It effectively leaches the soil of nutrients as it exceeds evapotranspiration and causes soil erosion by its characteristically torrential nature. The extensive savanna areas of other parts of the tropics all face these problems to a greater or lesser degree (Kenworthy and Glover 1958; Kowal and Kassam 1978; Agnew 1982).

Levels of agricultural productivity, success and failure, do, however, vary within the same edapho-climatic environments since social, tenurial, economic and political systems themselves vary and strongly influence the response to the environment. The problems presented by tropical environments become questions of how to improve existing systems, and of how to introduce new systems, and of how to create new agrarian and rural economies within environments which are far from supportive.

INDIGENOUS AGRICULTURAL SYSTEMS AND THEIR DEVELOPMENT

At the least developed level of subsistence, man has practised collecting and hunting and developed a culture designed to abstract from the natural environment a harvest of plant and animal food without the intermediary function of agriculture. Such systems exist today in a few localities but most have gone (Forde 1934). Both activities are still widely practised to augment agricultural harvests in dry seasons and in droughts. The bulk of the people of the Developing World, however, feed themselves on farming produce. What is grown and when, and whether livestock form part of the economy, depends in large measure upon climate, culture and historical experience.

In the wide expanse of the Developing World there are many contrasting environments. These environmental variations have produced different responses. A range of agricultural systems has resulted with some adjusting to the environment and altering it little while others involve major environmental modifications. Within this variety, common themes, goals and methods are discernible and it is more relevant to development issues to consider the major farming systems functioning in the Developing World rather than attempt a comprehensive description of their many regional manifestations.

The main aim of the indigenous farming systems of the developing countries is to produce food, chiefly carbohydrates, in sufficient quantity and as securely as possible, to feed the population. The central, crucial purpose is to produce enough food to survive. Additional to the staple diet will be plants grown, or collected, to make the diet more palatable, more interesting and, indeed, often more nutritious. Goats, sheep, pigs, poultry and dogs are often kept to provide a limited amount of animal protein to which can be added the meat of rats, lizards, snakes, monkeys, birds and, importantly where possible, fish. The keeping of large livestock such as cattle and camels is commonly a separate activity associated with distinctive cultures. Where they are kept by cultivators it is generally for use as draught animals. While pure subsistence economies are becoming relatively rare, part-subsistence economies are common and within them the first aim is still that of securing the basic food supply. Beyond that, other crops will be grown for family consumption and partly for sale and these will include non-food crops such as coffee, cotton or sisal. The growth of a cash sector is the first stage in economic and social development leading to an interdependence of societies and areas, to the development of marketing, to the emergence of comparative advantage playing its role in regional specialization and trade, and to the development of a space economy incorporating both town and country. The dramatic impact of largely European contacts, first in trade and then in the form of colonial development, in many areas accelerated this transformation to part-subsistence or indeed wholly cash economies as well as, in some cases, introducing totally new and alien systems of farming linked to external markets.

SHIFTING CULTIVATION AND BUSH FALLOWING

Shifting cultivation and bush fallowing is a farming system and agricultural practice that attempts to meet food needs by a husbandry which recognizes and is adjusted to the fragility of tropical ecosystems. Land is rotated rather than crops, acknowledging the limited resource of the soil without the application of inputs to replenish the drain upon it, though some crop rotation may be practised. Throughout the humid tropics with heavy rainfall and high temperature the weathering of rocks is rapid and the leaching of exposed soils is pronounced. Soils quickly lose their productive capacity and yields equally rapidly diminish. In seasonally arid areas yields also fall off quickly under continuous cropping. William Allan in his authoritative work, *The African Husbandman*, on the basis of field evidence from Zambia, has produced a set of hypothetical fertility gradients on a range of Zambian soils (Allan 1965). His graph of the yield performance under maize in normal years is the basis of Fig.

18. The minimum yield on which the subsistence cultivators of these soils can exist is of the order of 560–670 kg per hectare. The poorest soil, 3c, therefore cannot be cultivated; 3b is abandoned after two or three years, and 3a after four years, while class 2 soils can be cultivated for up to six or seven years. The class 1 soils are commonly tilled for only five years, though their yields would not fall to subsistence limits for some fifteen years, and are put into a natural grass fallow before being recultivated. The speed of fertility decline thus varies according to the original nutritional status of the soil and to its textural and structural characteristics since these control the degree of leaching. When yields approach the critical minimum level the land is abandoned to be recolonized by natural vegetation.

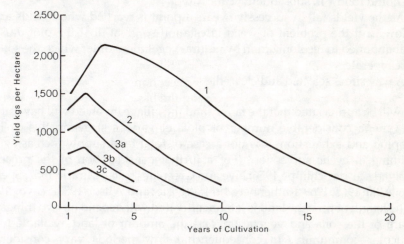

Figure 18   Decline in crop yields under shifting cultivation: Zambia

Throughout the tropical world, shifting cultivation is still widespread and is practised by both subsistence and semi-subsistence farmers in all its main essentials though elaborations of the system occur and the name, of course, varies. In Brazil it is known as *roca*, in Mexico *milpa*, as *masole* in Zaire, *chena* in Sri-Lanka, *kumri* in India, *ladang* in Malaysia, and the term *swidden* has been applied to its practice in Sarawak and the Philippines. It should be remembered that bush fallowing is not peculiar to the tropics. It was practised in Britain up until medieval times and survived in other parts of Europe longer. It is a simple and land extravagant method of land rotation which can be very successful. The basic procedure and sequence of events is as follows:

1.   The existing vegetation is cleared. If this is woodland it commonly means that some trees are left standing since, on the one hand, there is the consideration of costs and benefits in terms of the time and labour available for more complete clearance and, on the other, some trees may need to be preserved either because of the fruits they yield or the shade they may afford

to some crops. The clearing is generally done by a combination of felling and burning though on grassland and light scrub burning alone may suffice.

2. Crops are planted in the ash layers of the cleared sites. The sites themselves will be scattered and not contiguous, will often have no clearly defined outline and will contain surviving trees. Rarely will large areas be dominated by a single crop. More commonly a considerable mix is planted. To the European eye, accustomed to the fenced, demarcated field characterized by single crops grown in rows, the cultivation clearings of shifting cultivation do not look like the fields of an organized agriculture.

3. A succession of crops is grown on the cleared ground. Consideration is given to which crops should be planted first in the newly cleared soil and which should follow in subsequent years.

4. As the yields fall in successive years a point is reached when yields are so low, and the problem of weed infestation so great, that the plot must be abandoned to recolonization by natural vegetation under which the soil can recuperate.

5. A new site is selected and cleared.

It will be appreciated that the family and the village involved will have, at any one time, a considerable number of plots at various stages in this cycle of cropping and exhaustion. It is not a case of total food supply increasing and declining in cyclic succession. (For a further consideration of the processes involved see: de Schlippe 1956; Nye and Greenland 1960; Miracle 1957; Spencer 1966; Allan 1965, 1969; Ruthenberg 1971; Webster and Wilson 1980, Seavoy 1987.)

The length of time the land is left uncultivated will depend on two things: the nature of the soils and vegetation, and the amount of land available to the particular community. In consequence fallow periods vary considerably tending to be longer in woodland areas and shorter on grasslands. It is clear from evidence drawn from many parts of the world that under shifting cultivation only a small proportion of the land used for cultivation is in production at any one time. It is a system which requires a very large input of land; indeed land is used as the input rather than fertilizer and intensity of cultivation. Laborious though the initial clearing may be and time-demanding the subsequent weeding, total labour input in systems of shifting cultivation is less than in sedentary forms of all-year agricultural production.

The vegetation cleared may be virgin woodland or grassland or some intermediate vegetational association. It may be, and commonly is, a regrowth of vegetation after fallowing; tall secondary forest will develop if time is allowed, shorter fallows will yield an immature forest or indeed allow regrowth only to an early succession phase of scrub with grass and bushes rather than trees. On grassland, short fallows will revegetate with grassland of a different flora than in the complete vegetational succession. It is commonly reported that in woodland areas the farmers prefer to clear secondary rather than virgin forest because it is easier. The shorter the fallow, the more intensive the use of land and the closer the system approximates sedentary cultivation. Short fallows without the support of additional inputs such as fertilizer represent a heavy drain on soil resources. In the area which a village or tribe has at its disposal for cultivation, i.e. an area recognized by its nearest neighbours as its territory with rights of usufruct, the villagers will clear plots first in close proximity to the village. As these are abandoned to revert to fallow, sites more distant will be cleared and cultivated until the distance-decay of effective input is such that it is more reasonable to move the whole village to a central position within a new area. An intermediary stage with the setting up of a daughter settlement to initiate cultivation may precede the final move of the village.

In true shifting cultivation there is thus a combination of the rotation of land around the settlement, bush fallowing, and the rotation of the movement of the settlement through the land at its disposal. The result is a longer period of rest for any one area of land with the corollary that very large amounts of land are required for the production of crops.

In reality the practice is complex and sophisticated with many variations throughout the tropical world. Crops such as yams are planted in mounds or ridges incorporating compost or green manure. Sometimes pits are dug and the crops are planted in a mix of soil and vegetable remains. In other cases organic trash is piled into heaps, covered by soil, and then fired within the mound. Crops are planted in the resulting ash. All are attempts at fertilization to augment the soil and retard the drain on nutrients. The combination of bush fallowing with intensively cultivated gardens in and near the village is widespread. Because of their proximity they can receive not only food remains, domestic rubbish, animal dung and night soil, greatly to increase their nutrient status, they can also receive much greater attention resulting in better weeding and the scaring away of birds and other predators. Old village sites when abandoned commonly remain in cultivation for a period of years precisely because of the higher organic content of the soil. Shifting cultivators are able to identify by the natural flora the most suitable sites for clearing and within them the best areas for particular crops just as they can tell the stage of soil recovery by the vegetational succession on the abandoned sites. Cropping will reflect the need to secure a basic quantity of food but with some variety. The great intermixture of crops so often found reflects a knowledge of the micro-climate and edaphic needs of particular crops; it also minimizes the risk of diseases and pests, specific to particular crops, decimating a whole harvest. This mixture also ensures a near-continuous soil cover which inhibits soil erosion and leaching and results from the varying growth rates and manner of growth of individual plants. The variety of crops helps to spread out labour demands for planting and harvest while a more continuous yield reduces the need for storage capacity and the risk of destruction by insect and animal pests during storage. As the soil deteriorates, the less demanding crops will be planted. Cassava, though native to South America, is now widely grown throughout the tropics because it is not only tolerant of depleted soils but can be left in the ground without deterioration for long periods before use. It has become a significant component in shifting and other systems though its nutritional value is low. Far from being an elementary system, shifting cultivation and bush fallowing embodies a detailed and sophisticated knowledge of the environment and how to manage it with a minimum input, knowledge acquired through cumulative experience passed on from generation to generation but without scientific understanding.

Yields, per unit area actually cultivated, are very low. Measured against the total area used in the shifting system they are miniscule. The total output of the area can be increased by reducing the length of fallow but to do this without compensating for the shorter recuperative rest means that the initial yields on recultivation will be less, and those of subsequent years still lower. The only reasons for resorting to shorter fallows are if land is lost, for whatever reason, to the cultivating group, or if its population increases. Greater pressure of population on the land reduces the length of fallow, moving settlements are replaced by stationary settlements as more settlements occupy the land, long fallows are replaced by shorter and eventually each village comes to be surrounded by an area continuously cultivated with its more distant land rotated in bush fallowing. This is the infield–outfield system once so common in Europe. The Achilles heel of shifting cultivation is its inability to support increasing populations; it has to change. Change does come about perforce and change can be stimulated and planned. Boserup and others claim that it is an inevitable response to population pressure and that agrarian changes come about as a consequence of need (Boserup 1965).

As a system, shifting cultivation is not amenable to increased production and it is not an adequate basis for economic development unless it is considerably changed. It does not allow the establishment of the essential physical and social infrastructures which characterize permanent settlement nor is it an easy companion of road systems. Where there is no pressure of population it is possible to introduce cash crops into the system to give a monetary yield as for example groundnuts and cotton in northern Nigeria where shifting and bush fallowing systems exist, or bananas, dalo (*Colocasia esculenta*) and yaqona (*Piper mythisticum*) in Fiji. Where populations do increase markedly, shifting cultivation cannot support them, and distress and out-migration occur or resort is made to non-farm activities to a greater extent. The delicate ecological balance is upset.

There have been a number of attempts to systematize the essential principles of the system to make it more resilient or to guide its transformation. One considerable casualty of shifting cultivation is woodland; agriculture is the major destroyer of the world's forests. In the Indian sub-continent attempts have been made for shifting cultivators to work with foresters. They are used as forest clearers and crop the area for one year, after which the foresters take over and plant commercial timber and the area is restored to forest for seventy years or so (Ruthenberg 1971). In Fiji the Lomaivuna scheme in the 1960s attempted to resettle individual farmers from communal village communities in a forest area by allocating them 4 hectare blocks with 2.5 hectares to be devoted to the cash crop bananas, 0.8–1.2 hectares to subsistence crops with the rest in woodland. The bananas were intended for export, the project was supervised and the aim was to produce sedentary individual farms. Soil depletion and banana diseases caused the scheme to fail but the individual farms persisted with the farmers identifying cash crops for which there was a local demand. Supervision of the project became unnecessary. It highlights the need to appreciate that the indigenous farmer is often capable of responding to opportunities if circumstances allow. In the former Belgian Congo, now Zaire, attempts were made to systematize forest clearing in the so-called *couloir* system whereby corridors of cultivation were driven through woodland in surveyed plots at right angles to the direction of advance. Unfelled areas separated each strip; the latter being subject to a planned sequence of crop growth so that an individual cultivator had in any one year a sequence of strips with the allocated crop for the particular year of rotation. Each strip reverted to forest for some nineteen years (Dumont 1957). The benefits of the organization were rotations that could be practised using better varieties, with a more effective control over cultivation techniques, while the compact nature of the corridor of advance meant that roads could be used and marketing made more effective. Yields were increased but the system had all the land-extravagant characteristics of shifting cultivation and eventually administration declined and the system is no longer practised in Zaire.

The inability of shifting cultivation to sustain larger populations is obviously a Developing World problem. In many places it has changed and is others the process is under way. The first obvious transition is to the situation in which the village becomes permanent and with it the notion of a village area in which its inhabitants have rights of use, though land ownership in the European sense may not develop. Permanently cultivated land surrounds the village with fallow periods in a crop rotation, while further out longer fallows characterize bush fallowing on land which will commonly abut against that of neighbouring villages. Areas become demarcated and village territories characterized by a zonation of use (Prothero 1957; Blaikie 1971). Features of the transitional stage are the greater use of inputs to compensate for shorter rest periods by adding green and animal manure, greater care in planting, the sowing of row crops, more weeding and a greater tendency to grow blocks of the same crop. It can result in a more stable system supporting a higher population density, particularly if the soils are intrinsically of above-average fertility and if the climate is less debilitating.

Before considering the nature of the transition to permanent agricultural types, reference must be made to the livestock analogue of shifting cultivation, namely nomadism, which in some parts of the Developing World is the major form of livestock husbandry.

NOMADIC CATTLE-KEEPING

Nomadic herdsmen are essentially confined to Africa, the Middle East and central Asia. In Africa where cattle-keeping is restricted by the tsetse fly, the system is confined to areas too high or too dry for the tsetse fly to persist though in their journeys the nomads will cross infested country. The Fulani of West Africa exemplify the system. In the Sahel and Guinea Savanna zones the Fulani practise the animal equivalent of shifting cultivation. They use wide areas of poor territory, put nothing into the land and graze their cattle on natural and man-modified vegetation. The movement is a continual search for fresh vegetative growth and hence they follow the rains in their seasonal movement north and south across West Africa from the forest edge to the desert fringe. They occupy largely tsetse-free areas but are obliged to cross the Guinea Savanna areas which are notoriously tsetse infested, to market their cattle in the more densely populated south. In Nigeria, nomadic Fulani account for over 90 per cent of the country's cattle stock. Cattle provide food for the families, and milk and butter for cash sales to agriculturalists or for food by barter; the sale of cattle for slaughter is a further source of income.

Cattle are owned by individual families and their number is an indication of a person's prestige. They are transacted in social obligations such as dowry gifts and are the centre, indeed the *raison d'etre*, of the whole way of life. As a consequence, in many nomadic societies cattle are rarely sold, but among the Fulani selling has become a fundamental part of their economy. They may sell directly to the butcher in the towns of the South and receive a high price. In doing so they run the risk of their cattle dying or becoming diseased, and certainly of losing weight, as they move through the tsetse areas. The alternative, increasingly taken, is to sell to local merchants in the North for a lower price but without the losses of the long journey, and the cattle are transported by road or rail.

Grazing is carried out communally with whole clans moving together and sharing tasks. The system depends upon a delicate ecological balance between grazing available, intensity of grazing, and the need to move. The input into this system is what the land and climate offers. Commonly the group will be focused in one area, trekking through it in a form of grazing rotation, but it will also travel long distances in search of water and grazing. The parallel with the move–rotate–move sequence of shifting cultivation is strong. The Fulani family's wealth is stock not land. The animals must possess the ability to survive rather than the capability to yield heavily. Stocking densities necessitated by the poor quality of the vegetation grazed are, or should be, low. Generally they are of the order of one cow per eight or nine hectares, and in poor dry areas much lower densities. In northern Nigeria a symbiotic relationship has developed between the Fulani herdsmen and the Hausa cultivators. Their cattle are allowed to graze on the fallows, commonly the stubble of the Guinea corn (*Sorghum* spp.) which they improve with their manure and help soil consolidation by treading.

The Fulani practise a system which is vulnerable to economic and political change and to any developments which restrict mobility or the land available to the herdsman or which disrupt the symbiotic relationship. As more land is brought under cultivation, as forestry areas are demarcated, this resource base is eroded. Land developments can prevent Fulani movements and hence restrict access to grazing areas. Political boundaries lie across and hinder the free movement of a culture based on movement. As cattle keeping and mixed farming develops, the need to allow Fulani cattle to graze stubble diminishes. They become a nuisance rather than a benefit and a disease risk to sedentary livestock. As grazing is restricted so pressure on the remaining land increases, vegetational yield is less and soil erosion more. The pattern is repeated throughout West Africa and across to the Red Sea. Stenning's classic work has been augmented by later studies (Stenning 1959). Davies (1966), Wallach (1989) and Bascom (1990) have all written of the problems faced by nomadic pastoralists in the Sudan. In discussing the pastoralists of the semi-desert and low woodland savanna in northern Sudan, Wallach states that differing attempts to solve the problems have all failed because they did not address the basic problem of insecurity of land tenure. Burkina Faso and the Côte d'Ivoire present a situation similar to that in northern Nigeria and Niger. The characteristic seasonal movements of pastoralists while existing in a symbiotic partnership with peasant cultivators have created an uneasy relationship as crops are often damaged. The Fulani are obliged to move increasingly further south into the tsetse-infested but moister areas of the Ivorian savanna. A growing disease problem results which is currently being tackled by more frequent herd movements, by cross-breeding with more resistant stock and with the help of veterinary services. The question of restricting the movement of cattle in the area has arisen (Bassett 1986). The system is not compatible with the changes taking place. It can be converted to a form of ranching but in the conversion a way of life and a culture would disappear. The Fulani and their like run the real risk of being left out of development and they have little political cohesion or strength to state their case. The ranch may well emerge. It is a transformation with parallels in the American West where the cattle trail was essentially nomadic grazing over a vast unrestricted area. It was brought to an end by the spread of the cultivator and the invention of barbed wire. It was replaced by the fenced ranch and produces store cattle to be fattened on arable farms in the Corn Belt. The future of the Fulani and other nomadic grazers may be similar.

TRANSITION TO SEDENTARY AGRICULTURE

As shifting cultivation is unable to sustain large increases in population, the changes forced upon it may lead to a deterioration of the environment, to a situation of permanently low yields, to poverty and distress. Change for the better requires a reappraisal of cultivation techniques, of fertilizer practice, of plant and animal selective breeding. It demands innovations in labour use, the consideration of non-human sources of energy and a change both in the organization of production and the marketing of produce. In short it requires one good form of husbandry to be replaced by another, and not by an inferior form.

There is evidence which suggests that the pressures of population can lead to innovative changes by which farmers effectively transform their systems into sedentary forms capable of sustaining large populations. The dense rural populations of India and China bear witness to this. Esther Boserup in her book *The Conditions of Agricultural Growth* states simply 'One of the main contentions of this book is that the growth of population is a major determinant of technological change in agriculture' (Boserup 1965). If one puts aside, for a moment, the long-established rice cultivations of Asia, the question can be asked as to whether the process is discernible in the much more lightly populated Africa.

In what has become a classic study, Mortimore and Wilson examined the area around the ancient city of Kano in northern Nigeria which they describe as the Kano close-settled zone (Mortimore and Wilson 1965). Here the population is very dense by African standards. In the innermost zone, rural densities are greater than 190 per square kilometre while densities still exceed 55 per square kilometre at distances of 50–65 kilometres from the city. Mortimore stresses that these are rural and not suburban densities; by contrast, in the Guinea Savanna 'middle belt' of Nigeria the average density is 10 per square kilometre or less.

In the close-settled zone, shifting cultivation and bush fallowing have been replaced by permanent occupation of the land using methods developed over generations. The soils are essentially light and are physically capable of intensive working but, being sandy, are not intrinsically fertile. The rainy season is short (May to September at the maximum); the rainfall low (33 inches in conditions of high evapotranspiration) and highly variable. In winter the desiccating, dust-laden Harmattan blows across the area from the Sahara. Walled villages are characteristic though isolated compounds do occur and land is, obviously, in short supply. There are two distinctive categories. *Fadama* is the land along the streams with access to the stream water and, during the dry season, with access to the sub-alluvial water below the dried steam beds. It is the most valuable and productive land because of this. The much more extensive category is made up of the low interfluves between the streams beyond access to the stream water. On this land the staple grains, Guinea corn (*Sorghum* spp.) and millet (*Pennisetum* spp.) are grown together with cow peas (*Vigna sinensis*) and groundnuts (*Arachis hypogea*). The latter two are important sources of vegetable protein, and groundnuts are not only the main cash crop of the area but also of Nigeria. On these interfluves seasonal aridity prevents the year-round growth of crops though temperatures would allow it. The *Fadama* is intensively cultivated in tiny plots throughout the year producing a range of vegetables for the Kano market. Population pressure has resulted in the emergence of the idea of individual land owning. In northern Nigeria tenure has become a blend of traditional African forms of communal ownership, with individual use, and Islamic views on the laws of tenure and inheritance which stress order and organization. So totally is the land farmed that use and occupation have become coincident. Until the Nigerian Government nationalized all land, the land in the Kano Emirate belonged to the Emir. However, long before the government edict all plots were registered and the use of them could be inherited and indeed rights of use could be transferred.

Associated with these developments was the demarcation of plots by low hedges generally of useful shrubs or grasses. Land came to be used for such things as security for loans and since 1945 the sale of land has become common, indicating the strength of the idea of ownership. Progressive farmers can enlarge their holdings by buying and renting but as Islamic law requires a subdivision among all inheritors upon death, holdings go through a cycle of enlargement, fragmentation and consolidation. There is no spare land. As the population grows, enlargements can only take place at the expense of others. Holdings are approximately half the size of the average for northern Nigeria.

The dense populations are supported by an intensive and careful cultivation of fixed plots in which the users can invest, and which they come to know intimately. The land is manured, particularly the *Fadama* which in addition receives the benefits of water and silt. Crops are row planted and inter-cropping is characteristic on both *Fadama* and interfluves. The groundnuts are, however, mono-cropped since they are grown essentially for sale. Manure is the key to success. The holdings are too small, 1.5 to 2.5 hectares and less in solely *Fadama* plots, to make the keeping of draught cattle possible. Sheep and goats are kept and Fulani herds graze the interfluves in winter. But Kano is the important source of manure. The city (population 250,000 in 1962) has a very large animal population of donkeys, goats and sheep and is a source of town refuse. This assemblage of organic manures helps to replenish the drain on the light soils while the incorporation of legumes in the rotation further assists them. The town manure is the result of a reciprocal trade in firewood and manure carried on donkey back. From distances of eight to sixteen kilometres and more, strings of donkeys move in to the town each day throughout the year carrying fuelwood. During the season when manure is applied to the fields, from January to March, they return with manure sacks on their backs. Mortimore has calculated that some 27,800 tonnes of manure are carried in this way each season and spread onto the surrounding land. Areas outside Kano also supply manure. As increasingly donkeys are replaced by motor vehicles so the source of animal manure decreases and its important role in this intensive cultivation is diminished. Agriculture thus supports a large population but it is poor rather than affluent. Some farms are large enough to occupy fully the family labour resources. Others are too small and their cultivators commonly have non-agricultural additional employment, particularly hand-loom weaving. The *Fadama* areas more remote from Kano grow fewer vegetables and more sugar cane, rice and winter wheat on their irrigated plots. A fixed form of farming has thus evolved but further changes are difficult to introduce. Holdings are too small to develop mixed farming and it is difficult to create larger holdings unless a satisfactory alternative to farming is available for displaced farmers. Yields can be increased if new high yielding varieties can be developed, if artificial fertilizer can be applied, and insecticides used. Further technical development can allow a greater intensity of land use; without it regression may take place.

It is possible to hypothesize that some situations are more conducive to change and development than others. The Kano close-settled zone was favoured by the proximity of a market, a source of manure and by soils which could respond to intensive cultivation. The impact of these attributes decayed with distance from Kano. The Rural Economy Research Unit of Ahmadu Bello University, of which the author was a co-director when first it was established, has further explored this question around Zaria and near Sokoto, and demonstrated the importance of location and access in agricultural change. The greatest modification was evident in areas of urban proximity and next where road access was good. Areas remote in terms of accessibility, and not simply distance, exhibited the least change in agricultural systems from bush fallowing to sedentary types and in the extent to which non-farm activities were present in villages (Norman 1967; Goddard, et al. 1971; Norman 1972; Goddard et al. 1974; Norman et al. 1976). Gleave has reviewed the literature relating to population pressure and agriculture change in West Africa and concludes:

> The intensification of agriculture is not the panacea for agricultural development . . . . The difference between the most successful and least successful areas . . . is their position relative to markets both urban, national and international, in the level of their infrastructural provision in terms of factors such as roads and sources of inputs and in the willingness of their population to adopt new and improved methods
>
> (Gleave 1980).

The factors cited clearly have a bearing upon any attempt to promote rural and agricultural development.

SYSTEMS OF PERMANENT CULTIVATION

Permanent cultivation does not, of course, mean necessarily development in the sense of improved economic and social conditions or in well-being. It can simply be the end product of the pressure on land resources when no land is available to be brought into cultivation for shifting systems and less land becomes available per person. It can be characterized by rock-bottom yields and returns on labour input as, for example, in parts of India and Pakistan where population pressure has removed shifting cultivation and sedentary systems are insufficiently productive. Development should mean a permanent occupation of the land coincident with increasing yields to raise the food intake, to produce a surplus for sale and to enable additional labour inputs to become economically productive. It is important to stress that it is not a process insulated from other sectors of the economy.

There are two major categories of farming systems which permanently occupy the same land: those which rely on rainfall and those which are based on irrigation. The two practices are not mutually exclusive and indeed farmers will practise irrigation, whenever possible, if only on a small part of their land. Within the two categories there are many regional variations throughout the Developing World. Asian examples will be used to illustrate the types since Asia is overwhelmingly characterized by fixed rather than shifting systems, since population densities are high and because Asia provides examples of long-established and sophisticated irrigation systems.

On the Deccan Plateau of India where the black *regur* soils have developed on the Deccan lavas, and in adjacent parts of peninsular India where red sandier soils are associated with metamorphic rocks, small sedentary farms are characteristic. The soils present difficulties. The *regurs,* though intrinsically the more fertile, are clay soils difficult to cultivate and indeed almost impossible to till without a plough and draught animals. In the characteristic monsoonal oscillation between a very wet season and a very dry and hot season, the clays either bake hard or deflocculate into a plastic sticky mass. The red soils are easier to cultivate but lack plant nutrients. The population is numerous and holdings are very small. The average for India is 3 hectares but is much lower in the peninsula, 0.6 hectares being the average in Mysore. Most farmers in India, some 80 per cent, are owner occupiers. In the absence of land for bush fallowing the problem is of inputs to sustain output. Animal and vegetable manure can be added. Draught animals, zebu cattle *(Bos indicus)* and the water buffalo *(Bos bubalis)* are much more commonly used than in Africa. They can provide manure but they also require food to eat. They can graze on any scrub or waste land, they can be fed on the by-products of crops grown for human consumption, like grain-straw, or they can be fed on fodder grown for them. Fodder crops take land out of use for human consumption and there is therefore a reduction in human food supply even if the cattle are eaten or their milk consumed because of the energy used in the conversion. The calorific yield per hectare is less. In India, however, cattle are not eaten since it is prohibited by the Hindu faith and the slaughter of cattle is forbidden in the Indian constitution. Cattle supply milk and this is nutritionally important though yields are low due to inadequate feeding, the poor quality of peasant cattle, and energy lost in working fields. Overall they make a negative contribution to India's food output. On the Deccan laval soils where cattle are needed to plough, they consume up to 30 per cent of the farm output. Breeding cows serve only to supply new draught animals, and cattle surplus to this need and cattle too old to work are a further drain on very limited resources. In areas long deafforested, cattle dung is a major source of fuel and so does not find its way back to the soil to replenish it. The poor farmers on smaller plots cannot afford to keep cattle and if they cannot hire them are compelled to use hoes for cultivation. The dry central parts of the Deccan are characterized chiefly by millet, bulrush millet *(Pennisetum typhoideum)* and finger millet *(Eleusine coracana)* while wheat is the main crop further north. Maize is widely grown both as a fodder crop and a food crop. Despite manuring practices, crop rotation, the use of draught animals and steps taken to control soil erosion, yields are low. Nutritional levels are at a correspondingly low level and poverty is widespread. Cassava had been introduced to replace grains in many areas because of its high yields in difficult conditions. If a patch of land can be irrigated from 'tanks' (small reservoirs retaining the monsoonal rain) or from wells, the situation can be improved. Not only can a range of vegetable crops be grown but also, where possible, rice. Tanks are widespread in the

Indian sub-continent not only allowing irrigation downstream of the *bunds* (dams) but also the growth of crops on their beds after the surface water has been used. In addition to the staple grains the leguminous crops, chick peas *(Cicer arietinum)* and pidgeon peas *(Cajanus cajan)*, are grown on most holdings and are invaluable sources of protein.

India supports a very large population at a low level of nutrition. Wheat, millet, maize and rice supply between 70 and 90 per cent of her population's food. The need is to improve the output of these many small farms in the rain-fed areas. This requires a range of innovations easier to cite than to implement and India has made strenuous efforts to do so since her independence. First crop varieties can be improved by selective breeding. It can and has been done in research stations but the transfer to the farming community raises further problems. Likewise, selective breeding can be applied to livestock; in India's case this clearly, in relation to cattle, applies to milk production only, but dairy farming has much to recommend it as a form of Developing World farming. It provides a nutritious product with an important role to play in the improvement of health, and it yields a cash product with, in India's case, a ready market in towns. Importantly, also, it is a means of introducing an organized and demanding form of production which can do much to stimulate agrarian development in appropriate situations. Recent developments have indeed shown further progress in the organization of dairying in India. Rural producers in one area have been organized into cooperatives with the milk going directly to modern, hygienic, urban, processing plants thus cutting out the middlemen with their reputations for corruption. Aiming to be self-sufficient, this is as yet the only large-scale scheme operating in India but it may well be a pointer to others (Atkins 1989).

Improvements on the one hand require improvements on the other. Higher yields of new crop varieties can only be sustained by improvements in other aspects of farm practice. They require a greater input of fertilizer, greater care in weeding, an ability to control pests and diseases, and in some cases the use of improved agricultural implements. Livestock developments require improvements not only in livestock husbandry but also in fodder production, feeding skills and the availability of veterinary expertise. This all requires time, training and importantly capital, an assemblage of inputs to improve an unsatisfactory agrarian structure within a difficult environment.

The great concentrations of Asia's populations are not in the rain-fed areas but on the irrigated lands of the valley bottoms, the coastal plains and the deltas; indeed the population distribution of south and east Asia has been described as topographic. It is a valid description whether one applies it to the Indian sub-continent, to China or to south-east Asia. Even these areas are characterized by low material living standards though they have given rise to and supported some of the world's greatest cultures in the so-called 'hydraulic civilizations'. The obvious advantage of irrigation in tropical Asia is the coincidence of water availability throughout the year with all-year growing temperatures. More than one crop can be grown in the year and in some areas several. The area cropped is thus considerably greater than the area cultivated. It is precisely the reverse of the situation found in shifting cultivation. The production from a given area of land is thus much increased. In addition it is possible to obtain much higher yields of crops grown under irrigation since, other things being equal, water availability is the common restriction on plant growth. Fallowing is less necessary and sometimes unnecessary. It is possible to have a continuous, sometimes multi-sequential, schedule of cropping with the possibility of higher yields for each crop compared with the dry-land equivalent. Irrigation agriculture is much more responsive to increasing inputs be they of fertilizer or labour; in economic terms it is characterized by high marginal returns. There are additional and obvious advantages. Apart from exceptional situations of widespread and prolonged drought, dependency upon rainfall is much reduced. Variations in yields from year to year are on a small scale and this reliability lessens the need to divert resources into contingency strategies and so allows a fuller more productive use of them. Food is available throughout the year and so also are incomes from cash crops, while labour inputs are more evenly spread.

Central to Asian irrigation is, of course, rice. It has sustained large populations over centuries, allowing a permanent occupation of the land characterized by high yields and without soil fertility dropping to disastrously low levels. Whole holdings or only part holdings may be irrigated as topography and tenure patterns dictate. The explanation of irrigated, or paddy rice, productivity lies in its maintenance of soil fertility. The land is levelled and subdivided by low-banks, features which inhibit soil erosion, while the application of water provides nutrients in solution. On river flood-plains and in deltas, annual flooding may provide nutritious silt as well as water. These nutrient supplies do much to replenish the loss consequent upon harvest. In addition, stubble is not burned but ploughed into the puddled bed of the growing basin adding organic material. While fertility is thus maintained it can be increased by the addition of animal manure, household refuse, night soil, composted vegetable matter and, in some areas, by green-manure crops grown to be ploughed in. Where continuous cropping is practised with other crops alternating with rice, as in the most intensive systems, the detailed control of water, the great attention paid to crops and the large input of care and attention, all serve to suppress weeds and reduce pests. When water is impounded in paddy fields, nitrogen-fixing algae further enrich both water and soil (De and Mandal 1956; Grist 1975; Webster and Wilson 1980).

This level of care is reflected in crop rotations. Two crops of rice followed by a green crop may characterize the sequence found in a single year on ill-drained areas. On better land two crops of irrigated rice may be followed by a non-irrigated crop, wheat or maize or a root crop, in the cooler season, again within one year. A greater flow of cropping can be achieved by sequential planting in which a following crop is planted before its predecessor has been harvested. A sequence of harvests is thus possible and on land irrigated throughout the year as many as four crops, occasionally more, or as few as three, can be raised. On seasonally irrigated land two or three crops are more usual. With some crops, notably sugar cane and rice, and occasionally the sorghums, ratooning is possible, i.e. vegetative regrowth rather than replanting. This speeds up the sequence but yields successively fall and so it is not the most intensive form of cultivation. The use of nursery beds to grow the seedlings allows very intensive production and is widely practised with rice. In all these systems interplanting is used as it is on dry-land permanent farms. Row planting is characteristic, with each row given the particular attention it requires in a carefully organized system. In some areas inundated land yields fish, and ponds are rotated through the cycle of cultivation. The input of labour is high and the intensive nature of the system gives a large output which supports dense rural populations which because of their size, however, can only experience low material standards of living. Holdings are small and are subdivided as populations grow. The requirements of water management and civil engineering have been associated with the development of social, economic and political institutions but these have been essentially of a non-industrial nature. The pressure of population means that Asian irrigation agriculture continually needs to become more productive despite its intensity, as do all other farming systems in the Developing World. Within south and east Asia, the range of productivity and the degree of agricultural intensity are considerable. Even within such involved and sophisticated systems as irrigated multi-cropping there is room for increased efficiency and productivity. The challenge for future development is there.

INITIATING THE DEVELOPMENT OF AGRICULTURE

The development process in agriculture can be conceived of as incorporating the means to increase the production of crops and livestock, the ways by which the farmer may be drawn into the cash economy so that his products may be sold in exchange for non-agricultural commodities and services, the establishment of systems of marketing whereby he obtains a fair price for his product, and the creation of rural credit facilities which enable him to finance his developments without the evils of usury. Some techniques aim at an increased output from the land, others at increased production per worker; some involve completely new systems of farming, others modifications to the indigenous systems which have been described. The common goal is an increase in total production.

These various processes and techniques can touch upon every aspect of farming and agrarian society. So far-reaching and so intertwined are they, that it could be said that the only way forward is to change radically the whole rural scene. The tangle of tenurial constraints, fragmentation of holdings, absence of capital inputs, of plant and animal breeding, of fertilizer and insecticide use, the inadequate cultivation practices, and the rudimentary economic schedules could suggest that existing systems must be swept aside. In their place would be established large consolidated units, carefully planned, efficiently managed, soundly financed, embodying technical skills and using an appropriate mix of mechanization and paid-labour force. These large units, with a market identity, would be owned and operated by companies be they private, cooperative or state. It has been tried in one form in the Soviet Union with its state collective farms and in another and completely different way in China with her communes (Chapter 15). In both cases the entire agrarian economy has been changed. In Ghana where state farms and Workers' Brigades were established by President Nkrumah they met with an almost complete lack of success (Killick 1978). In some developing countries the plantation, either nationalized or left in private hands, is a model of the system. Government-implemented schemes of agricultural development may take the form of capital-intensive and large-scale environmental modifications, the object of which – and irrigation schemes are the obvious example – is to extend the cultivatable area. Indeed, it is only governments which can raise the capital required. There are notable examples in India and Pakistan. The Gezira scheme in the Sudan is another, where over three-quarters of a million hectares have been reclaimed from desert land by canal construction and irrigation. The land is divided into surveyed tenant plots, and all operations are determined by a central management including water supply and cropping rotations. Inputs of seed and fertilizer are provided and soil preparation is carried out by the central tractors organization. The major crops are cotton, sorghum and a fodder legume (Dolichos lablab). The tenants can keep cattle on the fallow stubbles. Marketing of cotton is carried out by the managerial body upon which tenants are represented. The sorghum harvest and that of other crops belongs to the tenant and he receives a share of cotton sales (see Gaitskell 1959). In many ways it is a successful scheme greatly increasing the secure output of cotton and grain in Sudan. Guy Hunter, however, feels that this scheme with its high degree of central control has inhibited local commercial and economic enterprise (Hunter 1969). It raises the wider issue of the role of the state in fostering or stifling human endeavour.

The alternative approach to agricultural development is to address the problems of the multitude of individual small farmers and to attempt to improve their conditions by increasing output and enhancing marketing opportunities, and to do this with a smaller injection of capital and a less involved technology than in the large modernization projects. This approach is made more difficult because it deals with a very large number of individuals and families embedded within the social and economic fabric of their traditional societies. It has to cope with ignorance, it has to be seen to be meaningful by the people for whom it is intended, it has to produce results in order to convince. This second approach puts to the fore the goal of increasing the income of the many peasant families which make up the majority of the Developing World's population and so of bringing benefits, albeit modest, directly and quickly to them. It is not the panacea for all the ills of the Developing World's rural communities but it is an approach which acknowledges that their condition cannot simply await the completion of some massive scheme or some long-term plan for structural changes in the national economy. It is an approach which everywhere has proved difficult partly because it deals with the disaggregated multitude and partly because many, perhaps most, governments have given it a low priority and viewed it with little enthusiasm. In many countries it requires radical changes in land ownership and tenurial systems – notably, but not exclusively, those of Latin America – and is therefore a development which impinges upon long-established relationships between hierarchical social structures and wealth.

As has been seen, almost all indigenous forms of agriculture are, without change and additional inputs, subject to limitations on their productivity. Increased production can take two forms, increased output from the land and increased output per worker. These increases can be the product of using improved methods, adopting new but appropriate technology, incorporating where possible livestock into cropping systems to provide animal protein, manure and draught power, of using improved breeds of livestock and varieties of crops, and of applying fertilizers and insecticides. Throughout this century agricultural research stations have been working to this end. In too many cases the innovations have not been adopted and in most it has been because the society for which they were intended has not been involved and its priorities and its economic and social structures not understood. Potential blockages in the adoption process have not been identified. All the evidence points to the need to involve the small farmer in the changes which are purported to be his benefit. G. Benneh, reporting upon a scheme to introduce mixed farming in northern Ghana, states 'illiterate small-scale farmers are responsive to agricultural innovations provided that the benefits of the recommended practices are demonstrated to them, the introduced technology is not too much above the level of that which is being replaced and an attempt is made to study the problems which the recommended practices aim at solving before they are introduced' (Benneh 1972).

In practice one innovation requires others to become fully effective; an improved variety or crop may demand a greater application of fertilizer and water, an improved breed of cow a more nutritious diet in order that its yield potential might be realized. The well-known high-yielding variety seed programme (HVP) introduced into the Indian sub-continent in 1966–67 exemplifies this situation. It was an attempt to increase production from the land in countries where land is limited and mostly fully used, and where the pressure of population upon resources is very great. This 'Great Revolution' Chakravarti describes as 'a significant breakthrough in feed grain production' and says that 'it appears to be a turning point in stagnating Indian agriculture' (Chakravarti 1973; Brown 1970; Frankel 1971; Johnson 1972; Nulty 1972; Farmer 1981; Bayliss-Smith and Sundhir Wanmali 1984; Goldsmith 1988). The traditional varieties of seeds used for the main crops, wheat, rice, maize, jowar (sorghum) and bajra (bullrush millet, *Pennisetum typhoideum*), were incapable of high yields as they had evolved over a long period of time as crops able to continue to yield under conditions of poor husbandry, drought and on infertile soils. The HVP introduced new hybrid varieties of these five cereals which were drought resistant (with the exception of rice), were very responsive to the application of fertilizers and had a shorter growing cycle which made multi-cropping more feasible. Importantly they were capable of yields two to four times greater than that of the traditional varieties. The success of these varieties depended, however, on the availability of other inputs, notably of water, fertilizer and insecticides. These required access to capital, the consequence of which was that the programme was most successful on the larger and more commercialized farms and least successful in the areas of poor small-holdings where most help was needed. It was not a matter of progressive farmers and conservative peasants but one of money. The implementation of the programme was hindered in its nationwide impact by the incomplete distribution of fertilizer centres and by the inadequacy of provision for credit and loans to adopting farmers. The sub-continent is, of course, cursed by the unreliability of rainfall over much of the area and the HVP came to have its greatest impact upon intensively cultivated irrigation areas and indeed led to an accentuation of regional disparities in agricultural production. The wheat growing areas most successfully adopted the new seed (a Mexican hybrid dwarf wheat). In the rice areas the new varieties were less successful, being very susceptible to pests and diseases and less acceptable in taste. Their shorter straw meant less fodder for cattle. As Chakravarti points out, disparities were enlarged not only at a regional level but also within village communities in which poor low-caste farmers unable to buy the inputs became increasingly disadvantaged compared with their wealthy neighbours who grew the new varieties. Indeed much has been written criticizing this differential effect of the Green Revolution not only in India and Pakistan but also in the Philippines. Not all are critical of the HVP programme and attempt to explain obstacles to its further progress. Rigg claims that the new technologies have been adopted by the poor as well as the rich and every class of farmer has benefited. He argues that increasing population pressure in India has made it difficult for new technology to augment the shortage of land at a sufficient rate (Rigg 1989). Dayal puts the case for the further adoption of HVP being conditioned by favourable factors such as greater commercialization, more irrigation and urban proximity; the last factor ensuring a better flow of information. The influence of commercialization has meant that districts which had already changed from purely subsistence agriculture to partly commercial responded more rapidly to the new technology with no evidence that small farms lagged behind. To promote further adoption of HVP rice, Dayal argues it is necessary to extend irrigation areas and to provide more information and institutional support to small farms (Dayal 1983). The HVP requires further work on the breeding side to produce more disease-resistant and palatable varieties. It has demonstrated the crucial need for ancillary facilities of fertilizer, water and financial help, and these, particularly the need for water, inhibit its application. It has, however, been a development which has increased the production of the staple food grains. To be critical of the high-yielding variety seed programmes, to condemn them because of their flaws, is too easy. A. H. Bunting has put the case in a different way.

Whatever the weaknesses, errors and difficulties which, with hindsight can now be identified, and which in future must be corrected or avoided, the initial decisions were taken by Ministers and governments. One may fairly ask the Western critic one single question: had he been Minister of Agriculture in India or Pakistan, or Turkey, in the mid-sixties, would he have refused to accept the new technology, and if so what reasons would he have given to the public and to the world? And what would the consequences have been?

(Hunter, Bunting and Bottrall 1976).

It is, of course, possible for developments on the macro-scale to be paralleled by efforts at the individual farmer level. Indeed it is important that this should be so. The work carried out by Bradley, Raynaut and Torrealba in Mauritania illustrates this situation and shows what can be done with minimum capital input (Bradley, Raynaut and Torrealba 1977; Bradley 1980). Mauritania is largely desert, has limited resources to expand food production and is obliged to import food. Areas of potential expansion are the alluvium of the Senegal Valley and its levees and a narrow band along the Malian border in the south and east. The valley alluvium is the more fertile and has access to more water for a larger period of the year. The dry land of the south and east can support only rain-fed agriculture of the dry-land type with poor yields on infertile soils. The Government, with financial aid from the World Bank and other agencies, has focused its development activities on the valley of the Senegal River in irrigation projects in which Mali, Senegal and Mauritania are involved. Bradley argues that the dry-land farming of cereals is totally neglected and yet could provide the greater part of the country's grain requirements. He and his colleagues chose to live among the Soninke people of the area and assess how the prevailing system functioned rather than to take in preconceived new methods to promote change. They discovered a system delicately balanced with the environment but disturbed by the emigration of males to work in French factories. Having been accepted into the community they identified constraints and bottle-necks in the system, tested out their ideas with the local community, modified them in the light of the feedback and were able to see them taken up by the local farmers and rapidly diffused throughout the area. Ensuing problems were again handled in the same way and by the end of four years an effective assemblage of new techniques had emerged. The result was an appreciable improvement in crop yields and the even wider adoption of the new techniques by surrounding villages which had not previously been interested. This work, involving the community in feeling its way to developments, is an example of success at the peasant community level. It has stimulated among the Soninke a dynamic attitude to agriculture which is evident in the continuing flow of new ideas and practices which have ensued. Alongside this work and in contrast to it, the Societe Nationale de Developpement Rural in Mauritania has set up pump-irrigated schemes in the river valley. Bradley found the targets set for production and financial return had not been met. He attributes this to the absence of infrastructural developments – physical, such as roads, and commercial, such as marketing arrangements – and to the lack of technical and mechanical support in a scheme where they were essential. All are deficiencies which could be remedied; as it is, this more capital-intensive scheme has proved less effective. Bradley argues that the irrigation scheme promotes too much dependence upon outside help and influences, instead of promoting self-reliance. It is a consideration which underlies all development issues and not solely those of the agricultural sector. Bradley is not alone in his criticism of large-scale irrigation projects; indeed his comments raise the issue of the questionable success of many massive interventions, often organized by the State, in the agricultural sector. The experience has been variable but sufficient examples exist to question the 'top-down' approach. In Mali, on the inland delta of the Niger, Thom and Wells describe a situation very similar to that encountered by Bradley. There flood-retreat agriculture is also practised and the active delta is the focus of dry season grazing. This economy is being disrupted by the schemes of the Niger Basin Authority which emphasize dams and navigation channels and in doing so not only disturb the indigenous agriculture but also will so diminish the dry season grazing as to lead to a reduction in the national herd which is currently the chief source of foreign exchange (Thom and Wells 1987). Adams, working in northern Nigeria, has described the way in which the traditional agriculture of the Sokoto valley, a sophisticated system, has been disrupted over a wide area by the State-constructed Bakolori dam. It is not a case of pleading for the retention of a long-established but inferior system; Adams makes the case that the traditional agriculture is far from static and fossilized and has very successfully incorporated a wide extension of small-scale farmer-managed irrigation. He argues that this was more likely to be able to raise output successfully than the big dam project if it had not been disrupted (Adams 1986; Kimmage and Adams 1990). This thesis is further developed in a comprehensive survey of all the large-scale irrigation projects in northern Nigeria (Adams 1991). With the exception of the Bacita sugar project, all the other six schemes are described by Adams as highly unsatisfactory in all aspects not only economically but also in their social and environmental impacts. The underlying cause would seem to be what Adams describes as 'irrigationism', in other words the emergence of an ideology which regards large-scale irrigation as a key element in agricultural and rural development and in modernization. It is characterized by over-optimistic assessments of the likely benefits and underestimates of difficulties such as unsustainable water supplies. All the schemes were badly conceived and managed but all had the magnetic political attraction of the 'technofix' solution; as with the modernity of the factory or power station, the big-dam project was physically manifest and up-to-date. The less spectacular progress at small-farm level was not only neglected but often considerably disrupted. All were in the thrall of 'irrigationism' whether they were politicians, bureaucrats, outside advisers or homegrown technocrats (Adams 1991). Nigeria's food problems are unlikely to be solved by high technology, centrally administered, expatriate managed and heavily capitalized projects (Okafor 1985). Such schemes in Okafor's view have had no impact on food production capacity or on the performance of peasants; the problem should be approached by giving financial aid to peasants and peasant co-operatives and directing advisory services specifically to their problems.

AGRICULTURAL DEVELOPMENTS AND SECURITY OF FOOD SUPPLIES

In theory an efficient agriculture can exist at all levels and so there is a place for the well-conceived fully costed and efficiently administered large project. Such a range of agricultural developments should aim to secure the nation's food supply by home production or by the trading of exported cash crops for imported food. It should also by exports be able to strengthen the country's economy. Yet many developing countries have failed to achieve this. Undernutrition despite improvements is still widespread. Clearly further agricultural improvements are necessary, particularly in the revitalization of peasant agriculture, so that it not only secures the food supply in rural areas but also contributes to and benefits from economic developments. This takes agricultural issues into the wider context of government policies, the national economy and the influence of the particular environment. The problem of food shortages, malnutrition and starvation which modern communications technology has put before the eyes of the well fed illustrates this.

Agricultural improvements at all scales and on all fronts may seem to be the solution to food scarcity problems but, while vital, they are but a part answer; the root cause as to why so many inhabitants of the Developing World do not have enough to eat is poverty. The elimination of poverty, both rural and urban, must be a priority concern in any attempts to promote social and economic development. The removal of poverty *is* such a development. At a world scale there is more than enough food to feed the world's population adequately and this is despite the fact that the global population is increasing more rapidly than ever before. Yet millions of people do not have access to an adequate or nutritious diet. The analysts describe this situation as lack of of *food security* and distinguish between chronic and transitory food insecurity (Reutlinger 1985). The former refers to a continuous shortage of food either because the resources to produce food or the means to buy food are lacking. Transitory food insecurity refers to a temporary decline in access to food due to such causes as changes in prices, family incomes or factors reducing food production. In its extreme form transitory food insecurity leads to famine. Chronic food insecurity is widespread and many millions experience it. In the 1980s half the people of south Asia, which includes the Indian sub-continent, and over 40 per cent of the population of sub-Saharan Africa experienced chronic food insecurity. In some of the countries experiencing a national chronic food deficit the total food available if it were distributed equitably could almost eliminate the food problem albeit at a low level. Therefore in some nations the problem is that some sections of the population lack the means to buy food, and not that the country cannot produce enough. In other countries, of course, the inability to grow enough food to feed their populations even if it were equitably distributed is the main cause. Bangladesh is one such country and there are many such countries in Africa. In these countries such is the size and growth of the population that an impossible expansion in agricultural production would be necessary to feed it; an increase in exports to purchase food imports would be equally unrealistic. There are no easy cures for this situation. Agricultural improvements to increase productivity from man and land are an obvious step but not everyone can be employed in this way. The parallel development must be a greater investment in rural areas to provide jobs and thus incomes in non-farming activities. Rural development and not solely agricultural development is needed. These twin approaches will take a long time to implement and their results will be slow. It is necessary therefore to put into operation other policies. One such is referred to as income transfer, that is the rationing of food to low income groups at subsidized low prices. This can help but there is always the risk of untitled people getting access to the supplies and reselling them on a black market. Another scheme which has been tried is to give public-sector employment to the poorest groups and therefore wages with which to buy food. A third stratagem is to subsidize the prices of all basic foods. This requires care since it can lead to lower prices being paid to the food producers and thus exacerbate farming poverty. In the case of transitory food insecurity, policies are designed to ensure stability. One approach is to achieve stability in food production. This has led to agricultural projects such as the irrigation schemes in Mauritania to which Bradley (1980) refers and those in northern Nigeria, in addition to agricultural developments at small scales. Stability in food prices is another goal which can be achieved to some extent if large buffer stocks of food are kept to even out fluctuations in supply. Large stocks are, however, notoriously expensive to maintain. Since transitory food insecurity is by definition temporary some economists recommend contingency funds from which to provide the means to buy food in times of scarcity and the institution of export and import levies designed to insulate food prices from international price fluctuations. Food insecurity persists; security can only be achieved by an equitable distribution of the benefits of economic growth.

Agriculture cannot be considered in isolation but must be regarded as an important part of overall economic and social development. The overview by FAO of the state of agriculture in the Developing World at the beginning of the 1990s shows how much still needs to be done (FAO 1991).

Africa is considered as particularly prone to climatic shock since it is so heavily dependent upon agriculture and has insufficient developed irrigation systems to act as a rain-deficit buffer. Population growth is such that pressure on a limited resource base is causing people to leave the land and enter the towns to become food consumers rather than producers. In many countries of the continent economic policies have often been to the detriment of agriculture and agricultural programmes have not been effectively implemented. The whole continent has been afflicted by political strife, coups and civil wars gravely disrupting economies. Within this dismal scene the FAO considers that government politics have often been wrong and many schemes have been characterized by such an overburden of costly administration as to be unworkable. There remains in many African countries a need for infrastructural improvements in transport and communications, for better agricultural extension services and research facilities and a more appropriate education for both children and adults. All these needs call for greater public expenditure which it is crucially necessary to implement effectively. The provision of credit to farmers, the creation of market services and the supply of imports can easily make unsustainable demands on government resources in poor countries. The private sector can help but government involvement in most African countries remains necessary and it is essential that their programmes are rationalized and targeted. There has been some success in implementing programmes through rural communities and non-government agencies. There is a trend in Africa towards market liberalization and this is regarded favourably by FAO provided that attention is paid to the risk of environmental degradation and the distribution of rural assets. Africa with the most rapidly growing of populations is experiencing pressures which not only lead to the fragmentation of holdings, the shortening of fallows without adequate compensatory inputs of fertilizer but also the bringing into cultivation of marginal land, often woodland, and an environmental degradation when sustainable systems are required. Food aid will remain important in sub-Saharan Africa but this does not solve underlying problems; agricultural improvements and the integration of agriculture in a more diversified rural economy are needed and international aid directed to this end would be well spent.

In Asia population growth involves the highest absolute numbers of additional mouths to be fed and hands to be employed. In its developing countries agriculture is already very intensive and further use of the land involves environmental problems. Poverty is widespread and the resource base is diminishing. Although many Asian countries are rapidly industrializing, manufacturing still cannot produce enough employment and again the need is evident for agricultural improvements to be embedded in wider rural development in which the poorer sector can take part. This need to shift investment into the rural economy must encompass not only increased efficiency in the use of agricultural resources including water but also incentive schemes, measures to stimulate agricultural exports including value-added processed agricultural products. Small-scale decentralized rural industries again have a vital role to play.

The countries of Latin America present distinctive agrarian problems. A plan of action was presented by the twentieth FAO regional conference at Recife, Brazil, in 1988 which identified the characteristic urban bias of Latin American development policies and therefore the need to achieve a balanced support of agriculture, industry and services and to do this in rural areas where all three sectors would interact to promote rural development; on the agricultural side particular support needs to be given to small and medium sized farms. All the participating nations endorsed this plan which was an attempt to change attitudes to agriculture which had been regarded in many South American countries as a provider of resources for industry and urban-based developments rather than a source of development itself (FAO 1991; see also Chapters 17 and 18).

The precise nature of the problems faced by agricultural developments and the rural communities of the Developing World varies from continent to continent but there is an underlying common difficulty which they all face: the low priority given to them.

The individual farmer approach which has been discussed has, however, its limitations. It can improve the conditions of the many, it can promote self-reliance and, perhaps, stimulate initiatives but it cannot put within reach the greater opportunities for further economic and social progress which exist. These require a much more profound change in society and economy. It can be regarded, however, as a stage, indeed an essential stage, which so many countries have attempted to pass by, that is preliminary to and lays the foundations for a more broadly based development. Much the same could be said of agricultural and rural development as a whole. The People's Republic of China identified the role of this stage and has used it as a means of egalitarian progress and a platform for further development.

In many countries of the Developing World, the over-reach for industrialization and the failure to conceive of the development of a rural economy, have placed an albatross around the neck of their development plans.

6 Agriculture: the European imprint

'The main influence upon the organisation of agriculture and the distribution of its products is the system of land tenure. Geography and climate are obviously of great importance, but different types of organisation can produce different consequences in the same natural conditions'

J Robinson
Aspects of Development and Underdevelopment 1979

The influence of Europeans, or peoples of European stock upon the agricultural systems and economies of developing countries has been great. It has taken three main forms. First, it created a market for the Developing World's agricultural produce external to the producing countries. Second, in some areas, it took the form of European settlement where Europeans acquired ownership of the land which has remained in the hands of their descendants. Thirdly, in some countries, the European impact resulted from the establishment of the plantation system. European land acquisition and settlement will be discussed in the context of Latin America and of the Kenyan White Highlands. Plantation agriculture will be considered in terms of its role in economic development.

LAND TENURE IN LATIN AMERICA

European settlement took place earliest, and had its most profound and lasting impact, in Central and South America where the Spanish and Portuguese conquered and destroyed pre-existing political systems, claimed title to the land and established colonies characterized by economies and systems of land-holding transplanted from Iberia. The independent countries of Latin America are heirs to these colonies and have shared an experience which has given the region its distinctive agricultural problems which bring to the fore the issues of land tenure in agrarian development (Barraclough and Domike 1966). Latin America is vast, its environments varied and its forms of agriculture many. As a continental area it is lightly populated. In pre-conquest times the small population was essentially peripheral with the greatest concentrations in the Andean valleys and plateaux and along the Pacific coast. The Spanish conquest, primarily interested in precious metals, focused upon this Andean area and made use of Indian labour. Interest in agriculture was secondary and rarely was it developed to produce crops or livestock products for export. Large land grants were made to Spanish settlers and were valued as much for the status they conferred on the recipients as for their productive capacity. The local Indian population was used as labour but the use of land on the *hacienda* or *estancia* was extensive rather than intensive except in proximity to the mining centres and administrative cities. The colonial economy continued to be based on mining rather than on agriculture. Plantations, in the modern sense, did not develop in Spanish America until the mid-nineteenth century when its countries were independent of Spain. The Portuguese in their Brazilian territory were far more concerned with land settlement and agricultural production and cultivated crops, such as sugar, for export. As in Spanish territory, large land grants were made by the Crown.

The conquests in Latin America markedly reduced the indigenous populations and led to the bringing in of negro slaves from West and Central Africa to work on plantations (Odell and Preston 1973). The large estates worked by labourers yielded their owners a satisfactory living because of their size even when worked inefficiently and at a low tempo. There was little incentive to develop an efficient and productive agriculture. In some cases the estate owners rented their land to workers in share-cropping or cash-rent arrangements and, as time went on, a proportion of landowners became absentee landowners. The surviving Indian population was relegated to the more difficult and less rewarding territory where they continued to practise their largely subsistence farming by traditional methods. Indians, Indian–Hispanic people (the *mestizo*) and the mixed negro groups (the *mulatto*) came to be the poor agriculturalists farming small rented or share-cropped plots, or tiny holdings on infertile soils or steep slopes, or became the forest clearers nibbling away at the woodland edge. The division of the agricultural land of so many Latin American countries between a few, wealthy landowners, owning most of the best land, and a large number of small impoverished farmers came about. To the poor farmers must be added the landless agricultural labourers who worked on estates, large and small, and on plantations. Many of these hired labourers lost all knowledge of how to operate a farm on their own account. The history of colonization in Latin America has left, though there are variations and exceptions, a continental area whose agrarian structure is characterized by an inequitable distribution of land, wealth and opportunity for development. Land tenure has become a major obstacle to the progress of agrarian society and the development of more productive farming systems in a region where land is in abundance. It is small wonder that many of the countries of Latin America are characterized by a most skewed distribution of income and by a rural population in a state of unrest (Paige 1975).

The first report of the Independent Commission on International Development Issues, popularly known as the Brandt Report, stated:

> In many countries, there are sharp disparities in land ownership: a minority of landowners and large farmers, often 5 to 10 per cent of the rural households, may own 40 to 60 per cent of arable land. The rest of the rural

populations is crowded on small, often fragmented pieces of land; many own no land at all. In many cases, a high proportion of rural land is held on a tenant or share-cropping basis with the landlord appropriating large shares of the total crop. Such agrarian structures are unjust and inefficient. In some countries, the large holdings are under utilized and output per acre is lower than on small holdings. To reduce poverty and increase food production, agrarian reform and the promotion of farmers' and workers' organisations are priority issues

(Brandt 1980).

It is a comment which appropriately describes the situation in many Latin American countries where much land exists in very large estates, the *latifundia*, worked in the manner described, with the remaining farmed land sub-divided into the tiny holdings, the *minifundia*. In Brazil in 1950 almost 60 per cent of the agricultural land was accounted for by only 4.7 per cent of the land holdings. In Chile in 1955 some 7 per cent of farm units farmed over 80 per cent of the land while in Peru in 1961 just over 1 per cent of farm holdings accounted for three-quarters of the farmed land. This inequitable distribution of land ownership and production was similarly evident in other Central and South American countries though to a less pronounced degree. The converse is the very large number of small farms which account for a very small proportion of the total agricultural land.

Clearly such situations represent obstacles in the way of economic and social development since on the one hand neither the large estates nor the small farms represent an efficient use of resources and advanced farming techniques, while on the other the egalitarian objectives of social development are frustrated. In several countries of Latin America attempts have been made to modify the agrarian structures. It is a difficult process since the major landowners are both wealthy and politically powerful with every reason to favour the status quo. As S. L. Barraclough states, 'most Latin American countries have favoured large landowners and relatively well-off urban groups instead of the small farmers and workers . . . Traditionally-oriented Latin American governments will not normally permit any type of *campensino* (small farmers and farm workers) movement that would seriously threaten established power relationships' (Barraclough 1976).

In Mexico after the revolution of 1910, large *estancias* were confiscated and re-allocated to peasant farmers in communities known as *ejidos*. Most of the *ejidos* are cultivated by individuals in separate plots though the land is owned by the community. In a minority of cases, chiefly on irrigated lands, the *ejidos* are operated as very large farms with the *ejido* members working as labourers on them in what are essentially worker-owned collectives. In the Yaqui Valley, by 1940, about one-third of the irrigated land was worked by collectives which co-operatively owned cotton gins, machinery and service centres. Changes in the Government lessened the support of the *ejidos* and led to their breakdown. In the late 1970s a Government more anxious to promote agrarian reform resuscitated and modernized the *ejidos* (Barraclough 1976). Preston describes similar developments taking place in the Fuerte Valley in the 1950s (Odell and Preston 1973).

Other examples of the promotion of co-operatives to achieve a more equitable sharing both of land and the opportunity for economic advancement are to be found in Peru where, by the early 1970s, some 1700 co-operatives had been created, but Barraclough reports that decision-making by members remained minimal. In Chile the Christian Democrat Government during the 1960s had expropriated a large number of big estates, which covered 14 per cent of the nation's agricultural land, and reorganized them into a form of co-operative called *asentamientos*. These units were jointly administered by government officials and a committee representing the farmers. The *asentamientos* came to be advantaged groups when compared with the bulk of poor peasants and their members came to operate more for individual rather than corporate interest. The Allende regime carried out much more extensive land appropriation taking in 40 per cent of the agricultural area and in addition set up 100 state farms. When the Allende regime was overthrown many confiscated estates were returned to their former owners and on the surviving co-operatives, farmers were encouraged to work individual plots (Barraclough 1976). Similar sequences of events have taken place in other Latin American countries where reforming legislation has been passed but rarely have subsequent substantial modifications of land holding followed. The great exception is, of course, Cuba where the Marxist revolution of 1959 took over all means of production. By 1968, 70 per cent of Cuba's agricultural land was farmed in state-owned and managed units by wage-labourers. One form of estate had been replaced by another.

EUROPEAN SETTLEMENT IN THE HIGHLANDS OF KENYA

A second example of European agricultural settlement in the tropics may be taken from Africa. Over much of Africa, European colonialists did not settle because of climate and disease-risk. Where altitude and location ameliorate the extremes of climate, colonialism in the twentieth century was associated not only with administration, mineral extraction and trade, but also with permanent agricultural settlements, notably in what were then the Rhodesias and Kenya. In the latter, the high altitudes along the equator produced a climate attractive to Europeans and offering a range of agricultural possibilities, from the production of tropical cash crops such as tea and coffee to temperate crops such as wheat and potatoes as well as the rearing of livestock using both temperate and tropical breeds. The Crown Land Ordinance of 1902 gave the colonial government total ownership of all land. This allowed the alienation of specific areas of land for British settlers. By 1914 there were over 1000 settlers in the Kenya Highlands which came to be known as the White Highlands. In 1960 there were 3600 agricultural land holders in the Highlands, owning 548,400 hectares of arable land and 2,193,200 hectares of pasture. In terms of area, the main crop was wheat but other important crops included coffee, sisal, tea, wattle, maize and pyrethrum. Odingo, in a very full account of the agriculture of the Kenya Highlands, describes how the crop cultivation in the area evolved as two distinct systems. One was a large-scale plantation system producing coffee and sisal; the other was a system of smaller-scale arable farming growing wheat, maize and pyrethrum together with some coffee (Odingo 1971). In addition there was livestock farming, since almost all the area was tsetse free, and by 1960 over 1 million head of cattle and half a million sheep grazed on permanent grasslands and cultivated ley grasses. Dairy and beef cattle were kept both on specialist stock farms and on mixed arable–livestock holdings. The ownership and management was in the hands of European settlers. The labourers were hired Africans. There thus emerged an enclave of a productive, efficient, commercial and essentially European agriculture in a country otherwise characterized by traditional forms of African tillage and livestock economies.

In an Order in Council of 1938–39 non-Europeans had been excluded from land-ownership and farming in the Highlands. Subsequent to the Mau-Mau disturbances, this Order was rescinded in 1960 and from 1961 the area was opened to all races and the British Government agreed to subdivide the mixed farming areas in the Highlands and to redistribute the land among peasant farmers. In 1962 over 500,000 hectares were allocated for resettlement. Subsequently very large areas were purchased by Kenyan Africans and Asians from those European owners who left Kenya after independence. These changes in land-ownership were almost entirely focused upon the mixed farming areas; the plantations and ranching areas remaining unaffected by either the resettlement schemes or land purchase. The Government of independent Kenya maintained this emphasis by subdividing the mixed farming areas. Population pressure on the land could be released as these areas were taken up by part-subsistence African farmers. At the same time the large coffee and tea plantations, so vital to the export sector of Kenya's economy, were left, largely British managed and owned, as efficient producers (Paige 1975). As the transitional phase since independence has passed, the Government has encouraged the growing of both tea and coffee, particularly the latter, on small-holdings, so that gradually the dominance of British planters in the agrarian sector of the Kenyan economy has been diminished. It is a transformation which has taken place smoothly; its initiation was very largely the product of political independence.

PLANTATION AND THE PLANTATION ECONOMY

Reference has been made to plantations established in the Spanish and Portuguese conquests of Latin America and the British settlement of Kenya. They are a form of land holding, agricultural organization, production and marketing which has a long history and distinctive features (Goldthorpe 1987). Plantations have, among some, become an emotive topic since they have associated them with a foreign exploitation of land and resources. They still exist in developing countries and in some are of great importance not only within the agricultural sector but also to the national economy.

Prior to the nineteenth century plantations were essentially associated with the Americas. In consolidating their position on the coast of Brazil, and thereby protecting their trade routes, the Portuguese colonial nobility began to organize sugar plantations in north-east Brazil as early as the sixteenth century, coming to use negro slave labour from their African territories. The plantations were all small with several supplying a central cane mill. In the seventeenth century the Dutch acquired for a time the Portuguese territory around Pernambuco and, together with Britain and France, began to take possession of the islands of the West Indies. Subsistence activities by the settlers were replaced first by tobacco and then sugar as cash crops for the European market. Small-holdings were amalgamated by successful planters and large plantations worked by negro slaves became characteristic by the eighteenth century (Angier *et al.* 1960; Courtenay 1980). When slavery was abolished, in most cases by the 1860s, many of the smaller islands of the Caribbean became small-holder and largely subsistence communities. In the case of Trinidad and British Guiana, Chinese and, notably, Indians were brought in as indentured labourers to work on what became increasingly large sugar plantations. In Jamaica, sugar production declined and was overtaken by small-holder production of bananas for large marketing companies, notably those which merged eventually to form the United Fruit Company, which shipped the fruit to the United States. Cuba remained a colony of Spain where sugar was produced by negro slaves on large plantations, slavery not being abolished by the Spanish until 1885, and with large areas of suitable land experienced the greatest growth in plantation sugar production. The system remained after the emancipation of the slaves. Independent Cuba was the recipient of much American investment and its sugar output became tied to the American market by preferential tariff arrangements.

While the nineteenth century saw changes in the old plantation areas and in the scale of the plantation economy, it also witnessed the European initiation of plantation production in Africa, Asia, the Indian Ocean and the Pacific involving a wider range of crops produced for the growing European domestic and industrial market. Coffee, and later tea plantations were established in Ceylon (Chapter 14) and India by the British. The Dutch were involved in a variety of products in Java, particularly coffee, sugar and indigo but including tea, cinnamon, cloves and nutmegs. At a later date they began to produce rubber and cinchona. The British established rubber plantations in Ceylon and Malaya, sugar plantations in Mauritius, Fiji and South Africa, and coffee and tea in Nyasaland (Robequin 1954; Courtenay 1980; Chapters 11 and 14). Plantations were established in other colonial areas. The Belgians in the Congo set up large-scale oil palm plantations, the Americans in Hawaii grew sugar cane, the Germans and French in the Cameroons and Togo established plantations to produce rubber, palm oil and cocoa. In Brazil, where plantations began and where slavery was not abolished until 1888, sugar began to give way to coffee production which became the country's major export crop. Increasingly the coffee plantations were established in the more temperate southern parts of the country on plateaux in the State of Sao Paulo. The labour force was provided by European immigrants largely from Italy.

By the Second World War, plantations and plantation economies were to be found in all the continental areas of the Developing World and though their histories had been varied, their impact in colonial territories had been considerable and they came to share a number of common attributes. With some exceptions they overwhelmingly bore the stamp of the industrial nations which operated them and have, indeed, been described as industrial forms of agriculture. In countries otherwise characterized by small-scale farming, they were large units manned by wage-labourers. They tended to be set up in sparsely populated and little cultivated areas and represented, therefore, additions to the farmed land. In the majority of cases it is incorrect to claim that they took land out of peasant food production. Shortage of local labour was a characteristic problem which plantation estates faced. Local farmers were either too few or saw no advantage in working on the plantations or were considered, in some cases, unsuitable. Whatever the reasons the large plantations of the world were worked by immigrant labour. At first, as has been seen, they were slaves from Africa. From the mid-nineteenth century they were usually hired labourers often indentured for a period of years after which they could return home passage-paid. They came overwhelmingly from the densely populated countries of the Indian sub-continent and China. In Ceylon, Indian Tamils from south India were used on the tea-plantations. Indians were shipped to Mauritius, to Fiji, to Trinidad, to British Guiana and to East and South Africa. Indians and Chinese were used in Malaya, Africans in South America, the Caribbean and Mauritius. Chinese, Japanese and Filipinos were taken to Hawaii. In India itself the tea plantations of Assam were worked not by local inhabitants but by the peoples of the Ganges plain and peninsular India. The movement of workers in this way, most of whom remained in the plantation countries, was on a large scale and has had not inconsiderable consequences for both race-relations and economic development in subsequent years. The plantations were not only large-scale implants of new forms of economy, they also constituted enclaves of alien people.

All the plantations produced crops in demand in the industrialized countries and while the market was not entirely external it was overwhelmingly so. Mono-cropping was characteristic and scientific methods of cultivation were applied utilizing fertilizer and insecticides together with the selective breeding of plants to maximize production, disease resistance and ease of processing. None of the crops were basic and essential staple foods and though they included food crops such as bananas and sugar cane, the non-essential commodities such as tea, coffee and cocoa became important plantation products increasingly consumed in Europe and North America. With the introduction of the motor vehicle and the invention of the pneumatic tyre, rubber became a dominant plantation crop in south-east Asia in the twentieth century. Palm oil, copra and coconut oil (both foods and industrial materials), sisal, cinchona, cinnamon and pyrethrum have all been cultivated on plantations as well as small-holdings. Many of these products require one or more industrial processes before they are exported and the necessary industrial plant operates most efficiently if it can achieve economies of scale and a year-round rather than a seasonal harvest function. Tree products which yield throughout the year lend themselves to this requirement, notably tea, rubber, coconuts and oil palm. In short, plantations were, and many still are, large, foreign-owned and managed, funded by external capital and worked by paid labourers who were immigrants or descendants of immigrants. Their cropping is scientifically based and managed, is specialized and is commonly concerned with only one commodity; there are exceptions such as the estates in Malaya which grow both rubber and oil palm. The majority of plantation products are exported and often their profits repatriated. It is an economy which can be insulated from its host country. In newly independent countries this poses problems.

It is possible to argue a case for the positive contributions of plantations to the development process in developing countries. They introduce new ideas of how to manage soils and crops, of insecticides, fertilizer and cultivation techniques. They link the host country to an external market and hence can earn exchange revenue. The wage-labour they employ itself creates a local market for food and consumer products and exemplifies the ideas of the subdivision of labour and its organization. In its activities the plantation invests in infrastructural developments such as roads, railways, docks, warehouses and factories. It builds settlements to house its workers, schools for their children and medical facilities for all. It becomes a source of scientific advice on crop husbandry and of economic advice on business administration. These positive benefits are manifest in most regions of plantation agriculture. There is, however, another side to their operation and this has become more evident during the twentieth century as the companies operating plantations have become much larger and more complex.

G. L. Beckford has drawn attention to the 'plantation economy' practised by multinational companies, an economy which doubly divorces them from the national economies of the countries in which they operate (Beckford 1969, 1972). Plantations can be operated as part of the function of large, and often multinational, companies. Reference has been made to the associated manufacturing element such as drying, blending and packing in the case of tea, of milling, refining, re-refining and packing in the case of sugar, plus the processing of the by-product molasses into rum or other alcohols, while with the growing of coconuts there are the necessary processes of copra production and drying, of oil milling and of copra-cake production. This has led to these many activities becoming vertically integrated both technically and economically within a single company. In addition the products require transport and marketing and again these services have tended to be handled within the company; indeed in the case of some companies, of which the United Fruit

Company is an example, their initial involvement in the plantation economy was as a carrier rather than a producer. The integration has thus been backward as well as forward. These vertically organized companies link not only the plantation products of the tropics to the markets of the industrialized world but also integrate the economies of the producer countries into a wider multinational company economy. Beckford argues that even after political colonial control was relinquished and with it, in many cases, land-ownership, the economic relationships have persisted. He claims that the independent countries of Central America can be regarded as extensions of the economic system of the United States, and the sugar plantations of the Caribbean as part of an economy which absorbs them within that of the United States and the United Kingdom. In some countries the role of plantations constitutes such a major part of their national economy that this becomes very much influenced by the activities and fortunes of the plantation company. Where this company is large and multinational in its operations, the individual country's economy becomes but part of that larger operation (Beckford 1969, 1972; Heady 1952; Jones 1968). Large companies, such as the British-based Tate and Lyle, Lever Brothers and Booker Brothers, McConnell and Company and the United Fruit Company, are not only characterized by vertically organized functions but also by a geographical spread of operations with their farming activities taking place in several countries. Beckford quotes, for example, the activities of the firm Booker Bros, McConnell and Company as involving the growing of sugar cane, the manufacture of raw sugar, the distillation of rum and other spirits and alcohol, and the manufacture of machinery for sugar making. In addition the company provides bulk storage for sugar, owns wharfs, handles the ocean transport of sugar, acts as shipping agent and insurer, and as merchant for agricultural equipment. This company produced over 80 per cent of Guyana's sugar in the 1960s and accounted for the greater part of that country's exports. It was the biggest employer of labour and owned all the sugar storage and shipping facilities. Essentially this company dominated Guyana's economy; it operated also in Barbados, Canada, India, Jamaica, Malawi, Nigeria, St. Lucia, Trinidad and also Zambia though largely in a non-agricultural sense. The Tate and Lyle company is much bigger and more complex and is involved in sugar plantations in Jamaica, Honduras and Trinidad, with sugar refining in the United Kingdom, Canada, Zambia and what is now Zimbabwe, and with sugar storage and distribution, molasses products and other activities in Africa, Asia, the Caribbean, North America and Europe. In the 1960s, it possessed eleven sugar ships and handled 65 per cent of sugar refining in Britain and 30 per cent in Canada (Beckford 1969, 1972). The United Fruit Company is much concerned with the banana production of Central America and their import into the United States. In 1966 it handled 55 per cent of all bananas imported into the USA and controlled all the area devoted to growing bananas for export in Guatemala as well as 70 per cent of that in Costa Rica and Panama and 56 per cent of that in Honduras. Of the fruit it handled, 47 per cent was grown on its own plantations, the rest coming from other growers.

Each of these firms is thus involved, to a major degree, in the economies of the developing countries in which they operate. Their range of activities extends beyond these individual countries and in their total activities the agricultural section, the plantation, does not represent the major part of their investment. The overall economy of the company relates to all its activities and not simply to agriculture, and since the firm's decisions relate to the promotion of an economy which spans several countries, the company's view of its economic development may not necessarily be coincident with that of a particular country in which it operates. A plantation may be well-run and efficient and yet its contribution to the social and economic development of its host country may be limited. The company will look, naturally, at its allocation of resources of land, labour and capital in terms of the functioning of the company and, since it functions in several countries, it will view the separate components in its integrated operation in an international rather than individual country sense. It will not be concerned with the question, in terms of comparative advantage, of whether a particular country might better use the resources in growing rice rather than sugar or bananas, or directing investment to the manufacture of shoes or transistors rather than raw sugar or rum. Profit maximization within the company will be related to all its operations and not solely its agricultural sector. It can draw the raw materials from a range of sources and not only from its own several holdings but from other supplies. A low price for raw sugar may reduce the returns to the farming function but means a low cost for the raw materials of its refining and distilling activities. Thus in this hypothetical situation, while the company's profit may be up, the beneficial spin-offs of its plantation activities in its host countries may be reduced. The risk of harvest failure in an individual country is cushioned for the company by its involvement in several others, though the stricken country suffers, while in less dramatic but equally significant terms, the large multinational company expands or contracts its activities as it sees its own best interests served, which are not necessarily those of the host country.

These features are not peculiar to multinational plantation companies but can be the product of any transnational activity. What is particular to them is that their contribution to economic development, though it can be considerable, is constrained by the nature of their operations. Not only are they concerned with products which are not staple foods and hence do not directly contribute to alleviating food shortages, but with methods and processes which are very specific to the crops grown. The technical spin-off is limited whether it relates to agricultural machinery, plant breeding or industrial processes. The vertical integration reduces the opportunity for ancillary supporting industries to develop so again the multiplier-effect is constrained. Even the exchange revenue earned by exports may be offset by the need to import food and equipment to sustain the work force and the plantation. While they provide employment opportunities for the local population it is largely as unskilled labourers, the managerial, scientific and technical functions remaining mainly in the hands of expatriate employees.

The greater part of these disadvantages are the product of the multinational nature of the large foreign plantation companies such as those described by Beckford. As a farming system it can have much to recommend it in its role in agrarian development. It encompasses techniques and concepts which can do much to promote agricultural productivity. If it is considered in terms of the development of plant breeding, of fertilizer and insecticide practices, cultivation techniques, land conservation, labour organization, the quality control of output and processing, and the development of management and entrepreneurial skills, then the plantation by its scale and its semi-industrial nature has a place in the economy of developing countries. If it can attract supporting industries and services it can further galvanize developments. In some developing countries former foreign-owned plantations have been acquired and subdivided among small-holders or operated by state or private local companies. In other countries, agricultural development projects have been implemented which are, in everything but name, plantations. The plantation can play a part in the transformation of the agrarian economies of the Developing World.

7 Industrialization

'Your worship is your furnaces
which, like old idols, lost obscenes,
have molten bowels; your vision is
Machines for making more machines'

G Bottomley (1874–1948)
To Iron Founders and Others

'successful development in virtually all countries has been characterized by
an increase in the share of manufacturing in total output. This structural
change, is both a cause and effect of rising incomes.'

H B Chenery *et al.*
Structural Change and Development Policy 1979

The countries of the Developing World, many of them former colonial territories
recently independent, saw themselves in the years after the Second World War
as markedly different from the nations of Europe and North America in two
central ways: they were poorer and less industrialized. The way forward, in
planning their own destinies, seemed to many to be the machine, the factory and
the power station. Industrialization was the way to a secure and less dependent
future, to the creation of wealth and the removal of poverty. Both politicians
within, and economists without, the developing countries regarded
industrialization as having the key role to play in development planning.
Kwame Nkrumah, President of Ghana, wrote of a 'rapidly increasing
industrialization that will break the neo-colonial pattern which at present
operates' (Nkrumah 1965). In 1967 the economist D. W. Jorgenson was stating
'The process of economic development may be studied as an increase in income
per head or as an increase in the role of industrial activity to that in agriculture'
(Jorgenson 1967). Others have argued that industrialization is essential to absorb
the 'surplus' population of the underemployed and unemployed of the agrarian
economies of the tropics. These views have raised considerable debate and in the
light of experience and with the wisdom of hindsight they can be evaluated, but
first it is useful to examine the bases for industrial development present in the
mid-twentieth century.

There are strong parallels in indigenous industry with the agricultural sector.
Indigenous manufacturing fits into the same pattern of social and economic
relationships as those of farming. Capital input is minimal, technological
sophistication is absent, inanimate power is rarely used and the scale of
production is small. The major inputs are labour and skill, the skills once

93

possessed by the craftsmen of Europe and now largely lost. Some developing countries, and China and India are notable examples, represent the last great reservoirs of craft skills. The level of production is, in the main, geared to local demand, and local raw material resources too small to sustain a large modern manufacturing plant take on a special significance. The sporadic occurrence of rarer minerals or crops related to specific environments gives rise to regional specialization and trade but essentially, industrial production is concerned with disaggregated local economies. Demand is the other dimension and it is an expression of the way of life in terms of habitation, of domestic utensils, of clothing and foodstuffs, and of agricultural implements and craft tools. The demands made upon industry are limited by this way of life and industry has grown up and responded to it. Manufacture has been attuned to agriculture and integrated into society throughout the cultures of the Developing World. In such societies the demand for non-agricultural goods refers not only to family and household requirements in the home and the field but also to the non-essentials such as jewellery and ornaments which can identify status and give expression to individuality. Many of the raw materials are the products of agriculture such as oils, fibres and leather, or of the forest, while the rest are common minerals such as clay or lean ironstones.

In Africa, the least industrialized and urbanized of continents, these characteristics of indigenous local industry are still widely evident. The towns and villages of West Africa exhibit such a range of domestic industries. Hand spinning and weaving using cotton, camel hair or wool fibres produce cloths for clothing and bedding. Clothes are embroidered in traditional designs, some indicating class and status (Heathcote 1972). Fabrics are coloured with vegetable dyes by specialist dyers. The products based on agricultural materials serve a general demand with some degree of market sophistication. The cattle, goats and sheep, which are slaughtered for their meat, yield hides and skins which by elementary forms of tanning are converted to leather for saddles, harnesses, cushions, belts and shoes in a sequence from primary processing through to skilled making-up and the application of artistic talents. Lean, lateritic iron ores worked from shallow surface pits are smelted at low temperatures and beaten to a purity which enables iron worked by village smiths to produce tools for the home and agriculture including hoes of sophisticated designs. Brass and copper working produces prestige goods such as the brass pots given in marriage dowries; potters, either resident in large settlements or itinerant workers serving many small villages, produce more mundane vessels from local clays. Carpenters make furniture and vehicles from local, not imported, timbers. Food processing, producing commodities such as groundnut and palm oil, and yam flour, compose another group. The essential salt is manufactured wherever possible, and is traded over long distances where not possible; soap manufacture also uses an amalgam of local materials. Beads, bracelets and musical instruments further diversify production. Some of these activities are recognized artisan skills practised by families through which the skills are passed from parent to child. Some are protected by trade guilds. Many are part-time activities practised by farmers during slack periods in the agricultural year. These activities are as characteristic of towns as of villages. Mabogunje reported a similar spectrum in Ibadan, the largest indigenous city in sub-Saharan Africa (Mabogunje 1968).

THE COLONIAL IMPACT AND NASCENT INDUSTRIALIZATION

What has been described, though there will be differences according to environment, culture and historical experience, can be found throughout many countries of the Developing World. It still survives but it has been and is being profoundly altered by external influences. As in the case of agriculture, the colonial experience and the impact of the industrialized world have made their mark as the localized equilibrium of agriculture and industry, supply and demand has been shattered by competing supplies and demands. Industrialization outside the Developing World rather than within it has profoundly altered its economies and this took place before independence opened the door to planned developments. The increase in the capacity to manufacture which Europe experienced in the nineteenth century created a demand for raw materials from the tropics, as has been seen in the case of agricultural products but it, of course, also included minerals in short supply in Europe. An enlarged market for European manufacturers also became necessary. These manufactures were often superior to goods made by the domestic industries of the Developing World and increasingly replaced them so that what survives is but a relic of a former economy. In West Africa the smith turns from the manufacture of hoes and adzes to the repair of bicycles or other imported machines. The tinker uses as his materials the scrap of imported tin boxes and cans. The wood-carver becomes a furniture maker and repairer. The weaver buys European or Far Eastern fabrics, invests in a sewing machine and becomes a tailor. The non-agricultural spectrum is enlarged not so much by manufacturing as by service industries, as imported watches, vehicles, printing machines, electrical appliances and a host of other machines are serviced and repaired. In Ibadan thirty years ago there were only forty-seven industrial establishments employing over ten persons and of these in only nine did employment exceed one hundred. The larger firms included soft drink manufacturers, tyre treading and furniture makers (Mabogunje 1968).

Some suggest that the mass of small industrial enterprises, present not only in Africa but throughout the Developing World, represent the fertile, bubbling concoction out of which industrial society grows, arguing that the scene of domestic, artisan industry described was characteristic of pre-industrial Europe. Is this the breeding ground of industrialization in the Developing World? It may be in terms of individual entrepreneurial aptitudes but it cannot be as a system. The parallel with the birth of the mechanical factory industry in Europe, so dear to economic historians, omits one crucial component, namely the contemporaneous existence of the Industrialized World vastly more possessive of a present productive capacity let alone a future one. It is the need to encompass this situation which presents the industrialization of the Developing World with a major problem; it also presents an opportunity for inter-linked development.

The amount and character of industrialization varies throughout the range of developing countries. Some are characterized by industrial developments initiated by the industrial nations during the colonial period. These are most commonly associated with the extraction of minerals and represent the industrial equivalent of the plantation. Examples in Africa are the copper mines and smelters of Zaire and Zambia and the tin mines of Nigeria dating from the early years of this century. In Asia the placer deposits of tin in the Malay peninsula and Sumatra, and in South America the tin, lead and silver of Bolivia and Peru and the copper ores of Chile are further examples. It will be noted that these examples are characteristically of relatively high-value metals, the value of which could help offset the remoteness of their locations from the processing industries which would use them. They were also minerals that were in increasingly short supply in Europe. Like the plantation, the mining industries, their associated smelters and the settlements of their workers, became enclaves of modern industrial technology in otherwise non-industrialized countries. Their subsequent development has had, however, in many cases, profound repercussions upon the development paths of the countries in which they were located (Chapter 10).

Manufacturing industry in colonial territories was inhibited rather than encouraged since it could serve to reduce the market opportunities open to the industries of the colonial powers. None the less in some countries European techniques of manufacturing and factory methods of production were adopted and took root. In Brazil, where there had been a considerable influx of Europeans to establish and work coffee plantations in the late nineteenth century, the development of coffee production and marketing not only raised capital but provided commercial and managerial experience upon which other industries could be based. Characteristically they first produced consumer goods, notably textiles. The manufacturing activity was then extended into iron-making, using local raw materials but in a modern high-technology manner, and into engineering. In the Indian sub-continent domestic cottage industry dominated the industrial sector throughout the nineteenth and early twentieth century. In an attempt to encourage small industrial developments, protective measures were enacted in 1922 to allow Indian textile factories to develop, modernize and grow, and to allow the iron and steel industry, using the coal and iron deposits of the Damodar–Jamshedpur area of the peninsula, to begin to supply the local demand for iron and steel manufacture. On independence in 1947 both India and Pakistan were still, however, essentially agricultural countries.

The colonial episode did stimulate economic development, including that of manufacturing, in many countries though the effect varied considerably. Capital and expertise were injected and methods for the large-scale utilization of the physical resources of agriculture, timber and minerals introduced. Factory organized wage-labour systems were set up in some countries and enlarged the size of the domestic market while, importantly, concepts of economic administration and business management were implanted. Some of these introductions withered; others such as the infrastructural developments of road and rail systems and dock installations remained. But overall the economic developments of the colonial era bear the mark of their external origin and focus. The transport development was essentially designed to extract resources and products and transport them to the new ports for shipment to the markets of the industrialized nations. The railway map of Africa demonstrates this most clearly with its series of isolated railway systems connecting interior to the coast but rarely country with country. It is a physical expression of economic developments skewed by a particular intent (Taaffe *et al.* 1963). Capital formation took place outside the developing countries, and ancillary developments, as with plantations, were restricted if they had advanced technological requirements. Indeed, most industrial developments were manifestations of the plantation type of economy; they were designed to further and sustain the industrial economies of Europe, North America and, latterly, Japan.

INDEPENDENCE AND THE LURE OF INDUSTRIALIZATION

Manufacturing industry, the missing element which seems to distinguish the rich from the poor nations of the world, presents a development problem different from that faced by agriculture. In its scale, its technological requirements, in its organization and concept of management, it is something so very different from what has gone before that it can be regarded as something totally new. It is an addition to, rather than a modification of, pre-existing economies. For all these reasons it has been put to the fore in developing strategies by most governments either in state enterprises or in their encouragement of private endeavour. In the context of political economies, industrialization is the banner of modernization and to the new politicians of the recently independent countries it appeared to have four outstanding attributes which could be proclaimed in speech and writ large on banners. First, that it is the missing element which, when introduced, will disrupt the equilibrium between subsistence agriculture and cottage industry. It is a thing of the new future rather than the stagnant past. Second, it can provide goods either directly or indirectly for the enlargement of material possessions. It is the way to wealth. Third, it can provide employment for the unemployed and underemployed. Fourth, it is the way to free developing countries from the yoke of the Industrial World, from the 'neo-colonialism' of which Nkrumah (1965) wrote. These are powerful notions which catch the imagination and offer hope; their translation into practice has proved a difficult matter.

The countries of the Developing World possess in varying degrees the factors of industrial production, raw materials, labour, energy resources and capital. These relative factor endowments will condition the form which their development path, including industrialization, takes. These endowments will have been influenced by their particular historical experience, including colonialism. The many countries in the Developing World differ from each other not only in their natural resource endowments but also in their size, their access to capital from outside the country and in their political ideologies, social goals, and attitudes to economic and social planning. The consequent opportunities, and problems, will thus vary considerably for industrialization as for agricultural development and the past three to four decades have illustrated this. Despite its inauspicious beginnings amidst inadequate social and physical infrastructures, capital deficiencies and untrained work forces, industrialization has been taking place to some degree in all developing countries. Some have become fully industrialized nations, for example Hong Kong, Singapore, South Korea and Taiwan, for which the acronym NIC, Newly Industrialized Country, has been coined. Others have a large industrial component in their economy and can be described as semi-industrialized. Argentina, Brazil, Chile, Mexico and Uruguay would seem to fit this description. Both India and China despite their huge agrarian populations which define them as agricultural, have developed large industrial sectors incorporating all stages of manufacturing.

The economist, Hollis Chenery, has for over thirty years been involved in the investigation of structural changes in the economies of developing countries and has attempted to establish what could be regarded as 'normal growth functions'. By examining a range of data from a large number of countries using multiple regression analyses, he suggested that at per capita annual income levels of US $100 (1950s values) the contribution of manufacturing was 12 per cent but that rises in per capita incomes were associated with an increase in manufacturing so that at levels of US $1000 per capita the characteristic proportion of manufacturing in the GDP was 33 per cent. The contribution of primary production fell over the same income change from 45 to 15 per cent of GDP. Furthermore Chenery was able to correlate changes in the nature of the manufacturing industry with increases in average per capita incomes. In a country of some 10 million people with per capita incomes at the US $100 level most manufacturing (65 per cent of output) would be concerned with consumer goods and only 20 per cent with intermediate goods, that is goods used to make other goods such as rubber or chemicals. Manufacturing devoted to the production of investment goods such as the making of machines to manufacture products or generate energy or the production of metals such as iron and steel, would account for the remainder. As per capita income rose this would be associated with a much smaller proportion of manufacturing concentrating on consumer goods, 43 per cent, while the basic investment goods would account for 35 per cent and intermediate goods for 23 per cent (Chenery 1960). Chenery and Taylor then took the analysis further and categorized countries into three types: large countries whose populations exceeded 15 million, small countries whose export trade was largely concerned with manufactured goods and small countries where exports were of the primary products of farm, forest and mine (Chenery and Taylor 1968). In both the large countries and the small industrialized countries the patterns of industrialization were similar, with the industrial share of GNP rising rapidly from 16 per cent of GNP at incomes of US $100 to 32 per cent at US $400. Thereafter, the rise in the share of industry was much slower, reaching 37 per cent at US $1200. Primary production's contribution fell from 45 per cent at the US $100 level to 12 per cent at US $1200. In the third category of country, the small primary export nation, the changes were different. Primary production remained more important than that of manufacturing up to the US $800 income level, falling from 50 per cent at US $100 to 30 per cent at US $500 while industry rose from some 17 to 19 per cent. Chenery and Taylor's analysis covered fifty-seven countries of which thirty-nine were developing nations. Fourteen were of the small primary export type in which industrialization is likely to present the greatest problems and twenty-five were developing countries in whom, it would seem, industrialization has a very positive contribution to make to development.

Chenery has extended his work still further and in 1975 published a wide-ranging statistical analysis of development patterns and processes over the period 1950 to 1970 (Chenery and Syrquin 1975). He and Syrquin were attempting, by examining the experience of over one hundred countries, to identify universal patterns of economic structural change which are associated with rises in per capita income in what has been described as 'modern economic growth'. The development of the industrial sector is an important part of that structural change. Its precise character reflects not only the abundance or scarcity of the factor endowments of raw materials, energy resources, and labour supplies in particular countries but also the development path which they have identified. Chenery has subdivided the three categories of country he defined in 1960, the large, the small industrialized and the small primary producer, according to the development strategy the particular country has adopted. These are primary specialization, import-substitution, balanced production and trade and lastly industrial specialization. The categories are based on coincidences with or deviations from normal average patterns of production and exports with the data relating to 1965 (Chenery *et al.* 1979). It will be apparent that involvement in trade is an essential characteristic of these strategies though the amount and the nature varies considerably. It reflects the alternative decisions of nations either to choose an absolutely, or relatively, self-contained existence, a choice normally open to large countries with a rich and varied resource base, or to opt for participation in the world economy with all that that implies in terms of imports, export opportunities, loans and foreign investments. In the political context of newly acquired independence, the majority of developing countries sought to achieve some degree of economic independence in terms of industrial goods from the industrialized countries. Some with the essential resource base attempted to lay the foundation of a vertically-integrated manufacturing structure and established capital-intensive basic industries such as iron and steel, heavy engineering and heavy chemicals. This was similar to the pattern used by the Soviet Union and India followed it (Chapter 16). Others, the majority, sought to establish import-substituting industries, producing consumer goods essentially for the home market but also generating capital which could be used, where possible, in a backward extension of linkages through intermediate to basic industries; the reverse of the Soviet model. Import-substitution appeared both from political and economic viewpoints to be the way to spearhead industrialization.

IMPORT-SUBSTITUTION, INDUSTRIALIZATION AND PROTECTIONISM

A degree of economic independence from the industrial nations was regarded as a means of both demonstrating the new political independence and of reducing the drain on scarce resources. Most developing countries were primary producers and the export of primary products was their major, often only, source of exchange needed to energize economic development. The import of manufactured goods from abroad not only limited this opportunity but also could, and often did, lead to crises in the balance of payments. Two concerted actions were open to countries in such situations. First, industries could be set up to produce goods at home to replace some of those imported and, second, tariffs, quotas and other measures, could be implemented to keep out a wide range of 'non-essential' manufactured goods and at the same time protect the infant home industries. This need for protection has often been increased in developing countries because they have over-valued their currencies. This meant the imports appeared as particularly low-cost competitors to home products. It continued to be the case of all imported commodities not covered by the protective measures. These could include raw materials and components used in the manufacture of 'inessential industries' and so their use would be encouraged. Griffin and Enos, who take a jaundiced view of the efficacy of import-substitution, point out that in such situations home industries develop production techniques based upon imported materials with the result that a greater use is made of imported inputs. Protecting the balance of payments may result in both the development of 'inessential industries' (i.e. industries which are considered of low priority either in laying the foundations of economic development or characterized by a low industrial linkage propensity) and a greater use of imports. So while the output of home-produced goods increases, being based upon imports it uses exchange revenue and does much to defeat the purpose of import-substitution. They argue that most new industrial ventures in developing countries are markedly less efficient than their counterparts in the Industrial World. In consequence the ratio of inputs to outputs is greater and their use of the imported raw materials much more extravagant. The net result is that, in money terms, the import-substitution effect is very low (Griffin and Enos 1970; Bernstein 1973). There is another possible consequence. Devoting a great amount of investment capital to the import-substituting consumer goods industries means less is available to establish capital goods industries. If these basic industries do not exist, or cannot expand, they are able neither to serve the import-substitute industries nor to initiate the establishment of any others. Capital goods have, therefore, to be imported and increase import costs still further. It is true, of course, that this investment in imported capital goods will, in the long run, help sustain output. Indeed it can be argued that is preferable to invest *ab initio* in capital goods industries and in manufacturing designed to produce competitive exports rather than to encourage a large domestic demand for consumer goods which, since it cannot be met adequately from home industries, will maintain a taste for imported items.

In the 1950s a large proportion of developing countries chose the path of import-substituting industrialization and protectionism. Behind tariff barriers they further assisted the new home industries by granting them tax exemption and providing capital loans at low interest rates. In some countries the state was directly involved, in others private enterprises, and in most a mix of the two. Protectionism it was felt would allow the new industries, with their inexperienced management and workforce, to establish themselves securely, iron out difficulties and become increasingly efficient so that they could eventually withstand the competition. This approach was adopted both by countries following the consumer-goods approach and by those taking the basic-industries road. Its effectiveness in allowing the development of efficiency proved to be a far from unqualified success. Griffin and Enos quote examples from Ghana where in 1964 twenty-two out of thirty-one state industries not only failed to make a profit but achieved massive losses, from Chile where the privately owned motor vehicle industry in the 1960s was grossly inefficient in its use of resources, and from Pakistan where inefficiency was present in a large proportion of industries (Griffin and Enos 1970; Islam 1967; Johnson, H. G. 1964; Johnson, L. L. 1967; Soligo and Stern 1965). The evidence appears to indicate a greater tendency to inefficiency in the consumer-goods sectors. It should be pointed out, however, that initial inefficiencies are to be expected; the question is the extent to which they are unduly prolonged by excessive protection. Continuing inefficiency and hence high unit costs together with an over-valued currency act as positive disincentives to the export of manufactures.

DEVELOPMENT PATHS OF INDUSTRIALIZATION

Chenery's classification of development paths, to which reference has been made, provides a convenient summary of the progress of industrialization and structural transformation in developing countries (Chenery *et al.* 1979). He points out that most countries begin by specializing in the export of primary products which result from their natural resources. Of the countries he analysed, those who remained in this category from 1960 until 1975 were either small countries or those with a particularly valuable mineral to export. The exceptions were the three oil exporters, Nigeria, Iran and Indonesia. In other countries the phase of primary export specialization had been succeeded by other developments; indeed Chenery describes primary specialization as a 'strategy of deferred industrialization', i.e. it is used to accumulate investments and to raise income levels so as subsequently to allow industrial developments. Clearly the rate at which this takes place will depend upon the size and value of the primary resource, the world demand for it and the ability to re-invest the derived income effectively. Malaysia and the Cote d'Ivoire are examples of countries who have been particularly successful in stimulating industrial growth by investing income from agricultural exports.

The second of Chenery's categories was that of balanced development. This essentially represents a combination of primary export and a manufacturing development which has not resulted in industrial specialization. Within this group Chenery distinguishes a sub-category of nations which have opted for a deliberate policy of import-substitution. He writes:

Over time these policies lead to relatively low levels of exports, diversion of resources from agriculture, and ultimately a slowdown in the growth of industry and of GNP as the possibilities for import substitution are progressively exhausted. Because of this market limit, the strategy of inward-looking development usually succeeds in eliminating the specialization in primary production but not in achieving manufactured exports

(Chenery *et al.* 1979).

Large countries, that is those with populations exceeding 15 million in 1965, have tended to be those which have persisted with import-substitution and to have been more successful with it because of their larger home markets which offer scale-economies to scale-sensitive industries. For other countries it does not appear to have been a self-recommending policy. The other balanced-development countries though engaging, at least initially, in some import-substitution activities, have retained a more open economy and with it a much greater incentive to export manufactured goods.

In the third category of industrial specialization are found countries which, having initially in some instances been concerned with some primary exports and indeed with the home market, have sought to manufacture goods for the international market and to seek their salvation in trade. A limited resource base has, of course, stimulated this course of action in a number of instances. This path requires the assemblage of entrepreneurial abilities, industrial skills and adaptive labour force which characterizes the industrial nation. It is not a package which is easily put together. Hong Kong and Singapore, notably deficient in natural resources, together with Taiwan have been equally notably successful.

Fifty countries in Chenery's analysis are depicted in Table 2. It is important to note that this classification is of policies and not of attainment. It therefore does not mean that Kenya, for example, is to be regarded as an industrial country. It is suggestive of a ladder of structural change and development though it does not purport to cover all dimensions of development. The majority of developing countries by the end of the 1980s still employed less than a quarter of their work-force in manufacturing; in many cases much less. The proportion has, however, continued to increase though the rate varies greatly geographically. The east Asian and western Pacific regions stand out as the most rapidly industrializing areas of recent decades and sub-Saharan Africa as the slowest. The proportion

Table 2   The Chenery classification of development strategies

Primary specialization	Balanced development Import substitution	Other balanced	Industrial specialization
Algeria	Argentina*	Costa Rica	Egypt*
Bolivia	Brazil	El Salvador	Hong Kong
Dominican Rep.	Chile	Greece	Israel†
Indonesia*	Colombia*	Guatemala	Kenya
Iran*	Ecuador	Ireland	Lebanon
Iraq	Ghana	Jamaica	Pakistan*†
Ivory Coast	India*	Morocco	Portugal†
Malaysia	Mexico*	Peru	Singapore
Nicaragua	Turkey*	Philippines*	South Korea†
Nigeria*	Uruguay	South Africa*	Taiwan
Saudi Arabia		Spain*	Tunisia†
Sri Lanka		Syria	Yugoslavia*
Tanzania		Thailand*	
Venezuela			
Zambia			

Source: Chenery et al. 1979
*Population exceeding 15 million in 1965
†Countries in which the net resource inflow as a share of gross domestic product was high, i.e. a high capital inflow

of GDP derived from manufacturing confirms this pattern and reveals Africa as the least industrialized of the continents (Fig. 19). The proportion of export revenue earned by manufacturers is another index and also reveals the increasing industrialization of the Developing World. Contrasts remain; in the high-income countries 81 per cent of exports are of manufactured goods compared with only 32 per cent in the low-income economies, excluding India and China (see Table 4). It is of interest that in the high-income economies, thought of as the world's industrial nations, the service industries employ more than manufacturing industries and contribute more to the GDP. This is suggestive of a further stage of structural transformation which is in part a result of the efficiency of capital-intensive means of production and raises the question of appropriate technology if manufacturing is to be regarded as a means of providing more employment and thus reducing poverty in developing countries.

In a large number of developing countries manufacturing remains a little-developed sector. As Hughes has written 'the most rapidly growing developing countries have been those that *without neglecting other sectors* [present author's italics] have achieved the most efficient and rapid growth in manufacturing industries' (Cody, Hughes and Wall 1980). Countries which have been most successful in their industrialization have been so for a variety of reasons. The existence of the group of rich industrialized nations has been described as a problem because the markets served by the nations of Europe, North America and Japan are difficult to penetrate. However, these same industrial areas also

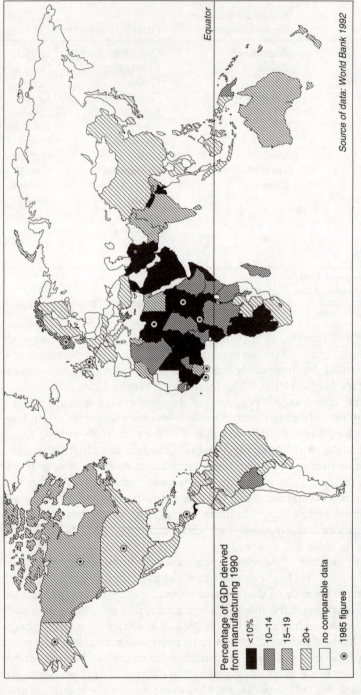

Figure 19 Percentage of Gross Domestic Product derived from manufacturing 1990

Equator

Source of data: World Bank 1992

Percentage of GDP derived
from manufacturing 1990

<10%
10–14
15–19
20+
no comparable data

⊙ 1985 figures

present an opportunity for interlinked development. Lawrence Brown in his thought-provoking paper on Third World development argues the case that exogenous factors and circumstances have the most profound effect upon development within the Developing World and that the effect they have will vary according to the particular circumstances in individual countries (Brown 1988). Industrialization provides a good example of this interplay. Changes within the industrial nations have meant that the economic and product ground they occupied has changed, allowing other nations to take over functions which were once characteristically theirs. The ability to take up these opportunities will vary from one developing nation to another. The sophisticated, wealthy, industrial nations trade largely with each other and are beginning to view the rest of the world, including their former colonies, in a new way. Once the sources of raw materials and captive markets, the former colonies, and other developing nations, have become locations for manufacturing wherever their labour supply is appropriate, infrastructure adequate and government amenable to foreign investment. These countries now present opportunities not so much for the mining of minerals and the harvesting of exotic crops but for revenue derived from manufacturing investment. The stimulus for this development is exogenous; the opportunities presented for industrializing in this way have been seen by governments as contrasting as those of China, Malaysia and South Korea. It is not solely a matter of inward investment; inputs of technological and managerial expertise together with the use of sophisticated machines have contributed to profound changes in the recipient nations' economies. Products are manufactured in partnerships, under licence or in other arrangements, with multinational companies playing an important role. As communications become more efficient, as transport is greatly improved, as goods made and traded become of increasingly high value per unit dimension, so distance becomes less an impediment to movement; the world network of manufacturing, trade and finance comes to involve the developing economies. These developments have had particular geographical foci, the most striking of which is that of the so-called Pacific Rim embracing Japan, coastal China, South Korea, Taiwan, Hong Kong, Singapore, Malaysia, Thailand, Indonesia, the Philippines and Australia. Together with Canada and the United States, this ring of industrial and industrializing nations may well emerge as the most important industrialized and high-income zone of the twenty-first century. For the majority of the countries of the Developing World manufacturing remains an undeveloped sector which must be further stimulated to energize the development process and fulfil as yet unfulfilled hopes.

Clearly to be considered adequately, the role of industrialization, the addition of large-scale manufacturing to essentially agrarian economies, must be viewed in the context of overall development policies. These policies embrace all sectors of the economy and all aspects of society since a development in one will have consequences for the others. Issues of population have been discussed and their relevance to social and economic development stressed. The nature of agricultural systems, and their potential for change, has been reviewed.

Traditional indigenous industries have been described and the question of industrialization raised. These separate considerations are parts of one entity and it is with the development of this, the political economy, that the peoples and the governments of developing countries are concerned. Many of these issues are exemplified in the case studies discussed in the second part of this book. The apparent underdevelopment of the economies of the many nations becoming independent after the Second World War raised the question as to the possibility of stimulating and guiding such economies so as to promote their growth and the well-being of their people. Development planning was conceived and this can now be considered.

8 The development process and development planning

'I have a Vision of the Future, chum,
The workers' flats in fields of soya beans
Tower up like silver pencils, score on score:
And Surging Millions hear the Challenge come
From microphones in communal canteens
"No Right! No Wrong! All's perfect, evermore"

J Betjeman
The Planster's Vision 1945

'What is the essential of planning economic development? I would say that the essential consists in assuring an amount of productive investment which is sufficient to provide for a rise of national income substantially in excess of the rise in population so that per capita income increases.'

O Lange
Essays on Economic Planning 1960

'One must avoid the persistent confusion of growth with development'

Brandt Report 1980

DEVELOPMENT PLANNING

Development planning has itself grown out of other developments. These include the study and increasing understanding of economic processes; the increasing internal integration of national economies which have replaced relatively discrete, disaggregated local economies; the increasing acceptance of the role of the state in the welfare of society; the development of administrative abilities and professional and technical skills to manage large and complex structures. These developments have all been encapsulated in the concept of the nation state. They became essentially characteristic of the industrialized countries of North America, Europe and Japan. In the Soviet Union, incorporated in the Marxist ideology, these trends were brought together into a totally planned economy, the command economy which was implemented after 1918. They appeared, for different reasons and purposes, in the fascist states of Germany and Italy in the interwar period, while at the same time increasing state intervention in the economy, in an attempt to mitigate the effects of the economic slump and to restore economic growth, became characteristic of the democratic countries of western Europe. The prosecution of the Second World

War itself forged instruments of national, concerted economic effort in all the belligerent nations from the United States in the West to Japan in the East. The countries of the Developing World were not part of these developments quite simply because most were not independent. In those that were, notably in Latin America, neither their organs of government, nor their political systems nor yet their economic structures were conducive to the emergence of development planning. In the greater part of the Developing World, up until the end of the Second World War, independent countries and distinctive national economies did not exist. National plans, where formulated, consisted of accounting exercises by the colonial governments.

The twenty-five years between 1945 and 1970 saw the vast majority of colonial territories achieve independence. Abruptly new sovereign states came into existence; new governments were faced with the task, for the first time, of administering their countries. They came into existence at a time when the acceptance of some form of economic planning was commonplace and when belief in the efficacy of planning was widespread. The populations of the developing countries looked to their new rulers, now their own kind and not foreign, for a promise of better things and for the means to obtain them. In retrospect it is possible to see how daunting, how formidable, this prospect was for the newly established and often inexperienced politicians – inexperienced, that is, in governing rather than political activism. Some were supported by their ignorance of the problems faced, others by the strength of their ideological beliefs; others, conscious of their inadequacies, not of leadership but in technical competence, sought help from international agencies such as the World Bank, from the former colonial power which had ruled them and from economic advisers both imported and home-grown. Their aim was to diminish rapidly the difference in the levels of economic attainment, standard of living, technical sophistication and economic independence between their country and that of the 'western' industrialized world. This was regarded as a goal which was not only attainable but as one which could be reached quickly. Development planning was to be the answer.

While these events were taking place in country after country in the Caribbean, Latin America, Africa, Asia and Oceania, scholars in the affluent world were increasingly turning their attention to this issue of development. Was there a discernible and inevitable process which transformed a technically backward, impoverished society into a more productive affluent organism? Was not the sequence of events which had taken place in Western Europe the apposite analogy? Could components of the European economic transition be isolated, examined, understood and re-assembled into a model which would not only describe the process but give indications of how it should be initiated in the new nation states of the 'underdeveloped world'?

The intellectual interest of economists had long been in the mechanism of growth which had brought about the transformation of the economies of the industrial countries. It now embraced the question as to why so many nations had not experienced this transformation. The theories of growth and nongrowth were the obvious source-material for economic development planning.

It can be argued that poor countries are poor because they are poor, that poverty is a self-perpetuating condition. At its simplest, the argument would take the form that low productivity means an insufficient food supply, that this leads to malnutrition and ill health, which in turn reduces the ability to work effectively resulting in a maintenance of low productivity. On the one hand this low productivity leaves no surplus of production which can be diverted to other and additional forms of production, and on the other it means that the producer has no means of purchase and so the total market is small and locked into a situation of a low-level equilibrium between supply and demand. The conditions of subsistence agriculture and domestic industry which have been described bear some of the marks of this condition. Nurkse developed this idea in his book *Problems of Capital Formation in Underdeveloped Countries* (Nurkse 1953; Mountjoy 1971). He writes:

> Perhaps the most important circular relationships of this kind are those that afflict the accumulation of capital in economically backward countries. The supply of capital is governed by the ability and willingness to save; the demand for capital is governed by the incentives to invest. A circular relationship exists on both sides of the problem of capital formation in the poverty-ridden areas of the world
>
> (Nurkse 1953).

He makes the point that some countries are poor because they lack mineral resources and because their environments are hostile to agriculture but that the lack of capital applies to their condition also. Leibenstein and Nelson have been associated with a further development of this circular relationship in the formulation of the theory of the low-level equilibrium trap (Leibenstein 1954,1957; Nelson 1956). This theory bears upon the issues raised in considering the economic implications of population dynamics referred to in Chapters 2 and 3.

It states that as incomes rise, populations will grow; the Malthusian proposition that populations grow and press up against the subsistence level. Malthus, it will be recalled, argued that this will then result in a reduction in per capita income, distress and a fall in population. Leibenstein's view is that since there is a physiological limit to the rate of population expansion (in the absence of immigration) to certainly not above 4 per cent per annum, this identifies the level of increase of per capita income below which the low-level equilibrium trap will operate. Increases up to this level will mean that population will expand in response to the income increase and, by absorbing the national income, again produce a situation in which savings approach zero and once more hold development down. The low-level equilibrium trap is shown graphically in Fig. 20 and is explained in the caption. Leibenstein was, however, not only concerned with the role of capital in a financial and physical sense but also as applied to the concept of the human resource, its skills, aptitudes and innovative energy.

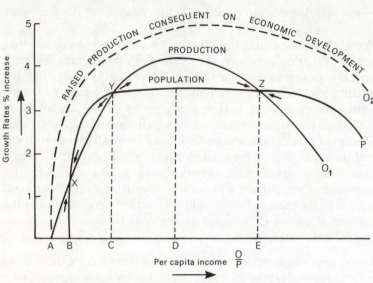

Figure 20 The low-level equilibrium trap

In this diagram the rate of increase of production (O_1) is shown as demonstrating characteristic diminishing returns. The rate of increase of population (P) is plotted against per capita income. The curve assumes that as per capita income increases so the population will grow until it reaches the physiological limit around 3.5 per cent. The two curves interact with each other. As per capita incomes grow from a very low level at A to a slightly higher level at B, greater affluence allows more to survive and the rate of population increase rises sharply in Malthusian fashion. At X the rate of population increase exceeds that of production and prevents a further rise in per capita incomes. As population continues to expand, increasing poverty operates its Malthusian checks, cutting back population so that it and production fall to a low level. Output and population increases oscillate, therefore, around X in a low-level equilibrium trap. If, for some reason, it is possible to continue to increase production until population increase reaches its physiological limit at Y then per capita income can reach C and continue to increase until D. Thereafter because of the operation of diminishing returns the production curve declines and crosses the curve of stable maximum increase in population at Z. At this point, per capita incomes begin to decline and the rate of population increase likewise falls until a further equilibrium is reached oscillating around Z, a second equilibrium trap. To break this containment of population–output equilibrium it is necessary to be able to harness the means of production so as to raise the production curve (O_2) so that it always exceeds that of population with the result that as population declines output continues to exceed it and per capita income increases. Continuing aggregated developments, when the development process is fully launched, will ensure that the diminution in returns on inputs are continually postponed as one form and means of production shifts into another. The model assumes, wrongly, that average per capita income, derived from the division of national output by national population, is the influence upon the size of family and hence the population curve. It is the actual family income which produces conditions favourable or contrary to the formation of small families (see Chapter 3). The distribution of family income therefore modifies considerably the relationships depicted in this figure.

THE ROSTOWIAN MODEL

These views are all centred around the difficulty of accumulating capital for investment for further development in countries which were initially characterized by poverty. The issue became one of how to break through this containment. Nurkse and Lewis could not see how a subsistence community could ever generate capital and therefore it must be an activity associated with some other sector, cash agriculture or mineral export, either privately or state operated. This in turn implied that one sector would become more affluent. Arthur Lewis regarded rates of saving as central to the issue of economic development. In his paper 'Economic development with unlimited supplies of labour' (Lewis 1954) he states that economic development is characterized by rapid capital accumulation. He associates this with increases in the rate of domestic savings. It therefore follows that if it can be explained why countries with previous low levels of saving and investment are transformed into nations with much higher levels (the levels he refers to are from 4 or 5 per cent or less of the national income to 12 to 15 per cent or more), then a greater understanding of the process of economic development can be achieved.

There are two points to note in Lewis's views. First, there is stress not only on saving as a source of investment but also the need for a rapid accumulation of capital. Second, Lewis was referring to the experience of the capitalist industrial countries and their structural transformation as possible guides to the explanation of 'development'. As applied to the less developed countries it implies that the opportunities for worthwhile investment exist; it is the capital which is lacking. When these views are tied to Leibenstein's exposition of the low-level equilibrium trap, possible explanations begin to emerge. What is needed is a massive investment over a short period to raise income levels above the containment rate. Thereafter, since it was believed that savings levels and hence opportunities for capital accumulation were determined by per capita income, this would allow a continuing growth of the economy so that it always exceeded any population expansion. Indeed, as we saw in Chapter 3, the rate of population increase would tend to decrease as its members became more affluent.

The ideas incorporated in theories of growth therefore suggested a succession of stages in what was conceived of as 'development' analogous to what had happened in Europe and it is with this concept that the name of Walt Whitman Rostow is most closely identified. Rostow progressively formulated his thesis in a number of works (Rostow 1952,1956,1960,1963). Approaching the problem as an economic historian and examining the development process in Europe, North America and Japan, Rostow put forward the hypothesis that:

> The process of economic growth can usefully be regarded as centring on a relatively brief time interval of two or three decades when the economy and the society of which it is a part transform themselves in such ways that economic growth is, subsequently, more or less automatic. This decisive transformation is here called the take-off
>
> (Rostow 1956).

Rostow is concerned with rates of investment rising and being employed effectively so that per capita output in real terms rises and that this crucial level of investment is such that it leads to fundamental changes in production and in the disposition of the resulting income, which in turn will lead to further investment, further increases of income and so on in a self-perpetuating manner. He says:

Initial changes in method require that some new group in the society have the will and the authority to install and diffuse new production techniques; and a perpetuation of the growth process requires that such a leading group expand in authority and that the society as a whole respond to the impulses set up by the initial changes....

(Rostow 1956).

In the course of a sophisticated thesis, Rostow examines historical analogies and identifies a series of stages of development of which the first two, the 'pre-conditions' and the 'take-off', he considered particularly relevant to the Developing World. The first referred to slow economic and social developments of a century or more when conditions for 'take-off' were established; the second to the twenty or thirty years referred to above. The third stage was the subsequent period of self-sustaining growth. According to Rostow, 'take-off' requires three conditions:

(a) a rise in the rate of productive investment from (say) 5 per cent or less to over 10 per cent of national income (or net national product).
(b) the development of one or more substantial manufacturing sectors with a high rate of growth.
(c) the existence or quick emergence of a political, social and institutional framework which exploits the impulses to expansion in the modern sector and the potential external economy effects of the 'take-off' and gives to growth an on-going character. The third condition implies a considerable capability to mobilise capital from domestic sources

(Rostow 1956).

Rostow makes the last point because he believes that while foreign capital can help to establish the pre-conditions for 'take-off', it cannot institute 'take-off' itself. This is the product of an internal transformation which subsequently allows a high marginal rate of savings. Rostow concludes:

This hypothesis is, then, a return to a rather old-fashioned way of looking at economic development. The take-off is defined as an industrial revolution, tied directly to radical changes in methods of production, having their decisive consequences over relatively short periods of time.

This view would not deny the role of larger, slower change in the whole process of economic growth. On the contrary, take-off requires a massive set of pre-conditions going to the heart of a society's economic organisation and its effective scale of values. Moreover for the take-off to be successful, it must lead on progressively to sustained growth; and this implies further deep and often slow-moving changes in the economy and the society as a whole

(Rostow 1956).

It is important to recognize two issues in this quotation. One is the recognition that economic development can be regarded as a discontinuous process of structural transformation. It is not the same as economic growth which may be simply enlargement without change or development. It is a view expressed as early as 1911 by J. A. Schumpeter whose work was translated into English by R. Opie in 1934 (Schumpeter 1911, 1934). Secondly, and related, Rostow refers to a 'massive set of pre-conditions'. It is this requirement which was so often to be forgotten. Too frequently attention has been focused only on part of Rostow's carefully considered arguments which were very persuasive and influential. Too many saw only the 10 per cent savings rate as the key to success and remembered the conclusion to his article of 1956 which reads:

What this argument does assert is that the rapid growth of one or more new manufacturing sectors is a powerful and essential engine of economic transformation. Its power derives from the multiplicity of its forms of impact, when a society is prepared to respond positively from this impact. Growth in such sectors, with new production functions of high productivity, in itself tends to raise output per head; it places income in the hands of men who will not merely save a high proportion of an expanding income but who will plough it into highly productive investment; it sets up a chain of effective demand for other manufactured products; it sets up a requirement for enlarged urban areas, whose capital costs may be high but whose population and market organisation helps to make industrialization an ongoing process; and finally it opens up a range of external economy effects which, in the end, help to produce leading sectors when the initial impulse of the take-off's leading sectors begins to wane

(Rostow 1956).

The argument is very plausible, the sequence sounds inevitable, the historical analogy is strong. It is a logical presentation of events among which the 10 per cent saving rate and industrial sector development stand out strongly. The ideas were taken up and followed by many. The phrase 'take-off' evokes the parallel with aircraft. A critical power level or thrust has to be reached before an aircraft can take off. All manner of lower level inputs will not achieve this; the aircraft will merely taxi around but not become airborne and self-sustaining. But Rostow's hypothesis has vital provisos: first, the true nature of development, and second, the crucial pre-conditions. It is not how much can be saved. Countries with large raw material export resources can and do experience savings of over 10 per cent of GNP and yet not move into the take-off stage. They can experience growth without development. It is possible to mistake the appearance for the reality of development. Some countries are rapidly developing with savings rates of less than 10 per cent. What is important is how the savings are productivity invested. This, in part, depends upon 'the massive set of pre-conditions' in the pre-take-off stage. These have to be created. They involve favourable political, administrative and social organizations. They involve infrastructural developments but more importantly they involve changes in attitudes and values which create a new awareness of how to do things differently and result in an ability to do so. The creation of take-off conditions and the controlling of the aircraft of development in its ascent requires a complex and delicate balance of considerations. These are the essentials of development theory and practice. As Paul Streeten says:

> All thought presupposes implicit or explicit model building or model using. Rigorous abstraction, simplification and quantification are necessary conditions of analysis and policy. But models must be realistic, relevant and useful. The trouble with many current models is that they are shapely and elegant but lack the vital organs

(Streeten 1972).

The economists of the 1950s and 1960s came forward with their models of development designed to rupture that ceiling of containment, to create the conditions of development and achieve the state of take-off. The approaches varied as their authors laid stress on differing factors and emphasized different goals. They were intended to form the basis of national development plans.

PLANNING INVESTMENTS

Essentially development strategies are concerned with making a country more productive by making a fuller use of its resources with the aid of more efficient economic processes. The strategy implements a certain pattern of investment designed to initiate certain forms of economic activity, to stimulate particular sectors which, it is hoped, will create a favourable economic, social and psychological environment to bring about the 'quick emergence of a political, social and institutional framework which exploits the impulses to expansion' (Rostow 1956). The mechanics of the strategy are capital investments in land for agriculture, in forestry, mineral and energy resources to make a fuller use of these resources and also to make a fuller and more productive use of the resource of labour. The productivity of labour is thus increased by a considerable margin, be it in agriculture, mining or in manufacturing, to a level over and above that previously pertaining. These developments make infrastructural demands. Capital is required to be invested in the transport infrastructure in road, rail networks, rolling stock and vehicle resources, in ports and port facilities. Energy supplies of a scale vastly greater than those made available by wood fuel, become necessary as factory production is introduced and as transport modernizes. Power stations, transmission and distribution networks, appear in sectoral development plans. Beyond these directly productive investments come the demands of the social infrastructure for education, training and health facilities, and for administrative functions.

All these investments, these economic injections and activities, are interrelated. All are calls upon scarce resources. The issue becomes one of how to balance them and arrive at the right mix to promote true development. Such true development would not only encompass the definitions of Schumpeter but also the dimensions of equity and social considerations which would influence who receives the benefits of development. The considerations in the 1950s and early 1960s came to centre around three main ideas. First, that the break-out of the low-level equilibrium trap of Nurkse and Leibenstein required a concentrated major economic effort – the so-called 'big push'. Second, that the need to raise the level of savings relative to national income described by Lewis would be the means of financing the 'big push'. Third, that industrialization would be the major part of that thrust and that the output of industry would be essentially for the home market to further stimulate structural transformation. As has been noted Chenery's work has done much to identify the role of manufacturing in structural changes (Chenery 1960; Chenery and Taylor 1968; Chenery and Syrquin 1975; Chenery *et al.* 1979). It was felt that planning would more quickly and effectively initiate these developments than the operation of free market forces, indeed that it was necessary to initiate the process of developments.

It is important to return to the role of savings and capital investment since this was central to all theories of growth and development. The work of R. F. Harrod and E. Domar on the theory of growth, though carried out independently, covered such common ground that it became known as the Harrod–Domar growth model (Harrod 1948; Domar 1957). This model states that the national rate of growth, that is the rate of growth of the Gross National Product (GNP), is the product of the ratio of savings to national income and the national capital–output ratio. It appeared, therefore, to provide the rationale behind the savings–capital–investment argument. As Myint so pertinently comments, it was a model based upon the analysis of mature economies with stable capital–output ratios and therefore not necessarily applicable to the stage of rapid structural change during take-off which Rostow's model required (Myint 1964). None the less the concept of the capital–output ratio is of significance in the understanding of development strategies and can be applied to the take-off stage if a non-stable ratio is involved. Simply, the ratio refers to the amount of capital investment required to produce a unit of output. If one unit of investment produced one unit of output the ratio would be one. If it required five units of investment to produce one unit of output the ratio would be five. In other words activities which required a large amount of investment to produce a small output would be characterized by high capital–output ratios. From the concept it follows that, in theory, it would be possible to estimate the amount of national investment necessary to give a desired increase in the GNP and therefore an increase in per capita income. It could set the dimensions of investment. Investment added to investment would further increase output so as savings were re-invested the self-sustaining growth characteristic of Rostow's final stage could be achieved, indeed the capital–output ratio came to be described as the incremental capital–output ratio (ICOR).

It has been shown that the populations of almost all developing countries are growing rapidly; to achieve increases in per capita income this must be included in the calculation. In a situation where the population is growing by 2.5 per cent per year, as in Nigeria between 1970 and 1980, to keep average per capita incomes at their present level the national income must grow by 2.5 per cent, otherwise population increase outstrips income and there is less to share out. This leads to a deterioration in conditions, the reverse of development. Even maintaining an output commensurate with the increase in population means only an attainment of equilibrium – or stagnation. How much investment is required to maintain this state of equilibrium turns, therefore, on the national ICOR. If it is, for example, four, this would require 4 per cent of the national income to be invested to raise output by 1 per cent; to raise it by 2.5 per cent, that is to maintain the status quo with a 2.5 per cent population increase, would require an investment of 10 per cent of the national income. To move into a growth situation with a 3.5 per cent increase would require a 14 per cent investment ratio. Rostow's and Lewis's figures of 10 to 15 per cent of national income being required to be invested reflect this concept of ICOR.

The ICOR does, of course, vary from activity to activity. What has been referred to so far is a capital–output ratio in a gross or national sense, a figure arrived at from a wide range of values. The national ICOR will be the result of the particular mix of activities prevailing in the particular country. The ICOR is characteristically high for infrastructural developments in transport, for public utilities such as water supply, for the energy industries and for some of the heavy industries. It is high for social investments in education and health facilities. In the case of the agricultural sector it can be very low. Other instances occur between these extremes. Investments in the less directly productive enterprises tend to be characterized by slow returns in output resulting from their implementation. Clearly these differences in ICOR and in return times raise the crucial question of the balance or mix of investments incorporated in a nation's development plan. There is another matter. Between the developing countries and those of the technically and financially advanced countries of the industrialized world there are differences, many of which have been discussed. These differences include differences in capital–output ratios. Not only do capital–output ratios differ from sector to sector but they are also subject to change. As technologies change, so does the ratio. The most profound changes take place precisely during the take-off stage of major structural transformation and the emergence of new values, attitudes and institutions. As a country changes from elementary subsistence-cash agriculture to commercial farming, from handicraft cottage industries to factories supplied by power stations, as transport changes from head loads to railroads, so the ICOR for each of its activities will change and they will not all change at the same time and in the same way. The nature of change will depend upon the nature of the activity. In the case of a manufacturing and assembly industry, such as motor-vehicle manufacturing, economies of scale, with increasing size consequent upon additional investment, will result in a continuing benefit from capital injections. It will result in the ICOR becoming lower. If the activity is one of primary production, such as agriculture using the resource of land, or forestry or mining developing mineral resources, the law of diminishing returns applies much more forcibly. Increasing investments produce diminishing returns so that the ICOR increases as the process goes on. As many developing countries emerged as independent nations with the colonial legacy of a focus upon primary production, this factor is important. In the advanced industrialized economies capital investments can be re-allocated into continuously arising new opportunities so that specific capital–output ratios do not go up and, in consequence, neither does the national ICOR. In such situations the self-sustaining and continuous development is characteristic and is a manifestation of this flexibility. Indeed it is of interest that inflexibility is not only characteristic of primary industry economies of the Developing World but also of the least successful countries and areas among the industrial nations of Europe in the continuing industrial revolution. The situation is made particularly difficult in developing countries where populations are not only increasing but where the resource base is diminishing as soils are over-utilized, forests cleared and mineral resources exhausted by primary export activities. There would seem, therefore, to be a case for instituting developments which would broaden the range of economic activities away from those characterized by diminishing returns to those where inputs of capital and labour were more significant than natural resources and where scarce capital could be used more effectively.

The study of the ICOR, the Harrod–Domar model of growth, the ideas of Lewis, Rostow, Leibenstein and others, all involve the effective use of capital and indeed in being able to generate capital for investment. The theory says this is derived from earnings. The problem occurs where national earnings are too low to allow enough to be put aside for investment to start the process going. The process needs to be started since the consensus opinion is that development is a discontinuous process of transformation. One side of the problem is therefore the generation of capital or access to it. The wealthy industrialized world is one possible source and this raises the issue of borrowing on economic terms, of receiving interest-free loans, of being given aid in money or capital equipment or in the form of technical assistance to save on capital which would otherwise be committed to this need. Trade is another possible source of capital accumulation. Activities can be promoted specifically to achieve earnings which can be re-invested. Obvious examples are the encouragement of foreign enterprises to develop mineral resources, of which oil is currently the best example, so that the exchange revenue derived from their sale can be allocated to development investments. Other instances are the development of commodity exports, generally of primary produce, for sale under international agreements or on the competitive open market; sugar, tea and coffee are examples. In these cases, use is often being made of earlier capital investment of the colonial period which, as has been seen, was essentially concerned with primary rather than manufacturing industries. As a consequence initial export earning has rarely involved manufacturing unless the capital for its installation has come from foreign and multinational investment. A further source of earnings has come from the development of tourism, where the physical and social environment has been conducive to its development, since it is an activity with a relatively low ICOR. These issues will be returned to later and more fully discussed.

If capital acquisition is one side of the problem, the other is its effective disposition and investment. Importantly, as has been exemplified in the discussion of agriculture, it is not simply a question of investment allocation to separate sectors since none is truly independent. It is a question of the interaction of investments as commonly the full utility of investment in one sector requires development in others. It is an obvious point but one which has often gone unheeded by the governments of developing countries. It is also an issue which underlies much of the thinking that has been devoted to the question of appropriate development strategies. There was a general agreement that an initial, large-scale investment was necessary – the 'big push'. There was agreement also that industrializing action should be the major component in the investment programme. Where the views of the strategists of the 1950s differed most was in how sectoral developments should be designed to interact.

THE BROAD-FRONT BALANCED-GROWTH APPROACH

Nurkse, in his concern for the difficulties in generating capital in low-income countries, was of the view that incentives to invest were limited because of the small domestic market which developing countries offered. In most of them the national population was small and, being composed of poor people in economic terms, was a very small market but Nurkse argues that his view applies to large countries, like China and India, also. The small market is a product of low productivity. He is referring to the 'vicious circle of poverty' (Nurkse 1953). By enlarging the market the incentive to invest would increase and this could be achieved by stimulating development along a broad front of activities. To make his point Nurkse hypothesizes the case of the establishment of a single factory to produce one product, say shoes. The factory workers earn wages which give them the ability to buy shoes but their demand for shoes is far smaller than the factory's output and they need to spend their wages on items in addition to shoes. Unless other developments are simultaneously taking place, increasing the ability of other workers to buy shoes, these other workers will not divert their income to footwear purchase from other needs such as food or housing. As Nurske comments, at low income levels demands are inelastic. The solution he claims is to establish a wide range of industries since this will enlarge the market. 'People working with more and better tools in a number of complementary projects become each other's customers. Most industries catering for mass consumption are complementary in the sense that they provide a market for, and thus support, each other.' He quotes John Stuart Mill's formulation of the classical Say's Law of Markets: 'Every increase in production, if distributed without miscalculation among all kinds of produce in the proportion which private interest would dictate, creates, or rather constitutes, its own demand.' Nurkse continues, 'Here, in a nutshell, is the case for balanced growth. An increase in the production of shoes alone does not create its own demand. An increase in the production over a wider range of consumables, so proportional as to correspond with the pattern of consumer preferences, does create its own demand' (Nurkse 1953). A further supporting argument in the case Nurkse presents is that of 'external economies'. He is using the term in a special sense, namely that each industry, in the wide range he envisages, by enlarging the market is creating economies external to itself. The more conventional meaning of external economies such as derive from the joint use of transport facilities, banking, communication and pools of skilled labour also, of course, can be advanced as benefits of the broad-front approach.

At first sight Nurkse's views may seem somewhat divorced from the reality of development planning and indeed their implications have been questioned. Myint questions the ways in which the establishment of a wide range of essentially consumer industries work (Myint 1964). He points out that if the new consumer goods industries are more efficient in the use of their resources than the domestic or handicraft industries they replace, then the extent to which their products replace ones made in the traditional way, rather than augment them, will create some unemployment. If they use labour more efficiently then a smaller labour force can produce the same amount and since the factory labour force it envisages is itself part of the market, then the contribution of the new industry in enlarging the market will depend upon it being able to produce more and find a market for the increase. Nurkse's case is that the increased market is provided by the workers of other industries. However, if one product can be produced at unit costs well below pre-existing costs or if it has a particular attraction, it may siphon off the demand for other consumer goods, the effect depending upon price–demand elasticity and cross-elasticity relationships. The favoured industry could, of course, invest its income in other industries and so enlarge the consumer market and employment opportunities but only if it did so. New industries which are not markedly more cost-effective than their competitors will not put them out of business and in Nurkse's concept will enlarge the total market by the total numbers of their employees. This is the total market for all goods including food and in reality it is composed of a whole set of individual markets for individual groups of products. Even if the total market is enlarged it may still mean that the individual markets are too small for the efficient use of industrial plant though industries do vary in their sensitivity to scale economies. For some industries, therefore, it will be possible for them to operate successfully in the conditions of the Developing World. Others, because of their scale, technical needs and skill requirements, may be difficult to incorporate in this balanced growth.

In a broad front approach aiming at setting up a range of consumer goods industries, essentially substituting for imports, producing for and creating a market by enlarging the industrial wage-labour force, it is not possible to keep them advancing at the same rate. Adjustments and re-adjustments will have to take place continuously due to differing levels of efficiency and productivity resulting in differential movements of prices, which in turn will result in changes in expenditure and demand patterns. It is a form of industrially-based development which is subject to a number of constraints (Myint 1964). Although in Nurkse's argument it is a means of generating capital and investment, the initial availability of capital for a wide range of enterprises is an obvious difficulty. Secondly, it is not realistic to assume that there exists an instant, ready-made labour force. The work in the factory may well be of a completely different order from that in handicrafts and agriculture. Labour may not be readily forthcoming from the land. Suitable labour may require training, and particular skills and aptitudes may be in short supply so that competition for them may raise labour costs and send up manufacturing costs. The precise effect will vary from industry to industry. As with capital this is a constraint which is most forcibly felt at the beginning of the process. Over the longer term a pool of industrial labour would have been created which by its very existence would facilitate industrial changes. The broad-front approach envisaging a whole series of endeavours is particularly demanding of managerial resources and these tend to be scarce in developing countries because of their economic histories. Experienced expatriate managers have often been employed to overcome this constraint.

Much of the thinking behind this broad-front approach has been focused upon manufacturing. The balanced growth was thought of as essentially a balance between manufacturing industries. Their successful development obviously is conditioned by the extent of the supporting infrastructure. It is something which requires not only the creation of a new kind of labour force but also the physical infrastructure of transport facilities, appropriate supply of energy and water, the economic infrastructure of financial and information facilities and the administrative infrastructure of government. As industry's requirements become more sophisticated, the infrastructural need incorporates the dimension of education and training. The compass of balanced growth spreads. But this is not all. One crucial dimension has often been neglected – that of the agricultural economy. The new, non-agricultural labour force requires to buy food. If in the conditions of an expanding market agricultural production cannot increase without unit costs going up as diminishing returns apply, then food costs go up. The increased food prices may feed back to the farmer and be regarded as a beneficial spread effect of industrialization. If agricultural production cannot meet the needs of the enlarging industrial and urban population as well as those of the vast rural community, then food imports become necessary and the economic development, so earnestly desired, may fall behind. What is vitally necessary is a planned development which incorporates agriculture and the rural sector. It must be part of any balanced growth programme. It so rarely is.

Development planning therefore requires, first, a balance between components in the productive sectors of industry; second, a balance between industry and agriculture, between urban and rural communities; and third, a balance between investments in the directly productive sector and infrastructural investments. The last of these presents particular problems. It will be recalled that the ICOR is high and the return indirect and long-term for such projects. They demand large amounts of initial capital and this is made particularly difficult because, like some manufacturing plant, they are subject to scale requirements. A road to connect A to B is not a meaningful investment if it only goes half-way. A dam for a hydro-electric project cannot partially cross a valley or be below a certain height. Infrastructural investments tend to come in big blocks. All manner of sub-optimum compromises can, of course, be used such as the building of a single-tracked railway which can be upgraded to two tracks at a later date when more capital is available, or the construction of low-cost roads with a low axle-load capability with the intent of strengthening them at some later date. It is true also that in many developing countries colonial infrastructural investments will have taken place so the task is of augmenting them rather than starting from scratch. None the less infrastructural investment presents problems. Unavoidably it makes too great a demand on scarce capital at first without immediate returns but if it is neglected it creates bottle-necks which impede subsequent developments.

Balanced development planning presents a broad spectrum of developments and involves the concept of interrelated and consequential developments over time. The question becomes one of which existing facilities to develop and by how much, which productive enterprises to encourage, how to smooth out the calls upon investment capital over time and how to react to inevitably changing situations as the economy progresses. It is a matter of integrated planning based upon a design but with flexibility. The structure being erected will include the basic foundations, the infrastructure which enables other activities to develop more readily, such capital and intermediate product industries supplying basic goods or semi-manufactured as are deemed feasible, and finally the consumer goods industries whose role, at least initially, will be to produce goods previously imported. It is a very demanding programme and in developing countries one very difficult to implement. Data sources and data collecting services such as exist in most industrialized countries are sparse or non-existent. It therefore becomes difficult to monitor progress and yet the balanced growth programme requires this if it is to respond in a meaningful way to changing situations, for change there will be. Inefficiencies in some activities will upset balances; technical problems will disturb nicely balanced predictions. The management of national economies has proved equally difficult in the free-enterprise countries of Western Europe and the command economies of the communist states of Eastern Europe and the Soviet Union. Many developing countries do not have the personnel to mount a development programme across the whole range of activities, particularly if the state is attempting to implement it. It places too great a burden on the civil service. In practice broad-front balanced planning commonly involves both governmental and private enterprise as for example in India and Nigeria. In some countries the state is largely concerned in infrastructural projects; not only the obvious social investments in education but also in the major energy and irrigation schemes, land reclamation, large agricultural estates and in the transport provision. Some governments also set up state-run heavy industrial projects. The state is involved in these activities largely because they are characterized by a high ICOR and involve large sums though in some instances the government may assume such functions for ideological reasons. While the state is thus involved, private enterprise will concentrate upon intermediate and consumer goods industries. In such situations there arises not only the phasing of these several initiatives into development plans but also the relationship between government operated enterprises and private businesses since technical and economical vertical linkages may be involved.

Balanced-growth strategies have become the most widely adopted though they vary greatly in their degree of sophistication, the effectiveness of the control of development and the range of activities involved. There are also variations which reflect how governments identify their goals and the means of achieving them. There are those who see their investment priorities in basic industries and infrastructures, giving a low priority for consumer goods and present material well-being. Their vision is of laying secure foundations for future developments rather than of providing for current tastes. This has an element of what might be called the Soviet model. On the other hand, there are those who see a need for the population involved in the development process to experience some return for their efforts which benefits them, that all planning cannot be solely for one's descendants. Between the extremes of 'all for the future' and 'all for now' lies the way which most countries have taken and, indeed, in democratic societies it must be so.

UNBALANCED GROWTH

The difficulties of maintaining the balances which, when applied, the balanced-growth theory encountered gave rise to some scepticism as to its practicality. It was argued by some that it is precisely the wrong way to stimulate the economic energies and initiatives necessary to propel development forward. Hirschman makes the point that in any economy its component sectors, and indeed activities within each, will develop at different rates. They do not advance together in a continuing balance. He writes:

> Just as on the demand side the market can absorb 'unbalanced' advances in output because of cost-reducing innovations, new products and import substitution, so we can have isolated forward thrusts on the supply side as inputs are redistributed among users through price changes, and or the cost of some temporary shortages and disequilibria in the balance of payments or elsewhere. In fact, development has of course proceeded this way, with growth being communicated from the leading sectors of the economy to the followers, from one industry to another, from one firm to another.... The

advantage of this kind of seesaw advance, over 'balanced growth' where every activity expands perfectly in step with every other, is that it leaves considerable scope to induced investment decisions and therefore economises our principal scarce resource, namely, genuine decision-making (Hirschman 1958; Myint 1964; Mathur 1966).

By 'induced investment' Hirschman was referring to a chain reaction whereby investment and growth in one activity would stimulate development in others.

This view is not the complete rejection of development planning which it may seem to be at first sight. Essentially it says that since developing countries have neither state nor private capital to invest over a wide range of activities, nor yet the technical and managerial capacity to handle them, a selection of priorities becomes inevitable. In the absence of adequate capital and expertise, a balanced growth path will be impossible to follow, the programme will begin to break up and an unbalanced growth will result, so why not plan an unbalanced growth rather than let it emerge out of chaos? Why not select a group of industries which have significant horizontal and vertical linkages, where economies can be achieved because of them, both internal and external, where comparative advantage exists, and concentrate on that group, focus the investment there and apply to it the scarce entrepreneurial and managerial skills? This group of industries, or activities, are more likely to succeed and pull ahead of the rest of the economic structure, break away from the old balanced relationships, produce a state of disequilibrium and by doing so stimulate other activities and sectors to catch up and readjust. It is the concept of the lead sector and it is an attractive thesis. In reality it presumes too much. It relies upon the ability to select the most significant growth sectors and to an even greater extent it relies on the emergence of new 'stimulated' activities of development significance. It demonstrates too little concern with the time-lags involved in 'backward' sectors catching up with the leaders. It presumes that a 'surplus' labour force exists in agriculture which can be drawn upon and then fed by those remaining. This cannot be done unless agricultural developments also take place and keep pace. Attention therefore *must* be paid to agriculture even though it may not be regarded as a lead sector. Infrastructural investments will be necessary to facilitate the forward thrust of the manufacturing sector and will make a call upon investment. Elements of a balanced programme begin to appear, a balance which must incorporate agriculture which, as has been discussed, is the keeper of old values, traditional societies and cultural identity and is often the most difficult and slowest to change. This essential agricultural component will set some limit on the rate of change. Many programmes have neglected it and it has remained out of balance. The difficulty with the unbalanced approach is not that it embraces the reality of continual change but that some sectors *are* left behind and do not catch up. It is a feature which underlies the problems of many developing countries today.

Whatever particular strategy is adopted, any development programme must address a complex of considerations. First, there is that of the generation of capital internally and access to external capital. Second, there is the choice of investments in the situation of limited capital and managerial and technical ability. Third, there is the need to balance investments between infrastructural needs, production activities, and social investments in education and health. Fourth, it is necessary to pay particular attention to agricultural productivity and the role of the rural community in overall economic development. Fifth, there is the need to consider the balance between basic, intermediate and consumer goods industries. Sixth, in all activities the issue of capital-intensive or labour-intensive strategies will need to be evaluated and with it the question of appropriate technologies. Lastly, all these considerations must be viewed in the context of the relationship between the nation's economy and that of the rest of the world. In this must be weighed issues of comparative advantage, of export earnings and import substitution. All development plans also require a time dimension in which stages of attainment can be phased. It appears complex because it is.

The ideas presented have been essentially concerned with economics and the ways in which national economies can be made more productive since development was conceived as being a matter of economics. This approach has been described as the 'orthodox paradigm'. This view, these ideas and methods were adopted by many new independent nations and, as a number of cases studied in the second half of this book illustrate, formed the basis of the development plans they formulated. Consequently, incomplete and imperfect though they are, these views have had a profound effect and cannot be dismissed as irrelevant today. Yet as the Brandt Report states 'One must avoid the persistent confusion of growth with development' (Brandt 1980). In many cases the carefully considered plans have met with only limited success, some none at all, so much so that not only have their goals been challenged but also their very need. Economies have often been approached *en bloc* as a closed structure; many plans have been shown to be too abstract from reality and to ignore many of its dimensions. Few plans take note of their spatial or regional implications and the questions raised by such features as backward areas characterized by lagging sectors of the economy; too few appear to consider the people for whose benefit the plans are ostensibly designed. The peoples of developing countries are not an homogenous mass neither economically nor socially. For example tribal and religious differences may have a spatial identity and be expressed in political power groups, or racial duality may be associated with differences in economic opportunity. This is the reality, the stuff of development situations which armchair panaceas tend to neglect but with which governments in the implementation of their development plans have had to contend (Chapters 11, 12, 14 and 16).

Some disillusion with economic planning is now evident and not only in the Developing World. As the Iron Curtain has been drawn aside the classic command economies of the former Soviet Union and its East European Socialist satellites have been shown to be ill-conceived, ill-managed, economically inept and to have done little to improve the well-being of their people. The interventionist character of development planning is widely criticized and models such as those of Rostov are no longer viewed as an adequate explanation of the development process and development planning. The nature of the development process and the causes of underdevelopment have been subject to a rigorous, and in some cases a radical rethinking, taking the issue away from national problems and national plans and into the global context (Seers 1969; Brookfield 1975; Harris and Harris 1979; Mabogunge 1980a and 1980b; Rimmer 1981; Chisholm 1980, 1982; Peet 1983; Ross 1983; Unwin 1983; Brown 1988). The orthodox paradigm presented considers development as something which can be measured in economic terms since it is expressed in material outputs and the availability of goods and services. This view underlay the production of the *Atlas of Economic Development* edited by Ginsburg and the analysis of its data by Berry referred to in Chapter 1 (Berry 1960, 1961; Ginsburg 1961). It is a view that encompasses varying degrees of intervention but underlying it are the concepts which Wilber and Jameson refer to as 'parables of progress' namely the view that development is the progression from one stage to another, each progression involving a higher level of development (Rostow 1952, 1956, 1960; Brown 1988; Wilber and Jameson 1988).

The attitudes to development thus fall into two broad categories. One approach is radical, often Marxist, and seeks to understand the present functioning of the world economy and holds to the view, rightly, that a particular situation to be fully understood requires to be set in the world context (Harris and Harris 1979). This radical view is critical of the interventionist approach at national level and seeks solutions in the world system. The second approach is pragmatic and seeks to solve problems on the ground by positive action. It commonly operates at local or regional level in such ventures as integrated rural development projects and health provision; at national level it produces the development plans which are discussed in the case studies presented in this book. This latter approach seeks to bring about changes from within and argues that these are more quickly obtainable and more directly beneficial than the goal of others who seek to bring about the restructuring of the world order by revolution. The revolutionary approach is increasingly being rejected. Experience has shown that pragmatic policies can be highly successful and that the benefits of growth can be equitably spread.

The now favoured liberal, pragmatic and market-orientated approach does not mean that economic measures and economic planning are discarded; they remain part of attempts by governments to stimulate national economies. Experience has shown that an over-involvement of the state in economic matters can stifle rather than stimulate development. On the other hand a degree of involvement in social considerations such as education and health care and in provision for the less fortunate seems essential.

Since all nations are interlocked in an increasingly integrated global economy the international dimension is powerfully influential in the process of development. This will be discussed in the following chapter.

9 The external dimension

'Neither a borrower nor a lender be,
For loan oft loses both itself and friend,
And borrowing dulleth edge of husbandry'

Shakespeare
Hamlet

'Above all, we believe that a large-scale transfer of resources to the South can make a major impact on growth in both the South and the North and help to revive a flagging economy'

Brandt Report 1980

In the orthodox development paradigm much is made of the central role of investment in the promotion of economic development. It is important to draw the distinction between capital in the sense of finance and capital as a means of production. Finance can be obtained to cover debts or deficits but it does not remove them; only the creation of real wealth by production can do so. Finance can be invested in productive capital developments which can create real wealth. The ability to invest can be acquired by earning it. The ability to earn it can be achieved by the appropriate investment of available finance. The finance can be derived from domestic savings, from trade and by borrowing. To contribute to the development process it must be productively invested. Developing countries can earn exchange revenue from their export activities be they of primary commodities or manufactured goods. With this they can purchase goods from other countries both to satisfy consumer demand for goods not produced at home and, in the case of imported capital investment goods, to further strengthen their productive capacity. Trade can, therefore, generate finance for capital investment and promote growth. However, as has been seen, the creation of this situation is not without its difficulties. Time lags in returns on infrastructural investments may, for example, hold up other developments intended to generate productive capital. Trade may not be able to develop sufficiently, for external reasons, to generate the necessary income for investment. These and other bottle-necks which balanced-growth policies need to avoid are of common occurrence – so much so that the majority of developing countries have had to seek external finances to maintain their development programmes. The problems of the Developing World's countries and their development paths thus acquire an external or international dimension. They are part of the world economic community. This involvement has three components. One is participation in world trade and this will be discussed first.

The second is the use of foreign resources which can include gifts, loans on the commercial market, low-interest or interest-free loans as part of aid schemes and aid in the form of technical assistance. The third is direct foreign investment and the increasingly important role played by international corporations in investment and international trade.

WORLD TRADE AND THE DEVELOPING COUNTRIES

Long-range world trade of large dimensions is of comparatively recent origin. It essentially began and was associated with the industrial revolution in Europe. As it spread throughout Europe and crossed the Atlantic into North America, trade itself grew in volume and reach. The new industries created new resources, new demands for new materials and new markets for their products. New technologies enhanced the capacity of international transport and new areas were drawn into an ever-growing system of trading links. As has already been discussed, the great upsurge of colonial activity was very much associated with acquiring access to new raw material and controlling new markets. Much of the development, though its beginnings were earlier, dates from the mid-nineteenth century. Compared with today, trade was very small but it continued to grow rapidly until the First World War and it was essentially a European affair. In 1914 European nations accounted for around one-half of the world's export trade and some 58 per cent, by value, of all imports. Much of this trade was between European countries. The countries now called the developing countries played only a very small part in this world trade but it was a growing part. In 1900 they accounted for around 16 per cent and in 1913 for 19 per cent of world trade. Between the wars world trade declined because of the economic recession in the industrial nations, and did not begin to revive until the late 1930s. The trade of the developing countries had risen to 25 per cent of world exports by 1938 (Bairoch 1975).

After the Second World War, considerable changes took place in world trade. The need for recovery stimulated both manufacturing and trade but it was within a new situation. There began the, at first gradual, and then rapid, dismantling of colonial empires. By the 1950s a large number of countries had achieved their political independence, and by 1970 almost all had done so. These new developing countries could view trade in a different manner. Industrial developments in Europe, North America and Japan created a new demand for raw materials. Some of this was due to technological developments such as those which created new uses for aluminium and thus a greater demand for bauxite, and some due to the increasing exhaustion of European mineral resources. Importantly there developed a growing and voracious appetite for oil in the industrial nations of the world. These events involved those developing countries endowed with the resources in demand. The impact was obviously first, and greatest, in the oil-rich countries of the Middle East. The post-war recovery led to an enormous increase in production in the industrialized countries, the like of which the world had never seen before, and to a big upsurge in trade. Much of this trade was, however, concerned with manufactured products. The nature of manufacturing industry was changing. Processing was becoming a much larger part of input as products became more sophisticated. The relative contribution of raw materials fell away so that as total trade increased, trade in raw materials, though it rose, did so by a smaller amount. In this trade in manufactured goods the developing countries were little involved, so as the world trade grew, and with it that of developing countries, their share of the total fell. From being 25 per cent of world exports in 1938 and 31 per cent in 1950, it had fallen to 21.3 per cent by 1960 and 17.8 per cent in 1970. (These figures exclude China though her trade was very small despite her size.) In 1990 the developing countries' share of world exports was approximately one-fifth including China. From a situation which had existed in the 1930s of a positive balance of trade, the developing countries, as a group, moved into the situation of negative balances which has persisted since the early 1950s. In the developing countries as a whole, the rate of growth of their trade was less than that of their Gross Domestic Product. It is important to stress that this general description conceals important differences. Those countries which possessed valuable primary products in great demand did not experience this decline and adverse balances of trade. In the rest of the developing countries trade was thus even less buoyant. By 1970 Latin American countries demonstrated trade deficits and so did those of Asia; Africa was in slight credit and the Middle East showed a major credit balance.

The big growth in world trade has been, therefore, mostly in manufactured products and has been between the industrialized nations. At the end of the Second World War most developing countries were largely primary producers and many still are; their participation has therefore been of a different kind and on a smaller scale but has been changing. Four main trade groups have emerged: the industrialized nations, dominating world exports and imports; the communist Soviet bloc countries characterized by very low levels of trade, though the situation is now changing as the countries of the former communist bloc are entering international trade; the oil-producers of the Developing World, most notably the high-income oil producers whose exports of a highly marketable primary product have been massive and sustained until the recession of the 1980s; and lastly the developing countries. How, therefore, has trade evolved in the developing countries since 1945 and how has it been incorporated in their developing strategies?

The economies of the vast majority of developing countries are open economies, part and parcel of the world's economy and linked to it by trade. This offers possibilities for the pursuit of balanced growth with an unbalanced internal economy. One obvious path is to identify any comparative advantage which may exist and afford export opportunities; it could be the export of primary produce from farm, forest or mine or, indeed, it could be a manufactured product. These commodities could be produced by focused internal developments, exported and earn revenue which could then be used to buy in items not produced at home because of the unbalanced concentration on exports. These imports could be of consumer goods or of capital goods designed to broaden the basis of the home economy. If the exports were primary products the strategy would be that of Chenery's primary specialization category.

To follow the path of export-led development meant, for most developing countries, the export of primary products such as sugar from Guyana, cocoa from Ghana, tea from Ceylon, bauxite from Jamaica, copper from Zambia or rubber from Malaysia. This policy has faced a number of difficulties. First, over a long period of time the value of raw materials relative to manufactured products has declined (United Nations Organisation 1949). The effect has been for primary exports to buy increasingly less manufactured imports in exchange. It is a process tending to retard the rate of development. Second, the world market for primary products is notoriously volatile. In agricultural products it may be a reflection of weather-induced good years and bad years. Periods of shortage may influence new developments in other developing countries as they see emerging market opportunities and, as a result, similar environments in tropical South America, Africa and Asia will be competing with each other in commodities with inelastic demands. Supply shortages stimulate the search for substitutes, the development of synthetic rubber being an example. Technical developments in manufacturing may increase the demand for a particular raw material, say tin, to the disadvantage of another, say copper. The very nature of agricultural production, particularly when it involves tree or bush crops such as rubber, cocoa or coffee, means that it is difficult to react readily to changing market conditions. The alternative markets for cocoa, tea and coffee, for example, are singularly limited yet it is not possible in the short-term to grub up the tea plantations and plant cocoa trees even if it were environmentally feasible. In mineral production capital investment is commonly on a large scale and is very specific to the enterprise. A copper mine and its equipment is designed solely to produce copper ore. It is this lack of flexibility in alternative output which makes the primary producer especially vulnerable to the vicissitudes of the world market but it is particularly significant in the context of the characteristically undiversified economies of low-income developing countries. The primary products exported often account for a very high proportion of the exchange revenue which is earned. In 1960, for example, cocoa exports in Ghana accounted for 60 per cent of the country's foreign exchange revenue. In Zambia, copper represented 97 per cent in 1979, and in Nigeria in 1960, before the growth of oil exports, cotton, groundnuts, cocoa and other crops accounted for 89 per cent of such earnings. This means that if prices for these commodities fall, the impact upon the national economy is dramatic. Home markets are usually too small to absorb the surplus production and to act as a buffer; fluctuations in world trade are thus felt immediately.

In these circumstances, the need is to plan for a more diversified economy, and to establish manufacturing which can itself absorb some of the primary production, by import-substitution save on exchange currency and, subsequently, enter the market of industrial exports. Clearly the export of primary products must not be abandoned lightly since the ability to enter the market for manufactured goods will depend upon their competitiveness. To maintain an effort on all three fronts of primary exports, import-substitution and manufacturing for export, is demanding of capital investment and requires unceasing vigilance over standards of production and market opportunities. It is possible to lose, with some commodities (as for example has happened with sugar) even the possibilities of primary production to the industrialized nations but it is also possible to succeed in the face of such competition. When the development of synthetic rubber threatened the Malaysian rubber growers, Malaysia devoted much research to the growing of natural rubber, introduced technical improvements, and successfully maintained her competitive position as a rubber producer.

Enlarging the manufacturing sector is therefore a process of diversification which helps to reduce the extreme vulnerability to market oscillations which primary producers face. To become even more industrialized, and fuel development by manufacturing for export, is a further stage in the structural transformation of developing economies. It is difficult because it has to take place in the face of the competition of the large, technically and commercially sophisticated industrial nations. It has been done and new industrial nations have emerged from the Developing World.

The development strategies and trading endeavours have produced significant changes in the past three decades. Table 3 depicts the situation in 1991 in which the continuing dominance of the industrial world is shown. If all countries in the World Bank's classification of low- and middle-income countries

Table 3 World trade in merchandise 1991

	Exports		Imports	
Country group	Value (millions US$)	Percentage share	Value (millions US$)	Percentage share
Low-income economies	161,496	4.84	167,270	4.76
(India and China)*	90,539	2.71	84,209	2.40
Lower middle-income economies	214,977	6.44	243,207	6.93
Upper middle-income economies	309,972	9.29	309,050	8.80
High-income economies	2,650,106	79.42	2,788,686	79.49
World	3,336,550	100	3,508,214	100

Source: World Bank Development Report 1993
*Figures for India and China are included in the low-income economy totals

are equated with the Developing World, its share of exports in 1991 was 20.5 per cent of the world total. This figure includes the substantial contribution to exports of oil-producing countries in the Developing World. What, however, has changed is the relative importance of primary commodity and manufactured exports from developing countries. Though there is much variation in the importance of manufactured exports among these countries, in almost all they were very much more important in 1991 than in 1960, as Table 4 shows. This reflects both the relative fall in the demand for raw materials, with the exception of oil, compared with manufacturing growth, and the extent to which developing countries have succeeded in exporting manufactured products. There has been a further change. While most of the exports of the developing countries still go to the industrial market economies of the so-called 'western' nations, this proportion fell appreciably between 1960 and 1991. To an increasing extent, Developing World exports are being absorbed by developing countries.

Table 4   The structure of merchandise exports 1965 and 1991

| | Percentage of commodity exports | | | |
| | Primary commodities | | Manufactured goods | |
	1965	1991	1965	1991
Low-income economies excluding China and India	90	68	10	32
China and India	42	25	58	75
Lower middle-income economies	82	56	16	44
Upper middle-income economies	70	45*	57	57*
High-income economies	31	19	69	81

Source: World Bank Development Report 1993
*1990

Despite these developments primary commodities have remained the dominant exports most obviously in the least industrialized low-income and lower middle-income countries. The attendant disadvantages with the persistence of adverse trading balances has led many developing countries to feel that the international trading situation is working to their continuing disadvantage. The first meeting of the United Nations Conference on Trade and Development, UNCTAD, called in 1964, became very critical of the trading policies of the industrialized nations. The import-substitution policies protected by tariffs in so many developing countries were resulting in goods being manufactured at home at a much higher cost than if imported. Tariff barriers were being raised against their manufactured exports by the industrialized nations and as a result their ability to sustain development with the help of export earnings was much diminished. The argument presented by developing nations at the conference was for a restructuring of the international trading situation which would allow them a greater chance of growing through trade. The so-called North–South situation was developing. In 1976 the Conference on International Economic Co-operation also addressed this problem as it was increasingly realized that some transformation of the world structure of trade was desirable. Such a transformation will inevitably be slow and indeed it has been further retarded by the economic recession in the industrial countries which have been tempted to protect themselves by tariffs and quota allocations from the exports of the newly industrializing countries. The Uruguay Round of GATT talks which began in 1986 and has continued into the 1990s for the first time included developing nations as major participants. The talks aim at liberalizing trade by lowering tariffs, allowing improved access to markets and by agreeing and controlling subsidies. A successful outcome and continuance of these discussions is of great significance for developing countries.

Table 5   Sources of manufactured goods imported by OECD countries 1970 and 1991

Source	Percentage of total imports	
	1970	1991
Low-income countries excluding China and India	0.38	0.98
China and India	0.61	3.02
Lower middle-income economies	1.40	3.62
Upper middle-income economies	2.55	6.49
High-income economies including OECD nations	95.04	85.87

Source: World Bank Development Report 1993

There are those, however, who view the international dimension and the role of trade as sinister and indeed some who attribute to it a counterdeveloping function which produces underdevelopment or, at very least, holds developing countries in a permanently tributary position. Three attitudes to trade have thus emerged: first, the attitude of those who see it as a proven way of economic advancement as demonstrated by the industrial countries; second, the attitude of those who concede that it can be of help but who claim that this is limited because the developing countries were in a colonial status, are in an economic colonial position and will remain so; third, the attitude of those who take the extreme view and see current trading structures as one of the worst manifestations of economic imperialism and claim that developing countries can only develop by realizing their own potentials in a self-reliant, independent way. The last group recommend a withdrawal from world trade and the world economy, unless the world's economic structure, and with it trade, can be radically altered (Frank 1967, 1969; Dos Santos 1970, 1978; Amin 1972, 1974; Corbridge 1986).

Are the developing nations locked into a world economic structure and trade pattern which either limits or reverses their chance of development? Morton and Tulloch address this question in their study of trade and the developing countries and, indeed, query whether their terms of trade are deteriorating in the manner alleged by the more radical commentators (Morton and Tulloch 1977). Prebisch and Singer initiated the thesis of deteriorating terms of trade referred to earlier and related it to the differing labour conditions pertaining between the industrialized and the developing nations (Singer 1950; Prebisch

1959). It was argued that since unionized labour was powerful in the industrialized countries, revenues which resulted from improvements in productive efficiency accrued to the wage earners. As unit costs of production went down, prices did not, while wages went up. The consumer, therefore, did not benefit from continuing improvements in productivity and, if he happened to be a customer in a developing country, faced either sustained or increasing price levels without the benefit of higher income. In the developing countries, increases in efficiency in manufacturing had the reverse effect. Because labour was in plentiful supply, the thesis said that it could not raise its wages and any improvement in efficiency could thus be reflected in lower prices. Goods exported from developing countries thus became characterized by falling prices while goods imported by them featured rising prices. Prebisch was therefore of the view that the continued progress of the industrialized world seemed to make its own workers more prosperous but, at the same time, produced the reverse effect among the labour force of developing countries. This argument has been countered by the application of the classical model of supply and demand. In such a situation this predicts that the world demand would be satisfied by obtaining its goods from the cheapest sources so that where certain commodities were available cheaply in the Developing World demand for them would rise and prevent a deterioration in terms of trade. The developing countries would be able to take advantage where their low labour costs led to low priced but technically acceptable products. As comparative advantages came into play the industrialized countries would specialize in goods which were competitive despite their high labour costs because they would be produced by capital-intensive means of production. However, this state of equilibrium has not been reached. In developing countries primary products are still the major exports and manufactures the major imports with all the consequent difficulties which have been discussed. Morton and Tulloch claim, however, that the thesis of Prebisch and Singer remains unsubstantiated. Deteriorations in trade terms have occurred but so too have marked improvements. The case, they would argue, remains unproven.

Emmanuel's argument, to which Morton and Tulloch also refer, again questions the beneficial effects of trade (Emmanuel 1972). Like Prebisch and Singer he argues that it is the level of wages in a country which determines the prices of the goods it produces. These wage rates, he claims, are not related to productivity but to factors, such as the role of trade unions, which are peculiar to an individual country. The price of exported goods is, therefore, little to do with their supply and demand but instead reflects the supply and demand for labour within individual separate countries. The stress is on individual countries because labour is regarded as internationally immobile; if it were mobile its movement would eliminate differences in labour costs from country to country. Since labour is not freely mobile while capital is, it is labour-cost differentials which produce international price differences. Emmanuel argues that since developing countries are characterized by abundant cheap labour their goods will always be characterized by low prices and that this results in an unequal exchange in trade if they trade with the rich, industrialized nations. He sees only two remedies to this situation; either the withdrawal from world trade with development based entirely on internal self-sufficiency, or the development of trade with other developing countries in which similar labour and other conditions pertain. The two approaches can be modified so that they combine. There are a number of questionable issues in Emmanuel's argument. It is predicated upon the belief that differential wage rates are the product of institutional differences, such as union influence, and not of differences in worker productivity. All the evidence suggests that the latter is the explanation (Morton and Tulloch 1977). Low wage rates cannot be equated with low labour costs; they turn on productivity. Neither is labour, in the sense of suitable labour, as readily available in all developing countries as Emmanuel believes. The alternative of a self-sufficient economy which is compatible with prospects of economic development does not exist for any but the largest and best-endowed nations; for most countries their resource base and domestic markets are too small and restricted.

Still other anti-trade arguments have attributed to imported manufactures from the industrialized world another undesirable role. Not only are they regarded as usurping home market opportunities but also are accused of creating a taste for manufactured goods which, coming from the industrial countries, have been made by sophisticated technologies. There can thus be two results. First, this addiction will create a market within developing countries which they cannot themselves, at first, supply. As was described in Chapter 7, they become repairers of imported machines rather than manufacturers themselves. Second, so it is argued, in order to manufacture these goods themselves rather than remain dependent upon imports, the developing countries have to adopt the production methods, often capital-intensive and of high technology, used in the industrialized countries. These methods developed under very different conditions may be most inappropriate for the developing nations and may not coincide with their own natural endowments of cheap labour. But it is, of course, possible in the case of many products to adopt labour-intensive means of manufacturing the same goods. These can be used to supply the home-market (import-substitution) or, if their quality and specifications are acceptable, to compete in the export market. This has indeed taken place as the newly industrialized nations have demonstrated and as China has done to some extent and will certainly do to a greater extent as she enters into world trade.

The situation is not, therefore, as bleak and restrictive as the critics of growth and development through trade would claim. It would be equally unrealistic to claim that the process is without its difficulties and that the present structure of world trade is satisfactory. Since self-reliance is not a feasible proposition for most countries, some form of participation in trade would seem essential. Trade between developing countries is certainly a way forward and, as has been seen, this is growing. The development of appropriate technologies, which will be discussed in a later chapter, may be of help in reducing technological dependence. But also, to quote Morton and Tulloch, 'Some trade, at least, with developed countries would seem necessary unless the baby of economic development (including economic growth as one element) is to be thrown out with the bathwater of socio-economic inequality' (Morton and Tulloch 1977).

The restructuring of world trade relationships in a way which deliberately takes into account the problems of developing countries is widely considered as something which is desirable. The Report of the Independent Commission on International Development Issues, the Brandt Report, considers this issue and states:

Action for the stabilization of commodity prices at remunerative levels should be undertaken as a matter of urgency. Measures to facilitate the participation of developing countries in processing and marketing should include the removal of tariff and other trade barriers against developing countries' processed products, the establishment of fair and equitable international transport rates, the abolition of restrictive business practices, and improved financial arrangements for facilitating processing and marketing

(Brandt 1980).

In the case of manufactured products the Report states:

The industrialization of developing countries, as a means of their overall development efforts, will provide increasing opportunities for world trade and need not conflict with the long-term interests of developed countries. It should be facilitated as a matter of international policy.

Protectionism threatens the future of the world economy and is inimical to the long-term interests of developing and developed countries alike. Protectionism by industrialized countries against the exports of developing countries should be rolled back; this should be facilitated by improved institutional machinery and new trading rates and principles.

(Brandt 1980).

Alas it is all too easy to make recommendations; to achieve international agreement in a world of apparently conflicting interests is a different matter and the distinguished Nigerian geographer Akin L. Mabogunje views the matter of trade as being but part of the wider problem of the dominance of the Developing World by the wealthy industrial nations (Mabogunje 1980b).

THE USE OF FOREIGN RESOURCES

Financial aid from foreign countries which assists in the promotion of development has been a characteristic of the developing countries' economies in the post-war period. It may be said to have had its origins in the Marshall Plan under which American aid was given to assist the recovery of the war-ravaged economies of western Europe. Such assistance has been seen, and sought, by its recipients variously as a pump-priming resource to establish initial developments, or as a means of augmenting revenue from trade or, indeed, as a necessary and continuing transfer of assets to help counterbalance the imbalance between development opportunities seen to exist between the Industrial and the Developing Worlds. Aid becomes more necessary when earnings from trade decline; when insufficient aid is forthcoming the developing nations argue more strongly the case for a reorganization of world trade to allow them a more equal share of opportunities.

The oil-exporting countries have had little need to call upon external capital resources in the form of aid and have, indeed, become donors. The higher middle-income countries, which have most successfully industrialized and enjoyed a trade-led development based upon the export of manufactures, have been able to sustain growth with less help from aid programmes. None the less, almost all developing countries have since 1945 made use of foreign resources whether it be aid in the form of scientific, commercial and technical manpower, grants, low interest loans, or loans obtained on the commercial market or direct foreign investment in economic activities within their borders. The use of such resources has become an integral part of the economic and development plans of developing countries with few exceptions. The need is felt to utilize this externally derived capital in order to sustain the impetus of development until a self-sustaining situation can be reached where, perhaps, savings will allow internally derived capital to suffice. This need clearly becomes greater in periods of world economic recession when earnings from trade fall but at such times it is less forthcoming as all nations retrench and take refuge in protectionism.

The supply of foreign capital takes several forms. Funds which are multilateral can be channelled through international agencies of which the International Bank for Reconstruction and Development, commonly referred to as the World Bank, is the most important. Official funds can be made available in bilateral agreements with individual developing countries. Outside this official aid, individual countries can borrow on the international money markets at commercial rates. Capital investments can be made by private firms which set up manufacturing, mining or other economic activities within developing countries. Lastly, foreign direct investments by firms, institutions and individuals are made in developing countries, and indeed in 1993 the fastest growing form of external finance for developing countries was portfolio investment. This accounted for 20 per cent of total investment flows to the Developing World, much of it taking place in Latin America (Ahmed and Gooptu 1993).

Aid, as opposed to commercial loans and private investment, has been given for a number of reasons. Some is undoubtedly given for humanitarian reasons to assist poor countries to break out of the low-level equilibrium trap and to offer hope to their peoples. Some aid, notably by charitable organizations such as Oxfam and Christian Aid in Britain and similar bodies in other countries, is given specifically to help the diseased and most poverty-stricken and is often given to relieve emergency situations. It deals, of course, with the symptoms and not the cause. Other aid is given because it helps the donor. The cynical point to the loans given with one hand and the tariff barriers erected against exports from the Developing World with the other. Aid can be given for specific purposes such as the construction of irrigation schemes, the building of hydro-electric installations or a factory, but given on the condition that the necessary equipment is purchased from the donor country. Again the cynic would describe this as a form of market creation and a means of ensuring employment in industrial countries. Help is also given for political and strategic reasons, to buy friends, secure allies or placate the neutral. The democratic nations of the capitalist world have given aid for these purposes; so too have the totalitarian states of the Soviet bloc though in the case of the latter it has commonly taken the form of military aid. Politically aligned aid is a very uncertain source since political shifts in ideology within recipient countries can easily result in its withdrawal. Tarrant points out that in the case of food aid the action is often not a matter of need but of countries disposing of surpluses and if an opportunity to sell occurs this will take precedence over aid. Most food aid goes to politically nurtured countries and its availability can have unfavourable repercussions on the prices of local food crops (Tarrant 1980). While such motivations still exist, much aid is genuinely given to assist the recipient countries. A greater source of concern is not that of motives behind the assistance but of its effective use. In the early years of donor schemes too little attention was paid to assessing the proposed use of aid and grandiose status-seeking projects resulted and littered the capitals and airports of the Developing World. In more recent years financial aid has been associated with assessment and advice in project design and implementation, the World Bank making notable contributions in this respect. As developing countries have gained experience, often bitter, of the difficulties in effectively using investment, so also have they become more efficient.

Who should receive aid has been a subject of much debate. To suggest that all developing countries, regardless of size, resources, or political alignment, should receive an equal, presumably per capita, share may appear equitable but it raises the problem of the definition of development status and would result in such a thinly spread capital augmentation as to be useless. Should the very poorest nations receive all the aid since their need is the greatest? The effectiveness of such a policy would turn upon how efficiently they could absorb aid and use it for development. Not to be able to do so would convert aid into poor relief and while this is a good cause it is not to be confused with capital transfers to fuel the engine of growth. Andrew Shonfield suggested that aid be focused upon two or three countries which were within reach of continuing development, which could effectively use capital loans and which, being focused upon them, could be given in such quantity as to provide an externally derived 'big push' (Shonfield 1960). None of these suggestions have materialized but official aid giving has persisted.

Official Development Assistance (ODA), funded by members of the Organization for Economic Co-operation and Development (OECD) which is composed of the world's affluent nations, is distinguished from strictly commercial loans by its low interest rates and favourable terms of repayment. It can take the form of bilateral assistance between donor and recipient or be a multilateral activity. In the early 1990s some 60 per cent of the total ODA from all sources went to the lower-income countries and 30 per cent to the lower middle-income group. In the heavily populated countries of south Asia ODA represented only a small proportion of domestic investment whilst in African countries it forms a substantial component.

The United States has been one of the major sources of aid to developing countries since it is the world's wealthiest nation. Much of its aid has been tied to purchase of American goods and equipment. Most countries in western Europe have likewise given aid, while the members of OPEC, the Oil Producing and Exporting Countries, began to give aid generously after the escalation of oil prices in the early 1970s, most of it going to other Arab or Islamic nations. The aid from the Soviet bloc to developing countries has been very limited. The amount of ODA in selected years since 1960 and the proportion it forms of the GNP of donor countries is shown in Table 6. Few OECD countries have allocated more than 1 per cent of their GNP to aid programmes.

Access to foreign resources of capital in the form of aid can clearly be important and helpful to developing countries but there are obvious provisos. Aid freely given without restrictive conditions can be the most effective in promoting development provided that the recipient government and its economic planners can identify the most significant investment and implement it. Sound economic advice, accompanied by technical support, can play an important role in this regard. There is, however, always the risk, in the view of some, that the continued receipt of aid tends to make internally derived capital less forthcoming and that it is an experience which dulls the incentive to progress independently. As Jean de la Fontaine has it, 'Aide-toi, le ciel t'aidera'.

Capital borrowed in the world's money markets is a different matter; it is borrowed on strictly commercial terms. It is, of course, common practice in all lands and, in theory it can allow investment in productive enterprises which will, by increasing production and earning revenue, pay the interest, i.e. the debt service cost. As production aggregates it will allow the repayment of the capital over the agreed period. Crucially such borrowing is dependent upon the success of the investment in earning revenue. This in turn depends not only on the cost effectiveness of production but also on securing markets for the products. A situation can arise, and indeed, has arisen, when the prospective income does not materialize in sufficient amount even to service the debt. This may be the product of world recession which restricts market opportunities, or of production inefficiencies, or both. In such circumstances the debtor country may borrow more in order to service the debt and negotiate for extended repayment terms. This means more money is leaving the country in order to pay for imported capital and a stage may be reached when more is flowing out than coming in and an adverse balance of payments results. It is vital that export earnings are at a sufficient level, and that investment of borrowed capital is sufficiently productive as to reduce the need for imports financed by capital inflows; then borrowing can continue since it will not get out of control (Mikesell 1968; Scammell 1980; Edwards 1988).

Table 6 Official development assistance from OECD and OPEC members

	Millions US$			
	1965	*1975*	*1985*	*1991*
OECD				
Australia	119 (0.53)	552 (0.65)	749 (0.48)	1050 (0.38)
Austria	10 (0.11)	79 (0.21)	248 (0.38)	548 (0.34)
Belgium	102 (0.60)	378 (0.59)	440 (0.55)	831 (0.42)
Canada	96 (0.19)	880 (0.54)	1631 (0.49)	2604 (0.45)
Denmark	13 (0.13)	205 (0.58)	440 (0.80)	1300 (0.96)
Finland	2 (0.02)	48 (0.18)	211 (0.40)	930 (0.76)
France	752 (0.76)	2093 (0.62)	3995 (0.78)	7484 (0.62)
Germany†	456 (0.40)	1689 (0.40)	2942 (0.47)	6890 (0.41)
Ireland	0 (0.0)	8 (0.09)	39 (0.24)	72 (0.19)
Italy	60 (0.10)	182 (0.11)	1098 (0.26)	3352 (0.30)
Japan	244 (0.27)	1148 (0.23)	3797 (0.29)	10952 (0.32)
Netherlands	70 (0.36)	608 (0.75)	1136 (0.91)	2517 (0.88)
New Zealand	..(..)	66 (0.52)	54 (0.25)	100 (0.25)
Norway	11 (0.16)	184 (0.66)	574 (1.01)	1178 (1.14)
Sweden	38 (0.19)	566 (0.82)	840 (0.86)	2116 (0.92)
Switzerland	12 (0.09)	104 (0.19)	302 (0.31)	863 (0.36)
United Kingdom	472 (0.47)	904 (0.39)	1530 (0.33)	3348 (0.32)
United States	4023 (0.58)	4161 (0.27)	9403 (0.24)	11362 (0.20)
OPEC		*		
Algeria		11 (0.07)	54 (0.10)	5 (0.01)
Iran		751 (1.16)	–72 (–0.04)	..(..)
Iraq		123 (0.76)	–32 (–0.06)	0 (0)
Kuwait		706 (4.82)	771 (2.96)	387 (..)
Libya		98 (0.66)	57 (2.92)	25 (0.09)
Nigeria		80 (0.19)	45 (0.06)	..(..)
Qatar		180 (7.35)	8 (0.12)	1 (0.01)
Saudi Arabia		2791 (5.95)	2632 (2.92)	1704 (1.44)
United Arab Emirates		1028 (8.95)	122 (0.45)	558 (1.66)
Venezuela		109 (0.35)	32 (0.06)	..(..)

Source: World Bank Development Report 1993
(Figures in brackets express assistance as percentage of donor GNP)
†German Federal Republic before unification
*1976
..data not available

The recession in industrial countries in the early 1980s by reducing the export opportunities of developing countries produced such a crisis of debt servicing and payment balance in a number of developing nations (Edwards 1988). As the World Bank has indicated, whereas medium- and long-term debt incurred by developing countries increased by 20 per cent per year in the 1970s, the increase in their resources and ability to repay capital and service debt was adequate.

Interest rates on loans increased by a smaller amount than prices and were less than the rate of inflation in the industrialized countries. The Bank goes on to say:

> After 1980 the position changed. Although the rate of growth of debt halved to an estimated 11 per cent in 1982, the slowdown in export earnings was sharper. As a result the ratio of debt to exports rose from 76 per cent to 104 per cent between 1980 and 1982; . . . For oil-importing countries, the ratio was far higher than at any time since 1970; for oil exporters it was no lower than it had been before the 1973–74 oil price rise . . . Boosted by higher interest rates, the ratio of debt-service obligations rose sharply from 13.6 per cent in 1980 to 20.7 per cent in 1982 . . . had export earnings risen at 10 per cent a year in 1980–82 (about half the average increase in the 1970s), the debt-service ratio would have risen by less than three percentage points instead of the seven points it actually did
>
> (World Bank 1983).

As a result more than twenty developing countries have since 1980 attempted to reschedule their debts, and in some cases loans have been written off by the lenders. By 1992 most commercial banks had ceased making voluntary loans to developing countries. The effects of this debt crisis have continued and investment in developing countries declined throughout the 1980s. Many countries have as a result modified their economic policies. The international

Table 7 Total debt-service as a percentage of exports of goods and services

	1980	1991
Low-income economies	10.1	21.0
Lower middle-income economies	16.7	19.5
Upper middle-income economies	33.0	21.1
Sub-Saharan Africa	10.9	20.8
East Asia and Pacific	13.5	13.3
South Asia	11.9	26.0
Middle East and North Africa	16.1	25.9
Latin America and Caribbean	37.1	29.2

Source: World Bank Development Report 1993

dimension of developing countries' development dilemmas could hardly be more clearly indicated. It is further support of Brown's views on the crucial significance of exogenous forces interacting with local endogenous factors in determining the development prospects of individual nations (Brown 1988). Corbridge has reviewed the three major schools of thought which currently address the debt crisis and its international causes and ramifications (Corbridge 1988). One group attribute debt problems to liquidity crises in individual countries which must therefore be dealt with by those countries; sound money policies promoting economic growth resolving the issue. Others view the problem on the global scale and argue that the crisis is a product of the instability of the global economy which can only be rectified by urgent collective action by both the lending and indebted nations. They also point out the deleterious effect which indebtedness has upon the prospects for economic development. The third view is that of Marxists who put the debt crisis in the context of the failure of the capitalist system. The solution of the 1980s debt crisis has been largely in the hands of the individual countries. Success has been greatest in Latin America which is now accepting direct foreign investment, and least in some African nations where political instability and economic insecurity have deterred investors.

FOREIGN FIRMS, MULTINATIONALS AND DIRECT INVESTMENTS

Foreign companies which set up manufacturing or other plant in developing countries represent the third method of capital inflow. The outsider identifies a business opportunity either related to natural resources or a suitable labour force, and with market potential either domestic or export. He goes to the developing country to make money but his presence has a manifold effect. He brings in all or most of the necessary capital and so helps to offset this characteristic deficiency. Second, he establishes linkages with the Industrial World. Third, he brings in experience of both manufacturing and commerce. He can thus help in the transfer of technology and the process of modernization by being part of it. The potential benefits from this use of foreign resources are considerable, particularly in the early stage of the development process. There are disadvantages which have already been discussed in the context of industrialization. To contribute to economic development, the foreign enterprise must demonstrate a multiplier effect which stimulates other developments or its operation must allow capital which would otherwise have been invested in its function to be redeployed, or it must earn revenue which can be re-invested in local development. Preferably it must do all three. If the foreign company repatriates the greater part of its profit there is a loss of income to the home country. If it re-invests most of its earnings in the host country it serves to increase the ratio of foreign to indigenous capital and if sufficient foreign firms set up this could lead to a large part of productive capital investment becoming foreign owned. If for nationalist or political reasons or on grounds of prudence this is considered undesirable and is limited, then remittances must be allowed on a sufficient scale to persuade companies to set up and stay. So while benefits can result from the use of external capital in this way, export earnings on a large scale is not one of them. The potential for dispute between host governments fearful of foreign domination or neo-colonialism on the one hand and foreign companies wary of nationalization or a moratorium on remittances on the other, is considerable. Nkrumah claimed that foreign capital was used for 'the exploitation rather than for the development of the less developed parts of the world' (Nkrumah 1965). In reality there is often a coincidence of interest between the political power groups in developing countries and the foreign investor. Exploitation undoubtedly takes place in some instances but, on balance, adequately controlled by mutually agreeable terms, the foreign investor setting up and operating plant in a developing country has a beneficial effect.

The companies which have attracted most attention and aroused most suspicion in the Developing World are the so-called multinational or transnational corporations (Tugendhat 1973; United Nations Organisation 1973, 1978; Windstrand 1975; Lall and Streeten 1977; Vernon 1977; Siddayao 1978; Mabogunje 1980b). Having grown considerably since 1945 their activities are such that it has been estimated that around one-half of the world's trade, by value, was taking place within transnational companies in 1975 and the proportion is growing. They display all the characteristics of the 'plantation' type of economy which has been discussed and because of their economic autonomy they can insulate their activities from the economies in which they operate, but their activities are much more varied and their international linkages more complex than those of the plantation economies; this increases the impact upon developing economies. They extend into the Developing World to develop tropical crops, scarce minerals and to acquire new markets but it must be stressed that the greater part by far of their interests is *not* in the Developing World though the proportion is growing. Multinationals are now the major influence in shaping world industrialization. Between 20 and 30 per cent of the world's foreign direct investment is in developing countries with about 40 per cent of this entering the manufacturing sector. Foreign direct investment is the channel for multinationals' intervention in the Developing World. Attracted by the same features which attract other foreign investments, multinationals have sought out pools of semi-skilled, low-cost industrial labour, particularly in south-east Asia; they have invested in some countries to escape import restrictions on finally finished products and they have set up assembly plants within the host countries both to serve their domestic market and to contribute to their exports. The contribution of multinationals to the exports of South Korea, Mexico, Brazil and notably Singapore, is considerable. As they have developed and prospered so the range of activities and the complexity of their multinational linkages have grown. No longer are plants in developing countries solely assembly plants for components made by the parent company; today multinationals set up subsidiary plant, making products in which there exists a comparative advantage in that particular country. It might be, for example, a product which requires a large labour input and the low-cost labour in developing countries will be the reason for its location. Within the multinational, activities can be so rationalized as to minimize taxes, import and export duties and production costs by the careful choice of location of each manufacturing and marketing activity. This situation is referred to as international subcontracting and can involve collaboration with other companies. Modern telecommunications, transport and computer facilities enable widely spaced activities in a complex of linkages to be handled efficiently. The multinationals and other forms of direct investment are having a major effect not only on industrialization in the Developing World but on its development and its participation in the global economy. Trade is taking place increasingly within multinational and between multinational corporations. As Mabogunje says 'these corporations provide today the most important institutional framework within which relationships in the productive field between countries of the centre and the periphery are being developed' (Mabogunje 1980b). What is apparent is that multinational corporations are increasingly the vehicle for the dissemination of information on all aspects of technology, markets, management skills and trading opportunities. Information technology is bringing the component parts of the global economy closer together.

The incorporation of the functions of multinationals in national development strategies is far from easy (big fish are liable to eat small ones), but it can be done as countries as dissimilar as China, Singapore and Sri Lanka have shown. There are risks and benefits and many developing countries are appreciating the opportunities for economic development which direct foreign investments can provide. As Lall and Streeton say, 'The skill in framing a policy and in negotiating a contract with multinational companies is to maximise the surplus of beneficial over detrimental effects that is still acceptable to the TNC. The key policy then, is selectivity, not ideology' (Lall and Streeton 1977).

These features have been discussed in the belief that they contribute to economic progress and enhanced productivity but with an equal awareness that these goals in themselves do not constitute development. Mabogunje raises the issue of the dominance of the affluent world over the Developing World. Writing in core–periphery terms and viewing the situation from the periphery he points out that the greater bargaining power of the core underlies its dominance and that without a reduced dependence upon the core, true development cannot take place (Mabogunje 1980b). All national development strategies are therefore implemented within an international economic framework unless total isolation from the world community is chosen. National strategies also possess dimensions other than that of the economic growth which was often the sole concern of early development plans. These dimensions must also be examined.

10 Spatial aspects of development and development planning

'The spatial pattern will change with shifts in the structure of demand and production, in the level of technology, and in the social and political organisation of the nations'

J Friedmann and W Alonso
Regional Development and Planning 1964

Much of the thinking of development economists and much of what has been discussed so far has been concerned with national economies. Particular attention has been given to the manipulation of sectoral investments in 'big-push', 'minimum critical effort' investment schemes utilizing balanced or unbalanced approaches but with no mention of their spatial characteristics. Yet economic developments have a spatial dimension. Resource bases have one, the population for whom one is planning has one, economic activities possess one, existing levels of development display a spatial dimension. To analyse or to plan *en bloc* a nation's economy involves spatial dimensions and consequences whether they are anticipated or not. This raises the issue as to whether the resulting spatial patterns are seen as desirable and, if so, whether they can be promoted, or if they appear undesirable, whether development plans can incorporate measures to modify them.

The most casual inspection shows that in all countries levels of economic activity are unevenly distributed. It can be argued that this represents the optimum activity-pattern which corresponds with the pattern of production based on the disposition of resources of soil, climate and minerals, influenced by comparative advantage, infrastructural facilities, transport structures and distribution of population. It is, however, a pattern observed at one point of time; all the components in the matrix, including the economic significance of the seemingly immutable deposits of minerals or fertile soils, can change. The significance of locations changes, the dimensions of space change. Since the essence of development is change it must necessarily be concerned with changing spatial patterns of economic activity, with the space economy.

Societies and their economies change and develop at varying rates in the course of their history. There are periods of relatively little change when all economic activities are spatially attuned to each other and a state of near spatial equilibrium exists. One can hypothesize a pre-colonial, pre-industrial, situation in the Developing World characterized by a series of localized and almost discrete economies, though with some marketing activities over-spanning them and some centres possessing particular focus because they were centres of political and administrative power. The same could be said of medieval Europe. If this somnolescent period is disrupted by some significant new development which has spatial attributes, the equilibrium is broken and a new pattern of activity takes shape. In eighteenth-century Britain this took place with the invention of the steam-engine, the use of coal for iron-making and the emergence of a machine age which gave greatly enhanced significance to the coalfields, so much so that their location became coincident with the main and growing areas of economic activity. The industrial revolution possessed a spatial dimension and an activity-pattern quite different from that which it had displaced. In many of the developing countries, a similar disruption of a period of little change in the spatial equilibrium took place with the advent of colonial penetration. Particular areas, because of the crops they could grow or the minerals they possessed, took on a new economic significance and new spatial patterns were formed. Those who subscribe to the idea of spatial equilibrium would argue that such disruptions are inevitable if there is to be change but that the competitive market processes will, in time, balance resources, needs, markets and employment opportunities, so that a new spatial pattern of economic activity emerges which represents the optimum activity-pattern coincident with optimum economic efficiency. In reality in most economies, and certainly in countries actively stimulating economic development, the process is one of continual change. At any one time there will always be areas of a country which are less active than others because there is not the national need to utilize their resources since they are not competitive in terms of returns on investment compared with others. The rate of change is crucially important and its significance varies between political, social and economic components. For example, an economic shift to achieve the optimum mix in terms of the use of factors of production may take place rapidly, while the social or political response to the movement takes place slowly. The resulting mismatch can produce social distress and create a potential for political conflict. It is this situation which poses a dilemma for all concerned with the spatial consequences of development strategies. Its neglect has produced major problems in many developing countries and led to tragedies in some. If the readjustments to a new spatial equilibrium are natural ones which maximize the economic attributes of locations, the question is whether they should be tampered with. To do so would impede a process designed to achieve economic efficiency and with it the maximization of growth. Where maximum, unimpeded growth is the sole objective of development strategies there is no problem. If development embraces goals additional to growth, such as an equitable participation in growth, interference may become necessary. In short, regional development cannot be divorced from the objectives incorporated in a national development policy.

The emergence of distinctive economic regions requires explanation and it is one which involves movement. There is a need to explain why labour moves from one area to another, why capital is redeployed in a new location, why business enterprise shifts from one place to another. The so-called regional export or export-base model does this (Tiebout 1956, 1968; Hirschman 1958; Moses 1955; North 1964; Baldwin 1964; Perloff and Wingo 1964.) Put at its simplest, it presumes factors of production to be mobile and states than an area will grow by developing the production of a commodity or range of goods in which it specializes and in which it has a comparative advantage either because of the natural resources it possesses or because of its access to markets. These advantages, by allowing it to produce these goods efficiently and cheaply, increase its competitiveness and sustain further expansion. As it grows it begins to experience scale economies, to attract skilled personnel and becomes able to introduce a continuing series of innovations which further reduce its unit costs and enhance its ability to penetrate additional markets in surrounding regions. This increased market opportunity feeds back to stimulate further production and a chain of cumulative causation is established. The group of growing industries attract ancillary and supporting industries and services and the spin-off still further increases economic growth and employment opportunity. This process involves other regions because in order to maintain growth it not only requires access to markets in other regions but also access to their resources. The burgeoning growth may begin to draw in labour and attract capital from other parts of the country; it has an impact beyond its own bounds. Returns on investment in the growth region may become so attractive compared with others that it draws in capital which would otherwise have been invested elsewhere and its wage rates and employment opportunities tempt people to leave their home regions and seek a better future in its boom towns. According to their relative accessibility, resource endowment and the strength and growth of their own economies, the other regions of the country will be affected by the growth region's emergence. Few, if any, will be able to compete with its specialist products and will cease to make them and instead buy them in from it. These regions may possess advantages themselves over other more remote or less endowed areas and may be able to export products to them. Most regions will, however, furnish the dominant region with the resources and labour which it requires and will themselves decline, or at least remain static, requiring less capital and using a small work-force as these assets move away until, in theory, a balance is reached when production and employment are spatially arranged in the optimum economic manner and capital is invested accordingly. There can be, of course, more than one growth region and each can exhibit a wide range of sizes and shapes. For example, a growth region can be ultra-linear aligned along a transport axis, or semi-circular centred upon a port. The model, therefore, involves the concept of a regional hierarchy of major growing or core regions, secondary and tertiary tributary regions, and peripheral regions which are the most remote and backward in the sense that they are participating less actively in economic growth. The international parallel of world trade and development discussed in the previous chapter is clear. There exists, therefore, a space economy which interlinks activities within a country by strands of varying strengths. This economic space has also an expression in physical space. In the advanced economies of the highly urbanized, industrial nations, the disposition of towns and cities holds and encompasses this network of linkages and the bulk of the population.

It was Walter Isard who did much to develop a general theory of spatial equilibrium and the concept of the space economy (Isard 1956). In writing of him Friedmann has said 'His graphic representations of the topography of the space economy were particularly suggestive . . . For the first time, planners could actually "see" a regional landscape that had "structure" but no boundaries' (Friedmann and Weaver 1979). But over half a century earlier the perceptive Mackinder had described the evolving space economy of Britain though in terms of the physical manifestations which connect the abstract space economy to reality:

At first a number of small market towns . . . were scattered all over the more fertile parts of the country. They were local distribution centres at nodal points. Then a few of them, placed upon coalfields, grew as though to rival the metropolis, yet possessing no equivalent nodality. And now a certain number of the remainder are being selected for city growth, while the rest dwindle with the general loss of rural population and the improvement of communications, which tend everywhere to eliminate the middlemen. But it is characteristic of the rising places that, like the industrial towns of the coalfields, they obtain their renewed importance no longer as general distributors of the second or third grade but by specialisation of some different type . . . It follows that they are not self-sufficing after the manner of the old market towns but must supplement one another, or depend on some vast neighbouring city. Thus a continuous organisation, with clearly articulated parts, is beginning to spread over groups of counties, binding them together into great loosely-knit urban federations

(Mackinder 1902).

It is a lucid description of the evolution of a space economy.

These ideas have been developed in the context of mature integrated societies. What of their relevance to the emerging economic structures of developing societies? Three stages can be identified: first, the pre-colonial space economy with near equilibrium established; second, that of the colonial impact which disrupted this; third, the post-colonial situation with the persistence of the colonial spatial pattern of development and the consequences of that pattern for the new nation's progress.

Beginning with the second stage, it can be seen that new regions of development were identified by colonial merchants. These were the source areas of minerals or where particular crops could be grown. In addition there were contact areas, the interface between the economies of the colonial territories and those of Europe. Most commonly these were coastal contact areas. The space economies associated with these external initiatives were associated with production for export. Infrastructural developments took two main forms: first, and most importantly, the construction of transport facilities to move the products out; and second, in the case of mineral exploitation, the building of mining settlements. Both were specific to the new economic activities though transport facilities could be, and were, used for additional purposes. To a large extent, distinct, separate enclave activities were established. This new space economy was not linked into the pre-existing one except where, in the case of agriculture, peasant farmers were involved rather than plantations. Two space economies emerged, one developing and externally linked, the other changing but slowly and inward-looking; and two groups in the population became distinct, those participating in these new colonial activities and those who were not. Unequal development began to take place and this had spatial dimensions. In so much as the new economic activities gave better economic returns and often greater influence, regions of growth stood out like mountains on a plain of traditional harmonized economies. They were the nodal growth points on the development surface. Regional equilibrium theory would have it that by the movement of people, capital and technology, regional disparities in affluence, or poverty, would be diminished as the economy readjusted. But the colonial economy was geared to an external linkage rather than an internalized one so that the distinctions tended to persist.

This general statement on the emergence of the colonial space economy requires to be linked to geographical space, the situation described so eloquently by Mackinder, to alignments on the ground and the tangible manifestation of settlements. In this the role of transport is crucial. It was particularly important in this phase because the new transport network implanted by the colonial powers was so totally different from that otherwise prevailing. Railways, vastly greater in carrying capacity, replaced bullock carts and head loads. These traffic arteries possessed a great significance in the focusing of the new space economy. They created growth regions, growth alignments and growth nodes. Areas distant from them were given a new inaccessibility and became peripheral areas. The friction of space was recalibrated in a way which markedly distinguished those areas on or near the railway route from those which were not. In consequence, the existence, alignment and density of transport networks came to be related not only to the new alien space economy but also to the way in which the indigenous space economy could readjust. Being the product of colonial initiative concerned with exporting goods out of the country, the transport systems were not designed primarily to promote regional development in an articulated sense. Inevitably, however, an evolving space economy grew around them. In a classic paper, Taaffe, Morill and Gould have postulated a model which describes these developments and exemplifies them with reference to Ghana and Nigeria (Taaffe *et al.* 1963). Figure 21 depicts the essential components of their model of the evolution of transport networks.

As transport networks developed they created new aligned accessibility and low-cost axes; in economic terms, economic rent increased along their routes.

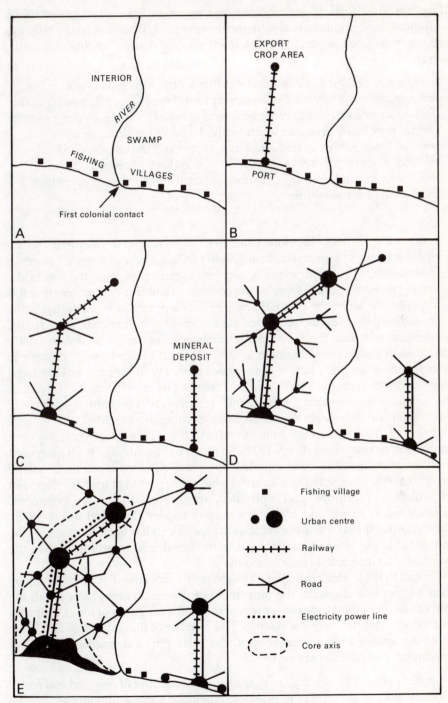

A

INTERIOR

RIVER

SWAMP

FISHING

VILLAGES

First colonial contact

B

EXPORT
CROP AREA

PORT

C

MINERAL
DEPOSIT

D

E

■ Fishing village

● ● Urban centre

┼┼┼┼┼ Railway

╳ Road

• • • • • Electricity power line

⬭ Core axis

Figure 21 Transport networks and economic development

The ports of their coastal termini took on a new significance and grew rapidly while other indigenous coastal settlements stagnated. Towns along the railways likewise prospered, especially at intersections. Mabogunje, writing of Nigeria, says:

> irrespective of their size, traditional urban centres which were not on the rail-line or on any other major routeway found themselves shunted into the backwater or economic decadence, losing many of their virile young men to centres now being favoured locationally. To make the situation more serious, the new type of trade based on export and import, exposed traditional manufacturers to competition from better finished articles. The result was the undermining of traditional craft and a worsening position in the decadent urban centres

(Mabogunje 1965).

Economic growth is, therefore, characterized by spatial development which is essentially one of regional differentiation marked by the emergence of urban centres which, as concentrations of economic activity, become the foci of the emerging growth-regions. Developing countries exhibit such a pattern and it is commonly one set by the colonial phase of development as is illustrated in the case studies presented in the second part of this book. It becomes the spatial pattern upon which the post-colonial developments must be subsequently based when goals of development are often different from those pertaining in colonial times. The very nature of the space economy in the industrialized and urbanized nations of the 'North' and the history of its growth has suggested that the urban concentrations of economic activity are the natural centres of innovation and change. In them new ideas develop, new technologies emerge and spread from one to the other involving the whole space economy in a continuing development (Berry 1972). So much so has this idea been promoted that 'urbanization' has been equated by some with 'modernization' where modernization is regarded as a comprehensive description of true development in which all attitudes and aptitudes are geared to a well-organized, technologically-based and highly productive society (Soja and Tobin 1975). Urban centres thus become the vehicles of change and it is but a short step to the concept of using them as spatial tools in the development process and in the creation of an integrated space economy.

Central to these ideas is that of movement: the movement of physical things such as people and goods; the movement of financial energy, of capital; the movement of methods; the movement of the vital intangibles of ideas, attitudes and values. Mobility is of the essence. This transfer of things, people and ideas gives dynamism to space economies and their physical manifestations. As Friedmann and Alonso express it:

> Spatial patterns will change with shifts in the structure of demand and/or production, in the level of technology; and in the social and political organisation of the nation. The economic and social development of the

nation is reflected in its patterns of settlement; its systems of flow and exchange of commodities, money and information; its patterns of commuting and migration; and its reticulation of areas of urban influence. And if there is a pattern corresponding to each 'stage' of development, it may further be suggested that there is an optimal strategy for spatial transformation from one stage to the next . . . At an advanced stage of development, the national economy will appear as a fully integrated hierarchy of functional areas, with most of the population and activities polarized in metropolitan areas

(Friedmann and Alonso 1964).

Emerging out of such statements is the spatial analogy of unbalanced growth strategies. Centres grow because they are the natural foci of development, their growth should be encouraged rather than restrained since this will maximize growth opportunities. If this means that some areas are left behind and regional inequalities appear, there is little cause for concern since the benefits of growth will spread out from the growth centres and eventually diminish the regional disparities. This argument, therefore, presumes flow and mobility which restore a balance. But do they? Are the factors of production freely mobile? Capital seeks out opportunities. It will move freely into growth centres. Will it as readily disperse into the surrounding areas where perhaps the certainty of return on investment is less? The labour force will see better incomes. The farmer will leave his subsistence holding for another occupation if he sees a worthwhile differential of income in prospect. The evidence shows that much labour movement is short-distance and localized and this is largely because in developing countries, even with radio for the illiterate and newspapers for the schooled, knowledge and awareness of opportunities is limited. But the big cities are known by repute. They can pull in labour from great distances, not by the reality of job-opportunity, but because they are perceived as places of betterment. Labour demand is an aggregate of differing aptitudes and skills and so opportunities are not coincident with gross numbers. Labour migration does not efficiently redistribute supply and demand and allow a balanced equilibrium to be reached. Material resources do move but overwhelmingly it is an inward movement. Indeed, it is the inward movement of capital, labour and materials which predominates because technical and commercial innovations are adopted more readily and are more meaningful in the complexity of interactions characteristic of the large centre rather than the surrounding areas.

In all these cases the issue centres upon the transfer of information; information about investment opportunities, job opportunities, information about markets for materials, information on the latest technological and commercial developments. Mobility is related to the diffusion process. The equilibrium theory and the views of those who mistook Perroux's abstract mathematical space economy with its growth poles for what he called *geonomic* space, or the space of reality, postulate that ideas and benefits will diffuse outwards from growth centres (Perroux 1950; Boudeville 1966; Darwent 1969).

This 'spread effect', as experience has shown, is greater in the immediate proximity of the growth region or city, the *core,* and diminishes rapidly with distance into the periphery. The 'backwash', the drawing into the core of people and resources from the periphery, has a longer reach but it too is subject to a distance-decay effect (Soja 1968; Riddell 1970; Gould 1970; Weinand 1973 writing on Kenya, Sierra Leone, Tanzania and Nigeria respectively; Robinson and Salih 1971; Parr 1973; Gilbert 1975).

Increasingly the evidence in the Developing World is that nicely balanced and integrated space economies have not emerged. It would appear that the backwash effect often does mean the sucking out of resources and that the movement to the centre is not compensated by the spread effect. Indeed it was Myrdal who as early as 1957 had coined the terms 'spread' and 'backwash', choosing the latter to stress the draining of resources. The surrounding areas lose, in Mabogunje's words, 'their virile young men', their business entrepreneurs, and their physical resources which move to the town for processing. Since essentially the comparison is between big cities on the one hand and the rural areas and small country towns on the other, the process is one in which the differences between rural and urban levels of earning widen appreciably. This stimulates a movement to the big cities where the rural migrant tries to find work and leads to a mushrooming urban growth, increasing urban unemployment and a failure of urban facilities to cope with the influx. On the one hand inward migration and rapid urban growth present problems, while on the other the benefits of the spread effect into the countryside are impeded by the need for radical changes in the rural economy and society to accommodate the spread benefits effectively. These changes, as has been discussed, cannot take place readily; the many blockages in the path of agricultural and rural development such as the need for land-tenure and land-consolidation reforms restrict the take-up of opportunities. As Friedmann puts it: 'for "spread" to occur, the structure of the rural economy must be thoroughly isomorphic with that of the city; it must be organized (or capable of being organized) along commercial, capitalistic lines. In addition, the periphery must be economically accessible to the core area' (Friedmann and Wulff 1976).

There are additional obstacles to the diffusion of ideas and economic transformation. Since spread effects are spatially limited, the process is most effective when it can take place around a series of towns which form part of a hierarchical network. This network of towns constitutes the paths which the diffusion process can take and it can, in the Mackinder fashion, bind them together 'into loosely-knit urban federations' (Mackinder 1902). Such networks in most developing countries are conspicuous by their absence. Their economic history has produced in most cases few urban centres. Often one single centre has grown so that it dominates all others. Urban primacy is common and forms a singularly inadequate mechanism for the diffusion of benefit and innovation in all but the territorially smallest countries.

Not only is the urban network unsatisfactory, but the slopes of the diffusion surface down which ideas, innovations and benefits can, as it were, slide, are neither regular nor continuous. In the abstract world they might be so; in the reality of the Developing World they are not. Distinctive divisions within the population are common. Tribal differences, religious divisions and ethnic plurality are frequent and significant. The importance of information flow has been stressed. This tends to occur within communities and within tribes, religious and ethnic groups; in other words, diffusion is often culture related. In so much as there may also be strata of society related to status and wealth this may further break up the diffusion surface. The spread effect is selective and travels at different speeds in the distinctive groups. Its impact will, therefore, tend to disaggregate the periphery rather than weld it into the fabric of the space economy since cross-cultural flows are impeded. Because these distinctive groupings of the population are often spatially specific this means that some parts of the peripheral zones will be quicker, and others slower, in adopting innovations. Even when they are not, it is possible to see in many parts of the Developing World evidence, albeit superficial, of the extent to which 'modernization' has varied between distinctive communities in the same area.

Throughout the Developing World the population is becoming increasingly urbanized. Between 1970 and 1981 the rate of growth of the urban population in the low-income economies averaged 4.4 per cent per year compared with 1.9 per cent for the total population. In the period 1980–91 the rate of growth of the urban population in low-income economies, excluding China and India, was twice that of the population as a whole. In the lower middle-income countries the rates were 3.3 per cent for the urban and 2.0 for the total population. Much of this urban population is accumulating in a few cities, often a port established in colonial times or the capital. In 1990 in half of the low- and lower middle-income countries over one-third of the urban population lived in the capital city, commonly the largest, and in 19 of the 84 countries the proportion was at least 50 per cent. Cities over one million accounted for 31 per cent of the urban population in low-income countries and 39 per cent in lower middle-income nations. This marked concentration has not produced an adequate involvement of the peripheral areas. Existing urban hierarchies have not proved to be the efficient machine of diffusion the model describes.

McGee indeed challenged the assumption that the large cities of the Developing World would generate development, writing 'in the context of the majority of the Third World Countries, it seems that a theoretical framework which regards the city as the prime catalyst of change must be disregarded' (McGee 1971). He regarded these cities as parasitic upon surrounding areas rather than the diffusion centres of development. Gottman, on the other hand, holds fast to the view that large cities are an essential component in development.

In the last 500 years no country in the world has achieved an advanced status without relatively large cities. In the twentieth century, urbanization has been the major characteristic of the acquisition of national wealth. No backward country pulled itself out of such status without large cities . . .

Large cities do not solve the problems. But cities are the instruments of development, lasting wealth and, with their turbulence, of social dynamics. There has been no known alternative for the masses

(Gottmann 1983).

Two problems have emerged. On the one hand a few cities in each developing country have grown so quickly and become so large that the expansion has outstripped the ability of the city to provide adequate housing, facilities and formal employment. The movement to the city has not been one of universal benefit and opportunity. On the other hand, the existence in the periphery of vast tracts of territory and people in a state of stagnation, or indeed decline as their resources drain away, presents a problem of human, economic and political dimensions. For many the promised new world of independent development and affluence is still but a dream; for even more it appears a cruel delusion.

To tackle these problems embedded in the spatial dimension of development, the issues have been reconsidered by scholars and by governments which have made attempts to remedy them. Three particular areas of concern have drawn attention. One is that of the spatial imbalance produced by primacy coupled to which has been the strategy of urban implantations in undeveloped areas to act as major growth poles. The second is the inadequacy of the urban hierarchy to diffuse developments. The third is the problem internal to the big cities themselves and this will be discussed first.

THIRD WORLD MILLIONAIRES?

Many in the Developing World still do not live in towns. The percentage of urban dwellers averages only 27 per cent in low-income countries excluding China and reaches 52 per cent in the lower middle-income group. The towns which do exist include some very large settlements. China and India, the world's most populous nations, have 78 and 36 cities, respectively, that are over 500,000 in size. A further 121 such cities exist in other low- and lower middle-income countries.

Cities in the developing countries are growing for two reasons. One is natural increase but the other is migration. This inward movement is part of Myrdal's 'backwash', the sucking in of resources from surrounding areas. The migrant moves because he perceives better opportunities than in his home village or rural town. Income differentials do exist but the job opportunities available do not match the scale of influx so urban unemployment and poverty are a major characteristic of developing countries. Some, the neo-Marxists, see this as a further manifestation of capitalist exploitation of the Developing World. In their eyes most of the large cities of the developing countries owe their economic character to the investment and activities of foreign firms of the kind discussed in the previous chapter. These firms, they argue, adopt capital-intensive forms of manufacturing brought in, unchanged, from their industrial homelands with the result that not enough jobs are provided. It can, of course, equally be argued that the excess of people over jobs is the product of false expectations on the one hand and the lack of opportunities and the poor quality of life in rural areas on the other. People may move because they feel it cannot be worse and may conceivably be better.

This in-migration takes several forms. Much of it is short distance and may be a direct rural to urban movement. This is true of the immediate surroundings of towns whether large or small. In other cases where a more comprehensive urban hierarchy exists, the movement may be from small town to larger and eventually to the giant city. The urban experience gives the taste for the urban life. Wilkinson, for example, has shown that in Lesotho migrant workers who have experienced wage-labour and urban life in South Africa, where, of course, they cannot stay permanently, upon return do not rejoin their old village communities but settle in Maseru the capital and largest city (Wilkinson 1983). Neither is all urban settlement permanent. In Africa particularly, but also in the Indian sub-continent, the migrant may return to his rural origins after a period of years whereas in Latin America, the move is commonly permanent. The explanation may lie in differences in agricultural practice. In Latin America the male is often a paid farm labourer whereas in Africa and the Indian sub-continent family small-holdings are more common and the women work the land while the men are away. The majority of migrants are young and in the child-bearing age groups so that they further increase the potential for growth of the cities.

Whatever the reason, and investigators of them have been many (Simmons *et al.* 1977), the city pulls in migrants in their thousands as individuals, in families and in kinship groups, all of whom identify better prospects there. They often join relatives or former neighbours and the feed-back of experience further stimulates migration. What then is the problem? It has two parts. First, immigration has not been produced by a demand to fill available vacancies in the industrial sector or indeed in other activities. The influx has far exceeded the capacity of manufacturing industry, located in the cities, to absorb it. Even the servicing sector, which in most cities in developing countries is large and is more labour intensive than manufacturing, cannot absorb all immigrants. Many immigrants have neither the training nor the aptitude for the jobs which do become available. Formal unemployment is thus a problem and can be a source of poverty. The second part of the problem relates to both the speed and the magnitude of the growth of city populations and to urban poverty. There is a mismatch between housing demand and housing supply. This raises property prices and rents. The low-paid and unemployed can neither rent nor buy. The city authorities cannot build sufficient public housing. The result is a spread of squatter settlements, that is settlements without legal title to the land used, around the cities, within it on sites too difficult for conventional housing and, indeed, on any vacant lots.

There has emerged, and this is characteristic of nearly all cities in the Developing World, a local, immigrant-produced and unconventional solution to these two problems. They create their own jobs and build their own housing.

They are solutions often viewed with suspicion, apprehension or outright opposition by the municipal authorities. The immigrants create 'informal', as opposed to conventional settlements and engage in an 'informal' economy to solve the employment problem. Two cultures and two economies operating within the same social and economic complex have thus emerged. One is made up of the salaried and wage-earners who derive regular incomes from conventional occupations be they bank managers or bus drivers, factory workers or company directors, civil servants or house servants. They live in private houses or apartments as owners or tenants or in municipal housing; they utilize and pay for public utilities and they pay local taxes. The other is composed of those with no conventional income and no accommodation other than that which they have built. Their huts are made of wood and scrap metal; any electricity used is pirated from illegal mains connections; sewage disposal is non-existent and water supplies at best will be the occasional stand-pipe; rather than on paved streets, their shacks stand on tracks dusty in drought and bemired in the rains. It is an assemblage characteristic of the shanty towns of Latin America, Africa and Asia. The inhabitants of these places engage in petty trading, in menial irregular services, in manufacturing at home in a putting-out system practised by small manufacturers, in salvage sorting and collection from refuse dumps and, in some instances, in prostitution and crime. They use their ingenuity, tenacity and capacity for hard work to survive in a situation which would otherwise crush them.

The scene depicted is, inevitably, full of generalizations. Between the formal and informal sectors the divide is not so sharp. The inhabitants of the shanty towns can, and do, get formal employment in shops, offices and factories though they may still be unable to buy into the conventional housing market. Others with entrepreneurial talents may be so successful in their informal activities as to be able so to improve their houses as to make them indistinguishable from those in the conventionally-built streets and some may increase their incomes by sub-letting. The shanty towns are not static sites of festering poverty. There is much evidence of their self-improvement, their consolidation, which encompasses better housing, laid-out and paved streets, electricity and water supplies and sometimes sewerage. They are converted into self-built and unplanned suburbs. Since this metamorphosis turns upon the acquisition of wealth it is by no means inevitable nor does it involve all and in the sifting process, the unsuccessful, the poor, drop out into a grim poverty. For those who stress this aspect it is, of course, useful to remember that the quality of life in, say, a poor Indian village on the Deccan is hardly one of comfort, affluence and rustic charm. None the less the vast areas of slum property and urban degradation which characterize such great cities as Calcutta and Bombay, or the large cities of Africa such as Kinshasa and Addis Ababa, can only be described as a problem, 'informal' solutions or not. As E. H. J Johnson puts it:

More than one Asian or African country is discovering that a planless drift of workers in the prime years of their potential productiveness to sprawling, slum-cursed cities, where huge manpower reserves already exist, may mean not only a tragic misuse of human resources but an equally prodigal wastage of scarce capital by reason of unwarranted pressure in all varieties of municipal facilities. Maintenance, repair, and depreciation costs accelerate; the costs of public surveillance, sanitation and welfare escalate at rates wholly out of proportion to the growth in the public revenue-yielding capacity of the urban economy. The presumed beneficence of urbanization is not very evident in Manila, Jakarta, or Calcutta!

(Johnson 1970).

The problem described is the problem of underdevelopment and the difficulty of developing. The fact that it exists in the proximity of modern factories, high-rise buildings, international banks and streets jammed with motor cars should not obscure this truth. The rapid rate of population increase in developing countries had been identified as an impediment to progress. In their great cities this rate of increase is much faster. Concentrating so many into so few places sharply focuses and exacerbates the problem and it becomes a microcosm of the national and international scene of underdevelopment and the development process. What are the solutions? Several lines of action have been taken and none have been wholly, or even largely, successful.

A possible approach is to focus all attention and resources on the city problem. But to build more accommodation, even at low cost, does not build jobs and it is the lack of adequately remunerated employment which is at the root of the problem. Attempts to create more jobs in productive industry, particularly if using labour-intensive methods, encounter all the problems of industrialization which have been discussed so it is not a quick and easy solution. The provision of non-productive jobs in the administrative and servicing sector, which is labour-intensive, can provide additional employment and has done so in many of the large cities of the Developing World where overmanning is so visually evident. But this increasing subdivision of employment does little to raise levels of income and it still leaves the question as to who is to pay the piper. Investment directed into the city still further deprives the non-urban areas and smaller towns of the opportunity for advancement and exacerbates the problems of urban primacy. Yet the problem of these big cities cannot remain untouched. The few major successes include the city states of Hong Kong and Singapore where essentially there is no periphery, it is all core, and where the ability continually to increase productive capacity has been established and investment is all urban-focused (Chapter 19). In the face of massive inflows of refugees Hong Kong has accommodated most of its dwellers in public and private accommodation though it has not totally eliminated squatter settlement. Singapore, without such immigration problems, has successfully housed its population and whereas Frazer in 1952 was able to report of thousands living in squalid conditions, the same is not true of the much larger city of today (Frazer 1952). Other cities in less favourable circumstances and less buoyant economies have met with little success. The building of low-cost housing has been widely practised though it does not always solve the problems of the most poor, of those who need to live near to where their formal or informal activities are practised, nor is it a solution which can be implemented on a wide scale (Gilbert and Gugler 1981). Some countries have sought to assist the squatters to help themselves by giving support at varying levels. It may be that title is given to existing squatter plots or that this, plus the provision of electricity, water supply and refuse collection, is arranged. In other words that squatting is accepted, legalized and incorporated in the city service and administrative structure. In some countries land is demarcated for squatting; in others to this is added the laying out of streets and plots with basic services provided; the incomer has only to build his house. Zambia, for example, has incorporated all these approaches in its attempt to accommodate the squatting influx into Lusaka. It has even built schools on the fringes of existing squatter settlements in anticipation of the inevitable further inflows. Most attempts by city authorities to ameliorate the 'squatter problem' are characterized, understandably, by a wish to create order and regulation where there appears to be chaos.

It is important to note two characteristics of the shanty towns and the squatting population. First, the population, because it is poor, is not necessarily incompetent, feckless and lacking in purpose or ingenuity. It is, after all, composed in part of Mabogunje's 'virile young men'. Second, both this population and the shanty settlements they inhabit are not static but in continual change. The initiatives demonstrated in devising employment and accommodation are evidence of this. The community will evolve its own evaluation of the relative importance of housing, its size, quality and cost, of proximity to a job, of educational opportunity, of entrepreneurial initiative, of a job in the town rather than life in the country. These evaluations may be totally different from those imagined by city officials, urban planners, or 'western' scholars. Furthermore, they will change as individuals' life patterns and prospects change. There are stages of evolution in the world of the city dwellers of the developing countries as in other worlds. The squatter shack, the shanty town, the informal sector could well be regarded as forms of appropriate technology. While other solutions but partially succeed or become pale palliatives, these autochthonous developments must be allowed to continue though they will never solve the problem completely.

An alternative course of action is to try to stop the big cities growing further by preventing in-migration. This can be done by prohibiting movement, an action only possible where a strongly authoritarian government can effectively enforce its laws by force or by a political indoctrination which produces compliance or by a combination of both. This degree of control and administrative efficiency is rare in developing countries, even those ruled by military regimes. Its requirements are most closely met in socialist totalitarian states of which the People's Republic of China is the most noteworthy. China has controlled, in large measure, internal migration but as part of a wider management of people and the space economy (Chapter 15). The Kampuchean policy of rustication of the mid-1970s, which emptied the cities, can only be regarded as a disaster produced by a political and malign bigotry, misguided and ruthlessly applied. For most countries an imposed control of internal migration is not a feasible solution. But there are further and obvious alternatives. These relate to the creation of more growth centres, either by stimulating the economies of existing but small towns or by building new towns; in other words, a policy distributing urban populations more widely and more thinly. This resurrects the wider spatial issues of the urban hierarchy and of counterbalancing poles of development.

CAPITAL IDEAS

As has been seen, in most developing countries, urban primacy is associated not only with urban squalor but also with marked differences in regional development. The primate city, often a port or the capital, frequently both, is the dominant in the space economy. To overcome the problems of the ever-enlarging city and regional disparities, a number of countries have selected sites in less-developed regions and built new towns to which in some cases the capital has been moved. Friedmann was involved in such a project in Venezuela. There, a new city, Ciudad Guayana, was to be built utilizing local undeveloped mineral resources of iron ore and bauxite in the industries which were to give it life. Energy was to be derived from hydro-electric installations. This new growth pole would help counterbalance the foci of urban and industrial development along Venezuela's north coast around Caracas and Maracaibo (Friedmann 1966). A new industrial city of over a third of a million people has emerged, contributing notably to the Venezuelan economy but it has not diminished the strength of the Caracas–Maracaibo axis or disseminated development into the surrounding countryside. The most famous, or notorious, example of a new city, Brasilia, was built to a grand architectural design by Lucio Costa on essentially virgin land in Brazil's interior and away from the great port of Rio de Janeiro and the industrial agglomeration of Sao Paulo. Its spectacular buildings contain the offices of government, and the houses and apartments of civil servants and others line its boulevards. This new capital, while it functions effectively as the seat of government, has done little to counterbalance the continuing growth of Brazil's great urban agglomerates or to open up Brazil's vast interior. The architectural showpiece made little provision for the work-force of its developing economy with the result that all the manifestations of the Latin American shanty towns are discernible around the periphery.

In Africa both the largest country, Nigeria, and one of the smallest, Malawi, have relocated their capitals in planned new towns. In Malawi the new capital of Lilongwe has been constructed alongside an old-established provincial town in the central part of the country in an attempt to initiate an urban growth-pole away from the major urban and industrial centre of Blantyre in the south. Spacious and well laid out, new Lilongwe is a beautiful city. Its houses, hotels and public buildings compare favourably with most countries and it is served by a new modern international airport. But it gives the impression of a city of residence rather than work, a place for car-owners rather than of workers using public transport, a place with even less industry and commerce than the old town alongside which it has grown. It is early to judge, but unless Lilongwe generates more industrial and commercial growth and houses the associated workers, its impact on the regional spread of development will be limited. Nigeria is vastly more well-endowed and populous than Malawi but despite its greater spread of large towns it too experiences a pronounced regional imbalance and exhibits in Lagos a huge port-capital development with all the ills of the Third World city. The new capital at Abuja in the underdeveloped central belt is an ambitious project designed precisely to counterbalance the political and economic concentration of power in Lagos. More thought, however, has gone into its own town plan than into the economic rationale and its completion is currently in doubt (Chapter 12). Elsewhere new city prospects have likewise been less successful. Those in the populous countries of India and China have essentially been over-spill, satellite towns rather than cities designed to give additional and alternative poles of development.

THE IMPERFECT HIERARCHY

Friedmann's experience in Venezuela led him to believe that the free operation of economic forces would continue to reinforce the strength of the cores at the expense of the periphery. He concluded that 'Economic growth tends to occur in the matrix of urban regions. It is through this matrix that the evolving space economy is organized' (Friedmann 1966). He was thinking in terms of a hierarchical central-place system in the Christaller mode. This represents the third of the three areas of interest and concern, that of the inadequacy of urban hierarchies in developing countries to serve as diffusers of development.

The idea of establishing new urban centres, or stimulating the growth of old smaller centres, the decadent towns, to make a more complete framework for the spread effect has much to commend it but its effectiveness depends entirely on how it is done. It is not simply a matter of establishing middle-sized settlements, which are missing in the rank-size distribution between the giant cities and the small quasi-village towns, and placing them at intervals at central points throughout the less-developed regions of a country. That is drawing-board geometry. It is a matter of creating centres of economic growth which, certainly after the initial phase, will be self-sustaining and will relate to the surrounding area, their hinterlands. They can create markets for hinterland-produce and offer services of health, education and commerce and while they will draw in resources and labour, they will none the less integrate their tributary areas into the development process because these new minor centres will more adequately divide up the modernization surface into smaller area units in which meaningful linkages can be established. Mehureta Assefa recommends such a solution to the problems of African development and attempts to produce a model of spatial reorganization which he hopes will correct the spatial distortions produced by colonial development (Mehureta

Assefa 1986). Jones likewise sees a strong interdependence between urban centres and the surrounding countryside as characteristic of his American experience and stresses this as a necessary component in the spatially dispersed systems of development he recommends in Kenya (Jones 1986). A similar argument in the Kenyan context is advanced by Gaile in which he refers to the Kenyan Programme for Rural Trade and Production Centres which calls for the setting up of 200 small towns by the end of the century (Gaile 1988). These towns will function not only as market centres and the sources of agricultural inputs but will also provide non-farm employment. Such towns will fill the gaps in the lower orders of the urban hierarchy and will, hopefully, be of a size to be able to achieve economies of scale in their functions. Crucial to this programme is, as has been stressed, the selection of sites which will maximize the opportunities to relate meaningfully to their hinterlands.

Since private capital tends to move to the cores and neglect the periphery, the role of the state, whether it is central or local government, is often seen to be crucial in this attempt to decentralize development. Public investment can be seen as less overtly economic, at least in the short term. It can be regarded as having a part to play in the laying of foundations and in the creation of a favourable economic environment which will make the new regional growth centres more attractive and persuade private capital to follow it. Public funds can, for example, be invested in infrastructural developments characterized, it will be recalled, by high ICOR and slow returns and often requiring large blocks of capital which are more difficult to generate privately. Transport facilities, energy supplies, public utilities, telecommunications, all sometimes encapsulated into industrial estates, represent such 'pump-priming' infrastructural investments. More directly productive state capital investments can also be made in precisely those areas where private capital is less likely to go. The private industrialist can be induced to set up plant in these new centres by grants, low-interest loans, tax concessions, low rents on sites and a whole range of temptations. It is a scene all too familiar to those acquainted with regional policies in the United Kingdom and western Europe, policies far from distinguished by unqualified success. As yet in the Developing World there is also little evidence of this policy's successful implementation with the exception of the special case of China discussed in the second part of this book.

What then have been the limitations placed upon this strategy of decentralized development? Economic science is the science of scarcities and hence of choosing between alternatives. The developing countries do not possess unlimited capital whether self-generated or borrowed; capital is scarce. Decentralized development may spread this limited resource so thinly as to make it ineffective in all places rather than effective in some. Secondly it could be argued that the periphery is by definition a sub-optimal area in the space economy and is indeed seen as such by private entrepreneurs. It is the area of slower, smaller and less secure returns. Therefore, the allocation of a proportion of the national investment to the development of these areas, it would seem, can only be done at the expense of retarding the rate of national development. In other words a distortion or manipulation of the optimum activity-pattern exacts a price. Any government must therefore balance the benefits of regional development policies against that of the overall national good. But what is this national good? Is it the GNP or does it embrace wider issues? This at once moves regional development into the political arena and the question of the goals of development policy. What then are the conceivable reasons for embarking upon a policy of regional decentralized development? First, it could be argued on grounds of equity, to improve the lot of the many and not the few. It should be noted, however, that spatial equity is not necessarily coincident with social and demographic equity though in the largely rural conditions of most developing countries that is more likely the case than not. Second, it could be said that regional development allows the use of otherwise un-utilized resources which might be of land, people or minerals. This would certainly be a valid reason in these countries with large areas of undeveloped but potentially developable territory. Outside South America and in some areas of tropical Africa, such as Zaire, few countries fall into this category. A third reason for promoting regional development would be to obtain political support. Reasons for not following such policies, in addition to their retardation of national growth, might be advanced by economic power groups within the country who regarded deviations from the apparent economic optimum as an unnecessary and too costly a philanthropy. Governments too might be influenced by the wish, or indeed the need, to see benefits go to their supportive tribal, religious or ethnic groups rather than be distributed solely in terms of impartial equity.

These factors both for and against may account for the lack of enthusiasm and lack of success of regional development policies in so many countries but there are more fundamental basic requirements if these policies are adopted and are to be successful. First, it is essential to identify accurately areas with a growth potential. Second, this growth must be a regional feeder rather than an enclave activity with metropolitan roots and connections. Branch-plant outliers of core companies are notoriously vulnerable to closure whenever the tide of prosperity recedes. Branch-plant management and government officials putting in their time in provincial outposts before returning to the core, are too centre-orientated to stimulate the essential interactions between the regional centre and its region effectively. Different attitudes are necessary. It would seem that the plugging of the gaps in the urban hierarchy has been an unsuccessful measure because alternative evaluations of development have maimed it, because most governments have neither the means nor the will to implement it, but most importantly because it is an insufficiently radical reorganization of the space economy in a manner appropriate for most developing countries. It produces the frame of the umbrella without the fabric.

Part 2 The practice

Let observation with extensive view
 Survey mankind from China to Peru
Remark each anxious toil, each eager strife,
And watch the busy scenes of crowded life.

Samuel Johnson
The Vanity of Human Wishes 1749

Dramatis Personae

Twelve nations have been chosen which exhibit many of the salient features of the Developing World and its geographical extent. The case studies begin with Fiji because within its small compass a wide range of development issues can be presented and comprehended. Fiji is thus an exemplar. The selection concludes with a consideration of the Far Eastern NICs, the new industrialized countries which have made the transition so many nations aspire to achieve; the manner of their transitions is instructive. Between beginning and end, Nigeria, Zambia, Sri Lanka, China, India, Peru and Brazil are presented. China and India, since they account for the great majority of the peoples of the Developing World, select themselves but each presents examples of significant issues. China has adopted a unique approach to the problems of development and pursued it within the framework of a Marxist command economy. India, which in so many ways invites comparison with China, has chosen a different path. In the case of both India and Sri Lanka the problem of population is emphasized and the difficulties arising from ethnic and religious plurality illustrated. In Africa, the poorest of continents, Nigeria, the largest and potentially the richest country, and Zambia, small and poor, together illustrate the significant importance of regional considerations in development and the difficulty of effectively investing export revenue productively. The distinctive characteristics of South America derive in part from her particular colonial history, a history which came to an end in a political sense long before most other areas were colonized. The two chosen countries illustrate this history and the way in which economic colonization has persisted. Peru is an example of such an extractive economy. Brazil, with the potential for becoming one of the world's major economies, illustrates many of the features characteristic of Latin America and in particular the legacy of cheap land and cheap labour.

The aim of this selection is to illustrate what has actually happened in a group of nations and in doing so illuminate the interplay of common features in particular circumstances. It is in the circumstances of this combination that the explanation of progress or stagnation lies.

11 Fiji and Pacific Island communities

"... it may be that for many of you the passage of time has dimmed the memory of the initial thrill of independence: our experience is that the feeling of independence is rather like that of leaving the cramped compartment of a jet liner. First the exhilarating and heady gulps of fresh air: then the cautious steps down to earth: and suddenly and immediately the need for direction, the offered hands to help with the burden – and then, I presume, a place in the rat race'

Ratu Sir Kamisese Mara,
Prime Minister of Fiji
Address to the General Assembly of the United Nations on the
occasion of Fiji's admission
21 October 1970

By the end of the eighteenth century, mercantile adventure, imperialism and the driving curiosity of exploration had all but tidied up the world into the 'known' and the 'unknown'. It was essentially a Eurocentric view and the 'unknown' was, in the main, the continental interiors of South America, Africa, Australia and Asia where climate, terrain and difficulties of overland transport made access less easy to the European. But there was another area little known, that of the Pacific whose very vastness meant that its scattered island communities could be as easily missed as found. Gradually the empty charts were filled and throughout the late eighteenth and the nineteenth centuries the flags of European empires together with that of the United States were raised on volcanic islands and coral atolls and beneath them was preached the Christian faith by missionaries each according to his sect. An alien culture, a sophisticated technology, new dimensions of a wider world were introduced into the small isolated communities and foreign names, New Caledonia, New Hebrides, Cook Islands, Gilbert and Ellice Islands, were intermixed with or replaced the old.

These islands of the South Seas, of all the islands of the world, possess the most romantic image. To the urbanized populations of Europe and North America they have an attraction made up of more than a physical beauty compounded of palm trees, coral sand, white surf and blue seas for many islands can boast as much. Few can, however, conjure up such images of the unspoilt life, of peaceful undisturbed remoteness, of mankind at ease with itself and nature, of life as it once was and, perhaps, still ought to be. Yet these islands are not the fictitious lands of some romantic tale but the homes of peoples with their distinctive cultures and traditions, people who are kindly, courteous and hospitable but people who are of today with today's problems, hopes and aspirations. They are not the inhabitants of some open-air museum simply to be looked at.

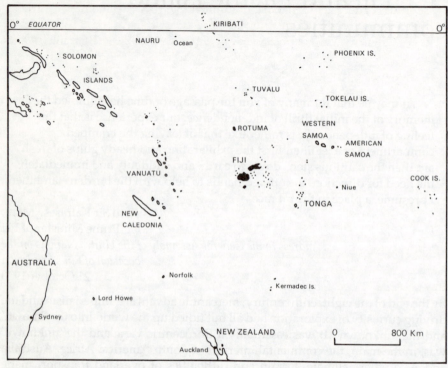

Figure 22   Fiji in the south-west Pacific

In the second half of the twentieth century these communities have become separate sovereign nations aware of other things and other places, of standards of living different from their own. They cannot revert to whatever former ways they had, though their cultures can and must remain alive and vibrant, nor can they remain static. The islands of the South Pacific present, literally, a microcosm of the problems and challenges of the Developing World. Their governments and people are planning and striving within the development process. Among the palm fronds and hibiscus flowers are prices and income policies, trade agreements and employment problems. They illustrate the ubiquitous Third World issues discussed in the first part of this book but they also highlight the special problems of small nations for all of them are small in area and population, and this is why they are here considered as an illustrative case study. All have very limited natural resources and in all cases these problems of size have been exacerbated by remoteness in the world's largest ocean where distance impresses itself even in the days of jet aircraft. The Pacific south of the equator is as large as Asia, Africa and Europe combined while its island communities total fewer than two million people. One country, Fiji, encapsulates the issues of development, is large enough and sufficiently endowed and compact to offer opportunities for substantial development, and yet faces the problems of scale. It is for these reasons that it has been selected as a case study. Other Pacific Island communities even smaller and more restricted by their geography and the paucity of their resources raise the issue of twentieth century nations whose circumstances are too limiting to allow their inhabitants hopes for a better future unless somehow they are linked in to the world at large. The ways in which this has been achieved in the South Pacific will be reviewed.

FIJI

Until the latter part of the nineteenth century when Fiji became a British colony, there was not a nation in the modern political sense but an archipelago of several hundred islands inhabited by peoples of Melanesian stock with a Polynesian component whom today are called Fijians. They had no written history but a rich oral tradition which attributed to them a common and divine ancestry and they had acquired a reputation as a fierce, war-like people hostile to outsiders. Their first contact with Europeans dates from the early years of the nineteenth century when traders came to the islands seeking sandalwood, and later bêche-de-mer, for the lucrative trade with China. The interest of the British Government was drawn to Fiji by the activities of its Australian colonists in the mid-nineteenth century but an expedition sent to survey the group's potential for colonial development reported unfavourably. It was not until 1874 that Fiji was ceded to Britain, at the request of the dominant Fijian leaders of the day, after a period of unsavoury, unofficial European speculation in Fiji. With Victorian precision an area of Pacific ocean almost the size of Texas was defined as the Colony of Fiji. Less than 3 per cent of it was land.

After ninety-six years of colonial rule Fiji became an independent nation in October 1970. In this period there had been significant changes. Fiji had acquired new peoples and a broader resource base as commercial agriculture, mining and tourism were added to the subsistence economies of former days. The future was now in the hands of the peoples of Fiji themselves and a programme for economic development was mapped out. A development plan some eighteen months in the making appeared within a month of independence and marked the watershed between the period of colonial rule and the independent future.

DEVELOPMENT ISSUES

Fiji has development problems much in common with other countries of the Developing World. There is a need to raise the material standard of living though in Fiji per capita incomes are on average much higher than in most of Asia and higher than in any country in Africa save South Africa and Libya. The average is weighted by a small but affluent commercial, professional and industrial sector whose higher earnings offset the low monetary income of the farming community which still contains a large subsistence element. The income disparity between the various sectors of the economy had considerable implications for any planned development policy. Secondly, Fiji's earnings of foreign exchange had been, and still are, closely associated with the export of a few basic commodities: sugar, coconut products and gold. Sugar alone accounted for two-thirds of the value of commodity exports in 1970. In none of these exports was Fiji's position secure. There was an obvious need for the diversification of the country's economic activities. Thirdly, Fiji's population was expanding rapidly. The need to provide increasing opportunities for employment clearly required a major effort.

In addition Fiji faced problems less ubiquitous. First, as a South Pacific community she was isolated in a vast ocean. Second, she was small. With an area of 18,332 square kilometres and a population of some 527,000 in 1970, Fiji was a small nation in which economies of scale would be difficult to achieve over a wide range of activities. The problem was aggravated by the country's archipelagic nature which breaks the country up into several still smaller units. Unlike other South Pacific countries, Fiji is a country of racial duality. Over half the population consists of Indians – descendants, in the main, of the indentured labourers brought in to work sugar plantations between 1870 and 1921. In so much as the original Fijian population and the introduced Indian group are distinguished from each other in terms of land tenure, participation in the cash economy and the extent of their urbanization, racial duality is a factor in development planning. Lastly, Fiji, in becoming a tourist centre since the Second World War, has not only developed a previously untapped resource which now contributes significantly to the country's invisible exports, but has also accentuated the problems of what might be called her 'threshold' situation. This has two facets. The first is that though small, Fiji is not too small to rule out any considerable prospect of development. She has hopes for the future which appear attainable and the earnings from tourism have helped to place her on this threshold. The second facet is sharply cut by tourism; it is that of subjective or comparative poverty. Into an island community of subsistence farmers, fishermen and peasant cane growers has come the epitome and caricature of twentieth-century affluence, the tourist. Coming from the more prosperous sections of the more prosperous countries, the tourist in his mood of holiday-spending highlights the contrast between the affluent society and that of the Third World. His presence demonstrates the comparative poverty of the peoples of Fiji and raises their level of expectations. In so much as these expectations may not be realizable in the short-term, development planning, if it is concerned solely with laying a secure foundation for sustained growth, may appear too restrictive and too slow. It must incorporate provision for the present as well as the future.

RESOURCES FOR DEVELOPMENT

Fiji is an agricultural country; 70 per cent of its population live in rural areas and these are almost entirely agriculturalists. Agricultural commodities have long dominated the export trade so clearly land is a basic and fundamental resource for Fiji's development, yet much of Fiji is steep and mountainous country and a tenth of the Dominion's area is in scattered groups of small islands. The small area of the country is an even smaller resource. Twyford and Wright classified 70 per cent of Fiji as difficult land including 40 per cent considered to be quite unsuitable for agriculture (Twyford and Wright 1965). The best quality land includes some areas in the mountainous interiors of the two largest islands, the use of which is restricted by their inaccessibility, and other areas in small islands too fragmented and remote to be significant. Of the total area under cultivation 82 per cent is on Viti Levu and Vanua Levu. The land at present cultivated represents only 8 per cent of the total area and about one-quarter of the good land. At the time of independence 85 per cent of cultivated land was under sugar cane, coconuts or rice, the remaining area being devoted to a wide variety of crops including village crops grown in systems of bush fallowing. Since the ratio of cultivated area to fallow is one to seven, the maximum area required for this latter range of crops is even greater than that under the three major cash crops. Even so this means that the total area devoted to crops in Fiji is only half the area of the better land. The area under pasture other than common grazing is only around 14 per cent of the size of the area under crops. In Fiji there was, and is, clearly considerable scope both for the extension and the intensification of agricultural production.

Much of the land unsuitable for agriculture on the steep slopes of the wetter areas is clothed in rain forest of a varied flora while secondary forest covers cut-over areas. Indigenous species include both valuable hard and soft woods, the commercial properties of which were imperfectly known until the 1960s. Approximately 248,500 hectares contains commercial species. Extraction without replanting had been common even after the Second World War but the Forestry Department began to establish a number of plantations, chiefly of the exotics Caribbean pine and mahogany, so that forest resources would not be prematurely and unnecessarily exhausted. The wet Fijian climate promotes rapid tree growth. It is estimated that pines can be grown on rotations as short as 25 years for timber and 10 to 15 years for pulp-wood. Some 810,000 hectares is considered as suitable for forestry without any encroachment being made on land of good agricultural quality. The land potential of Fiji is thus considerably increased by its ability to produce commercial timber.

Fiji's outstanding natural beauty is a further attribute of its land resources and one for which the word 'recreational' seems hardly adequate. From its coral coasts and sandy beaches to the rugged interiors of the larger islands the Fijian archipelago delights the eye with its natural beauty; the warm climate and long hours of sunshine, particularly in the northern and western parts of the larger islands, are a further attraction. These are not the only qualities which bring in tourists in considerable numbers but they are fundamental and represent a land resource which requires to be husbanded as carefully as those of farm or forest.

The development of these resources of the land is, however, conditioned by the character of land tenure. As a result of British policy designed to ensure that land ownership remained in the hands of the indigenous Fijian people, and because individual land ownership was unknown in Fijian society, some 84 per cent of the land area was at the time of independence communally owned by Fijian village communities and could not be sold. This Fijian owned land was administered by the Native Lands Trust Board which placed a large proportion into the category of Native Reserve. This land could not even be leased to anyone other than a member of the Fijian race; at least three-quarters of the total area of Fiji could only be developed by Fijians, who represented 42 per cent of the population. The government of independent Fiji did not change this situation. The remaining quarter of the country included Crown Land (now Fiji Government land), freehold and leasehold land, much of the last two categories being of high quality. It is upon this land that most of the sugar cane and rice is grown and beef and dairy cattle kept. Clearly, in realizing the full potential of Fiji's land resources particular attention must be paid to the communally held Fijian land.

Mineralization in Fiji has been on a small scale and scattered. Gold, because of its value, has been least affected by this situation and is still the only mineral of significance mined. First worked in Vanua Levu, the only gold mining remaining is at Vatukoula in northern Viti Levu. By 1970 the economic ore reserves were already approaching exhaustion. The static world price of gold and rising costs, made inevitable as increasingly leaner ores are worked, produced a financial crisis in the industry and the new government of independent Fiji in its first year of office gave the mining company a considerable subsidy to maintain operations. Other minerals, manganese and copper, have been mined on a small scale but neither has been significant.

THE PEOPLE IN THE DEVELOPMENT PROCESS

Central to the development goals and strategies of any country is its population. As has been discussed in Chapters 3 and 8, its size, distribution and demographic characteristics establish parameters for economic growth. Its health and education are both development goals and development resources. In Fiji's case racial duality has also its socio-economic and political implications.

At the time of the first census in 1881, 127,500 people were enumerated, of whom 115,000 were Fijian and 500 were Indians brought in under the indentured labour scheme. The growth of the Indian population up until 1921 when the scheme ended was largely the result of immigration. At the time of independence Indians totalled 263,500 (50 per cent of the population) and Fijians 220,000 (42 per cent) out of a total population of 527,000. The rapid population increase had produced a youthful population with 57.4 per cent under the age of 19 in 1966 and an estimated annual growth rate of 3.3 per cent. As a result of a birth control programme and other influences in fertility rates, birth rates have been reduced and the population in mid-1990 was put at 744,000. None the less the potential labour force has been growing faster than the population and the rate of population increase sets, of course, the lower limits to the rate at which national productivity must grow.

Differences in the socio-economic characteristics of the racial groups have a considerable significance for planning. Both major groups are overwhelmingly employed in agriculture. The Indians are concerned mainly with cane growing and to a lesser extent rice and vegetable production and they are commercial rather than subsistence farmers. The Fijians are predominantly subsistence or part-subsistence farmers within the traditional village economy but they produce about half coconuts exported from Fiji, almost all the bananas and cocoa and a range of other crops for the internal market. The Indian population is largely confined to the two main islands, particularly the cane areas, whereas the Fijians are more widely distributed and are the dominant group in all the smaller islands (Fig. 23). Since there was little additional land available to Indian cultivators, extra employment on the land could not be readily provided. This situation channelled many Indians into non-agricultural occupations for which they were suited by aptitude and inclination. Indians became active in commerce, industry and the professions, and became highly successful as small entrepreneurs. At the managerial level, Indians outnumbered Fijians by 3 to 1 in primary industry, 11 to 1 in construction, 8 to 1 in commerce and 7 to 1 in transport. At supervisory and clerical levels and among skilled workers there were twice as many Indians as Fijians. Most of these activities are urban-based and consequently Indians became the more urbanized group.

Education is easier and cheaper to provide in towns than in rural areas. As the 1969 Commission on Education pointed out, the scattered and rural character of the Fijian population presents severe logistic and scale problems for education (Fiji Education Commission 1970). There are 55 separate islands with Fijian schools and 130 Fijian schools can only be reached by sea. Fijian schools averaged only half the size of Indian schools and their isolated or rural character has often meant that both teacher and pupils are deprived of outside stimulus. Conditions for study in villages are most unsatisfactory and rural poverty makes difficult the maintenance of adequate standards. Secondary education for rural areas can be provided, in the main, only by boarding schools and increased expenditure. The Indian rural population is less scattered and though rural poverty presents problems, it is not to the same extent aggravated by the geography of the population of Fiji. The Commission stated that a disparity did exist in education, opportunity and achievement between Fijians and other races of the country (Fiji Education Commission 1970). This situation is one which bears directly upon the quality of manpower available as a resource for economic development and it began to receive urgent attention.

In contrast to the situation in many parts of the Developing World, the population of Fiji is healthy. There is no malaria; leprosy and filariasis are declining diseases, while tuberculosis which had long been a major disease, has been brought under control and is residual in a decreasing number of adults.

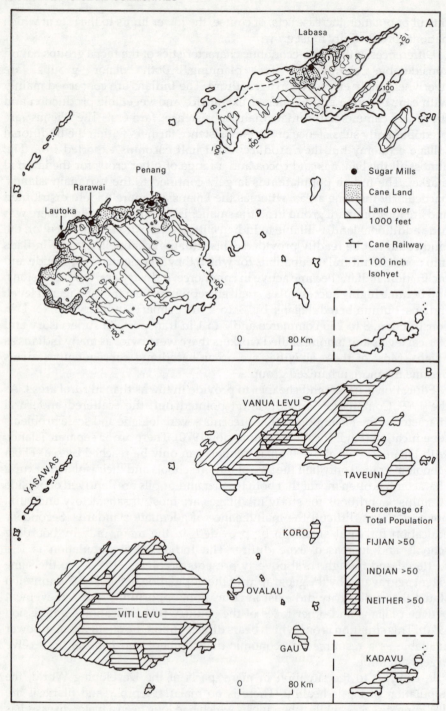

Figure 23   Fiji: (A) sugar areas; (B) ethnic groups

DEVELOPMENT STRATEGIES AND DEVELOPMENT PLANS

In 1966 the new Commonwealth Development and Welfare Act required a comprehensive development plan to be submitted by the United Kingdom's colonies requiring aid. Fiji's first such plan covered the period 1966–70. The plan produced by the newly independent government published in November 1970 was both far-reaching and sophisticated. It assessed the achievements under the modest pre-independence plan, considered the economic position at the time of independence and presented proposals which attended to both short-term problems and long-term development.

In making her proposals Fiji was very much conditioned by her particular circumstances, her resources, remoteness, scale, racial duality and the need to achieve fair and equitable opportunities for employment as her population expanded. The classical free-trade approach to Fiji's situation would be to stress her international comparative advantage in her export commodities and use the earnings from these to correct any domestic imbalances. Tourism could be viewed in a similar light. Over the years this, in fact, had been the situation. Earnings from sugar exports and to a lesser extent from coconut products and gold exports had fuelled the Fijian economy and paid for increasing quantities of imports. Good progress had been made in recent years which had witnessed both a considerable expansion of sugar exports and an increasing efficiency in its production but the situation had obvious weaknesses. Too much of the national revenue and foreign earnings was derived from one commodity and that a primary agricultural crop, subject to diminishing returns, and with a future market less certain than in the past. Secondly, as raw materials, consumer goods and indeed capital goods for investment programmes were to a large extent imported, their costs, including transport charges, were out of the control of the Fijian economy. The country was vulnerable on two fronts. It was essential, therefore, to diversify the economy: first, to extend the range of exports and so reduce the risk (tourism can be viewed as a beneficial development in this context); second, to reduce dependence upon imports by the creation of import-substitute activities in both industrial and agricultural sectors; third, by enlarging the field of activity to provide greater employment opportunity in the cash sector.

This was broad-front planning involving a balance between internal and export activities and industrial and agricultural sectors. It is the balance between those developments which calls for the most careful assessment and the most delicate of decisions. It was not possible to advance at all points. Capital availability and resources were restricted and priorities had to be allocated. Fiji was hoping for capital to be made available in two ways: first from overseas aid, essentially for government-controlled spending in developing the infrastructure of communications and public utilities but also for major agricultural projects; secondly from private investment, particularly in tourism and industry, for which the essential prerequisites were the creation of an infrastructure to make the capital productive and the maintenance of political stability. It can be argued that the execution of such a broad development plan makes calls upon administrative and organizational resources beyond the capability of a small and newly independent country. Indeed, for this reason the activity of private enterprise was to be encouraged since it would release the government from some of the burden. However, this also has its drawbacks. Local entrepreneurs and administrators are likely to be as much in short supply in industry as in government. Secondly, the private capital available was mainly foreign. In consequence both investors and management over wide areas of private enterprise could well be foreign and so raise fears of foreign domination of the economy. It will be recalled that in theory at least, broad-front development creates economic cross-stimuli between agriculture and industry and within industry allows the development of horizontal and vertical integration.

The alternative strategy to this broad-front approach, that of a focus upon one sector in which interrelations of activities are likely to be fruitful with the resulting rapid growth upsetting the economic equilibrium and stimulating further change, was an alternative not available to Fiji. The sector would have had to be manufacturing since its level of activity responds more quickly to growth stimulus and holds prospects of an import-substitution role and even of exports. As manufacturing is more flexible in adapting to market changes than primary production it would enable the country to cushion itself more effectively against changing world market conditions. Such a focus would have implied, however, a capital emphasis upon this one sector to the extent that others would be starved of it. In Fiji, manufacturing, excluding the processing of sugar and coconut products, contributed only 5 per cent to the GDP. Wage earners made up only 7.2 per cent of the employed population, the remainder being self-employed and essentially agriculturalist. Of the 48,100 farming mainly for cash, 15,600 were cane-growers. A further 29,000 were subsistence farmers. To concentrate upon a narrow-front development of industry would be to accentuate the income disparity between town and country, between main islands and minor and, to some extent, between Fijian and Indian. The development plan recommended a broader approach. It attempted to embrace the problems of racial and economic duality, of geographical dispersion and urban–rural contrast and of over-dependence upon the export of a primary product, sugar, whose growth potential was considered to be severely limited.

THE FIRST DEVELOPMENT PLAN OF INDEPENDENT FIJI

In discussing its objectives, the authors of the plan point out that economic development must encompass social development and is therefore concerned with the optimum use of resources both human and physical. They stated that the moderation of increasing income disparities was perhaps the most important single objective of the plan and from it stemmed many of the plan's proposals (Parliament of Fiji 1970). As the disparity was a rural–urban one and as agriculture was the predominant activity of the country and land one of its

Plates 1 and *2* **The future of the Developing World.**
Children in China (above) and Fiji.

Plates 3 and 4 **The problem of poverty.**
Villagers on the Jos Plateau, Nigeria. (above)
A village in Malawi.

Plates 5 and *6* **The problem of disease.**
An anti-filariasis poster in Fiji (above). Combating parasitic worms. Note the open sewer behind the notice. Port Harcourt, Nigeria. (opposite).

Plates 7 and ***8*** **The problem of population.**
Family-planning posters in the world's largest country, China (above) and one of the smallest, Tonga.

Plates 9 and *10* **Shifting cultivation.**
Clearance by burning in Guinea savanna forest (above) and cultivation mounds in a forest clearing. Central Nigeria.

Plates 11 and *12* **Nomadic livestock economies.**
Fulani cattle on the move (above) and grazing Guniea corn stubble on Hausa fields.
Northern Nigeria.

Plates 13 and *14* **Cash crops in indigenous agriculture.**
Cotton (above) and groundnuts in Northern Nigeria. The pyramids are of groundnut
sacks awaiting transport.

Plates 15 and *16* **Intensive sedentary agriculture.**
Irrigated vegetable crops, Kano close-settled zone Nigeria (above). Terraced vegetable small holdings, Sri Lanka.

Plates 17 and *18* **Rice the great food staple of Asia.**
Rice nursery beds in southern China (above). Paddy fields amidst coconut groves Sri Lanka.

Plates 19 and *20* **Small holder cash cropping.**
Coconut husking on a small holding Vanua Levu, Fiji (above). Sugar cane harvesting on an Indian 4 hectare holding Viti Levu, Fiji.

Plates 21 and *22* **Plantation Agriculture.**
A tea estate in Malawi (above). Within a rubber tree plantation Malaysia (below).

Plates 23 and
24 **Traditional indigenous manufacturing.**
Hand-spinning and hand-loom weaving by Hausa people in northern Nigeria.

Plates 25 and 26 **Contrasting technologies.**
Labour intensive and capital intensive tin mining on the Jos Plateau, Nigeria.

Plates 27 and 28 **Appropriate technologies.**
A mini hydro-electricity station in China serving a village of 300 people (above). A gas-turbine electricity generator in the Nigerian oilfields (below).

Plates 29 and *30* **Decentralized development: industry in the countryside.**
Rural brickworks (previous page) and a fan-making factory in the Chinese
countryside.

Plates 31 and *32* **Decentralized development: manufacturing in the rural economy.**
A rural tractor factory in China; the process and the product.

Plates 33 and *34* **The newly-industrialised nations.**
The Jurong industrial estate, Singapore (below). Flatted factories in Hong Kong
(opposite).

Plates 35 and **36 Dimensions of trade.**
The international dimension, the container terminal, Singapore (above). The domestic
dimension, a periodic street market in a rural town, China. (below).

Plates 37 and *38* **Urban attraction.**
A main street in Singapore (above). A main street in Nuku'alofa, Tonga (below).

Plates 39 and *40* **Townscapes in the Developing World.**
Vernacular domestic architecture, Zaria, northern Nigeria (above). Colonial
institutional architecture, Kuala Lumpur, Malaysia (below).

Plates 41 and *42* **Educational contrasts.**
A traditional Hausa school, Zaria, northern Nigeria (above). Ahmadu Bello University, Zaria, northern Nigeria (below).

Plates 43 and *44* **Natural beauty: the intangible but fragile resource.**
The river Gui in the limestone country of Guangxi province, South China
(above). Fijian shores: the peace of the South Pacific (below).

major resources, it followed that agriculture must be given special attention. A more prosperous rural population would mean a greater market for home-produced manufactures and a more secure food supply for the urban worker. It also implied the bringing into the cash economy of the subsistence farmer. As these farmers were located in the more inaccessible areas and in the outer islands, the effort for agricultural development embraced the need to improve transport and marketing conditions in these areas and so to tackle the spatial problems inherent in Fiji's geography. The agricultural effort was not considered to require putting more land in Fiji's most lucrative crop, sugar-cane. Indeed, as production per hectare was increasing as a result of the technological advances made in cane growing, the area under sugar was thought likely to decline. The need was for agricultural alternatives to sugar. As is so often the case, agriculture presented the greatest problems for resource development. The plan balanced this agricultural effort by envisaging industrial and service developments which would help to provide additional employment opportunities in urban areas.

Central to the whole agricultural scene was the position of the giant sugar industry. The plan assumed a 1975 sugar production of 368,000 tonnes based upon the maximum export quota allowed under the International Sugar Agreement and including 25,400 tonnes for home consumption. The sugar-milling company, which initiated the industry in Fiji and operated it for 90 years, decided to withdraw from Fiji because it considered the 'new contract apportioning sugar proceeds between itself and the growers, formulated under the Denning Award of 1970, to be unacceptable. After the plan was published the milling company sold its interests to the Fiji Government. The state thus became involved in a financial commitment of some F$10 million not allowed for in the plan and the owner of the country's major industry the uncertain future of which made all the more imperative agricultural expansion on other fronts.

Agricultural proposals fell into two categories. One was of small-scale investments characterized by low ICORs and the other of large infrastructural developments with higher ICORs. On the one hand the detailed proposals for agriculture make reference to training programmes to improve the small farmers' techniques, extension work to change 'value and behaviour patterns' and to facilitate the introduction of new ideas, to the provision of rural credit to finance small-scale investment in agriculture and to the need to continue and extend subsidy and price support schemes to provide financial incentives for cash farming and farm improvement. Such proposals would relate to the bulk of the farming population. On the other hand the plan incorporated a major commitment to the development of irrigation schemes for paddy rice in three of the nation's major river valleys. These schemes, far beyond the capacity of individual farmers, were designed to boost the production of rice and so reduce the need for imports.

The aim of the effort in agriculture was, therefore, to improve the income of the farming community and so reduce the drift to the towns and to prepare for a future in which the expansion of sugar production was thought unlikely and a contraction possible. The improvement in agricultural efficiency and the diversification of agriculture was not something to be achieved quickly as progress in the preceding plan's period had shown. The small scale of activities and the fragmentation of the land resource meant that more efficient marketing arrangements were imperative if improvement was to be encouraged. The plan described them as poor and uncoordinated and stated 'the creation of an efficient marketing system is the single most pressing requirement to stimulate agricultural production' (Parliament of Fiji 1970). Proposals for a national marketing authority and for the development of co-operatives were made. On land tenure there were no specific proposals but the plan said 'It is in the national interest that arrangements for the use of the limited agricultural lands of Fiji be designed to preserve and develop their productivity' (Parliament of Fiji 1970).

The plan's proposals for the development of Fiji's forest potential were muted because an FAO team was assessing it at the time of the plan. It was, of course, another primary resource but forestry development could offer a much needed diversification of land use, provide opportunities for employment and contribute to exports. Steps had already been taken in the pre-independence plan to further these ends including the extension of a plantation programme and the restriction of lumber exports so that the value added by sawing and processing could accrue to Fiji.

Manufacturing industry was described as 'not a key component'. It had, and has, however several important roles to play. For example, the gold industry was in difficulties and receiving government help. While the mine is producing gold the industry earns foreign currency so that governmental assistance may be regarded as a conversion exercise transforming Fiji's natural resources into exchangeable earnings. There is clearly an opportunity cost limit to such an operation. The Government had, however, another objective. It regarded its assistance as a short-term bridging operation to prolong the life of the mine so that alternative employment could be created for the mine workers by the time the ore was exhausted. It was important to maintain the Vatukoula community, which totalled some 5000 persons, with its infrastructure of housing, roads, power and public utilities and to prevent the dispersal of that rare commodity in so many developing countries, a trained and organized work-force. The mining company made land available for an industrial estate which could use the power, services and training facilities at Vatukoula.

Manufacturing clearly had a role in providing additional opportunities for employment at Vatukoula and in urban areas yet it was little developed. Like agriculture, it was dominated by the sugar industry which accounted for nearly 60 per cent of the net output and 40 per cent of employment in manufacturing industry. As has been seen, industrialization is so often thought of as synonymous with economic development that it may seem surprising that more was not made of it in the plan. The appraisal was, however, realistic and the problems of industrial development, certainly over the period of the plan, were recognized as considerable. It is in this field of activity where Fiji's small size hits hardest because it directly affects her international competitiveness. Fiji had not the manpower, energy resources or capital for large-scale production nor a large domestic market to support it. The possibility of import-substitute industries is often canvassed as a means of providing employment and as a saving in import costs. However, Fiji did not have a balance of payments problem and to produce goods at home more expensively than they could be imported would have been an extravagant way of providing employment in the circumstances. There were, however, other possibilities; if import-substitute industries could also command a local export market in the South Pacific this would recommend their development, particularly if they were of a nature in which scale-economies were not important. Secondly, industries processing the country's agricultural and forestry products would be viable projects if the requisite quality control were established. Such industries would add value to exports and provide employment. They would include fruit and meat canning, coconut desiccation, and, using timber, wood pulp manufacture, plywood and fibre-board making. The authors of the plan in outlining these possibilities stressed the need for a careful scrutiny of proposals and for government help by way of tax concessions to new establishments. Already government-developed trading estates, to assist industrial development had been established in Suva and other towns. Labour-intensive techniques were recommended where possible though this was not considered to be the major criterion for development.

The activity which was the most buoyant and required guidance rather than stimulus was tourism. The 1960s had seen a spectacular rise in tourist numbers, averaging 23 per cent per annum. Total spending by tourists rose from F$7.5 million in 1965 to F$26 million in 1970. However, tourist spending was on duty-free goods which were imported so that the net benefit to the Fiji economy was much less. None the less net tourist receipts served to offset the large adverse balance in commodity trading which had developed in Fiji and which was expected to increase further as imports necessary to support the proposals in the plan were brought in. Clearly tourism is a most welcome diversification of the economy. Its contribution was small (3.5 per cent of the GDP in 1970 compared with 20 per cent for sugar) but its growth potential was considerable. Furthermore, if carefully directed it could help in the development of the rural areas, including those of the smaller islands, both directly and by providing a local market for vegetables and craft goods. Like every silver lining tourism has its clouds. In the first place unless guided and controlled it could destroy the beauty and exoticism which gave it birth and detract from land development for other purposes. Land use planning is essential; this was proposed in the plan and technical assistance for an overall survey of tourism in Fiji was sought from the United Nations Organisation. Secondly, tourism must not be allowed to distort the use of scarce resources so that other essential developments suffer. The social repercussions of tourism are commonly great and some may create difficulties; the development of subjective poverty has been mentioned. Others may be undesirable and require control but given forethought and planning, tourism can be a valuable part of an economy. It is essential, however, that participation in this industry, and not merely as wage-earners but as partners, is open to the people of Fiji. Tourism, like commodity exports, is subject to external influences. The size of its clientele depends upon taste, fashion, and the changing prosperity of the countries from which the visitors come. It is also dependent upon transport technology. Most of Fiji's visitors come by air and stay for a few days en route for another destination because Fiji is a staging point on the trans-Pacific routes. Developments in aircraft technology could result in the overflying of Fiji to the detriment of its short-stay tourist traffic.

SUBSEQUENT PLANS AND DEVELOPMENTS

The first plan of independent Fiji thus addressed the crucial components in its development process. A well-conceived and balanced plan which proved to have laid secure foundations for Fiji's further development, it was not successful upon all fronts. Its implementation was facilitated by the maintenance of racial harmony though the disparities between the races were not removed. It initiated the diversification of the economy though in its short span it could hardly produce significant changes and it acknowledged the problems of the increasing movement from outer islands to the main island of Viti Levu and from the countryside to the towns but it did little to stop it. The distribution of land ownership was mentioned only in passing. External influence affected the plan considerably. While tourism did well, financed largely by foreign capital, the major sugar industry performed less well and the increase in oil prices contributed to a further enlargement of the import bill. The subsequent plan attempted to correct some of these trends. It recommended that foreign investment should not exceed 50 per cent of capital finance, that a hydro-electric station be developed to reduce dependency on oil imports, and that a greater proportion of the earnings from tourism be retained in the country. But as the international economy went into recession so the tourist trade suffered. Output from the newly acquired sugar industry declined and inflation pressures, part imported and part created by demands, increased rapidly. The trough in annual growth was reached in 1975 with GDP experiencing a decline in that year. Thereafter the economy became much more buoyant, largely due to improvements in the fortunes of the sugar industry.

This second plan, like its predecessor, aimed to increase employment opportunities, encourage a more equal participation of the races in the cash economy and to improve conditions in rural areas. It put much more stress on regional development, seeing it not simply in terms of rural–urban differences but in the income differentials which existed on a wider regional basis within and between the large islands and the many smaller ones. Importantly the idea of developing manufacturing industries within the rural areas began to emerge.

In addition to this concern with regional disparities the plan couched its proposals in such a way as to maintain or increase Fiji's ability to meet its own requirements. The reasons were obvious; the more that could be produced at home, the more employment could be created, imports could be reduced and the balance of payments situation improved. However, the very nature of the capital goods required to support development objectives meant that they could not be manufactured in Fiji. The same was true of a range of consumer goods. As oil prices further increased in the late 1970s the import bill continued to grow and helped further reduce national economic self-sufficiency. Fiji remained a country dependent upon the export of primary products with sugar strongly dominant. The growing numbers entering the labour market and consequent unemployment emerged as a major economic and social problem.

Subsequently Fiji's planners, while setting out objectives similar to those in their first plan, continued to put greater emphasis upon issues of the political economy and social justice. Later plans notably incorporate shrewd evaluations of the psychology of development. Plans became more ambitious, far reaching and more detailed in their proposals; they are difficult to fault. The third plan produced by independent Fiji, the so-called DP8, exemplified these features. There were six major components in its development strategy. It recognized the need to grow economically, to generate more true wealth so that there is wealth to be distributed. It recommended a reduction of the dependency on Fiji's two major sources of wealth, sugar and tourism, and this meant a diversification of the economy in all sectors. For example, within the sugar industry itself, it argued the case for the greater processing of sugar products within Fiji instead of a reliance on raw sugar exports and hence the production of sugar-based spirits such as rum, of ethanol for motor fuel, of particle board from the cane stalk, and of animal feed from molasses. Such developments were designed to increase self-reliance, create employment, substitute for imports and also widen the range of exports.

Secondly, the plan stressed the need for greater economic and social equity achieved by a wider participation in the development process. It drew attention to welfare provision for the poorest to cater for their basic needs. It stressed the significance of location policies in industrial development, agricultural investments and service provision which would stimulate new developments in the disadvantaged regions and islands. So significant was this issue regarded that the plan's authors devoted a second volume of their plan entirely to a comprehensive system of regional economic planning involving the deliberate planting of manufacturing and service units in identified growth centres throughout the main islands and the archipelago. These rural growth centres are the most recent and significant developments in economic planning in Fiji.

Sectoral change, regional developments and the goal of greater equity all have a bearing upon the increasingly grave unemployment problem which has been emerging since this manifestation is the outcome of the present inadequacies in Fiji's development path. The plan pointed out that increasing incomes among the employed created consumption tastes for goods which could only be imported, that too rapid an adoption of mechanized techniques in some industries had also reduced employment opportunity, that the attraction of the towns had pulled in young people from rural communities where they could be employed to the towns where they could not, and that the education system played a vital role in differential employment opportunity. These are features common to many developing countries and the plan sought to modify their effects and 'provide access to productive activities for individuals' (Parliament of Fiji 1980/81).

Thirdly, the plan reaffirmed the need for greater self-reliance, particularly by changing policies and attitudes. It stated:

> efforts will be concentrated upon increased community participation in development activities and decision-making; discouragement of reliance on non-essential imported goods and services; careful monitoring of Government assistance to ensure against reliance on hand-outs; increasing local people's ownership of assets and total investment using appropriate fiscal policies and institutional arrangements; careful screening of foreign investment proposals as to their social and economic benefits; assessment of the appropriateness of technologies imported and used and also greater local capacity to design and manufacture simple tools and implements; careful monitoring and evaluation of external assistance programmes especially their necessity and adaptability to local needs and conditions
> (Parliament of Fiji 1980/81).

Thus the plans formulated in Fiji came to embrace a wide range of proposals with social considerations and equity at their core. Rather than being solely concerned with economic growth they came nearer to the concepts of full development. Economic proposals, such as the introduction of cocoa, coffee and tea plantations, were there to further diversify agriculture and reduce the country's vulnerable dependence upon sugar cane. These and other economic initiations were implemented but the plan which was to carry Fiji into the mid-1980s stressed another need; the need to cultivate a stronger feeling of national unity in a country so divided into two distinctive major racial groups. Fijians and Indians had co-existed in harmony over many years, a conflict between them could only be regarded as disastrous for the nation. The plan advocated a greater understanding of the other's views and culture by each of the two groups and recommended steps to be taken in the fields of education, with the press and radio to further this end. In the event these hopes for continued harmony were dashed as, in 1987, the political framework within which the peoples of Fiji conduct their affairs was tested and broken.

Prior to independence, Fiji's Legislative Council was elected by three communal divisions representing the Fijians, the Fiji-Indians and the remaining minority racial groups. No one racial group had enough seats allocated to it to obtain overall control. The system was maintained after independence and subsequent elections produced an essentially Fijian government supported by much of the third communal group. Indians formed the opposition party. In 1987 the election put the governing party out of office as, unexpectedly, a Fijian splinter group voted to join with the Indian party. For the first time a non-Fijian dominated government came to power. Under the leadership of Colonel Rabuka, the army, composed entirely of Fijians, overthrew this government in a bloodless coup. A new government was set up and a new constitution enacted which ensured Fijians ruled the country as long as they voted for the same party. Many professional Indians and Indian businessmen left Fiji to the detriment of its economy. The tourist industry declined for a period and a strike by the predominantly Indian cane growers affected the harvest of the country's most important crop. After another election General Rabuka, having secured an alliance with Indian members of the Fiji Labour Party, became prime minister. The nation remains uneasy and the disruption on racial lines must be resolved if Fiji is to reap the benefits of her well-conceived development programmes.

THE MIRAB ECONOMIES OF THE SOUTH PACIFIC

Fiji exemplifies many of the features and problems facing developing countries; even its racial and political turmoil is, unfortunately, far from exceptional. Fiji has demonstrated a steady development and a shrewd use of her resources, has produced realistic and not overly optimistic plans and holds prospects for further development. Throughout the South Pacific there are, however, independent sovereign communities much smaller than Fiji and far less well endowed in natural resources. Remote, with economies based on a small agricultural production subject to the limitations of diminishing returns, such countries have but a limited prospect for material advancement. Their cultures are often rich, harmonious and sophisticated; they have a wholesomeness. But it is an equilibrium easily disturbed if their young people are unable to see a fulfilment before them and if they are aware of the affluence of others. There must be a way which allows those who are poor in the material resources necessary for the technological society but who are rich in culture to live in dignity and with self-respect. Such a way has emerged embracing many of the Polynesian communities of the South Pacific. It has been described as the MIRAB system (Bedford 1984; Bertram and Watters 1985, 1986).

The acronym MIRAB (migration, remittances, and bureaucracy) describes a system which connects isolated Pacific communities to the affluent economies of Australia and, more notably, New Zealand. This system dominates development in the participant island communities. First there is migration, but it is migration for work. Islanders who go to New Zealand are often employed as unskilled labour and send part of their earnings back to their families and communities. The remittances increase the expendable income of their families at home and the possibility of their savings. The island resources are enhanced both by the diminished drain on them and by remitted investment put into them. The migration and the remittances are part of a deliberate family decision. Migration is not regarded as permanent. There exists therefore an economy internalized within the family group which both participates economically within the modern rich community and in the indigenous home economy.

Bertram and Watters describe it as a 'transnational corporation of kin' (Bertram and Watters 1985). These islands annexed by New Zealand in the early twentieth century, such as the Cook Islands, Niue and Western Samoa, subsequently had access to New Zealand. It was possible for island families to settle in New Zealand over a long period and on retirement receive New Zealand pensions which they could take back to their home villages. Whatever the nature of the work migration, long term or short, the link with the home community was never severed and remittances remained an important inflow of income. Whenever the New Zealand economy was buoyant it required labour and this, together with population pressure on the islands, stimulated migration while air transport facilitated the move. By 1981, 44 per cent of the population of Cook Islands, 61 per cent of that of Niue and 45 per cent of Tokelau was living in New Zealand. The moves are characterized by a great deal of reciprocal visiting so the contacts are never lost. A flow of labour out and remittances in joins the islands of the South Pacific to their affluent neighbours and an internationalization of economic activities takes place as kinship groups seek out employment opportunities (Bertram and Watters 1985).

A further component in this internalized economy is aid. All the island communities receive it and it has become regarded as a permanent part of their budget. In its origin aid was regarded as an augmentation of domestic savings to be used for investment in developments. It has become in practice a form of income to be expended. Much of it is used by the island governments to finance their administrations. It is a very significant input since up to a half of government budgets are financed by aid and well over a half, in some cases up to 90 per cent, of paid employment is in government service. Aid thus creates employment, is a source of income for families and so stimulates consumption. Aid via government employment pays for a large part of imports and so helps reduce balance of payment deficits. Because some island communities have few resources in which to invest, both aid and remittances become incomes often used to buy imported goods and, indeed, in some cases diminish incentives to produce goods for export from the islands.

The net result of MIRAB is that a better standard of living can be enjoyed as part of the community maintains the traditional economies and culture and part derives benefit for the whole community by participation in the New Zealand economy. The maintenance of the island economy is important; it prevents whole communities from becoming wage-labourers completely dependent upon a distant and alien society. The system requires all its components. Continued aid is essential; so too is access to the New Zealand labour market, and both are likely to fluctuate over the years. What gives the system its strength and gives hope to the island communities is its twofold structure; on the one hand are indigenous economies never deserted and always maintained by kinship links and frequent visiting, and on the other the need in New Zealand for a continual refreshment of new immigrant workers. It is a development solution which has grown out of the particular circumstances and is one likely to continue successfully.

12 Nigeria

'. . . wonder is packed every morning into the birth of an African day. Heat and sweat and weariness come later . . .'

Elspeth Huxley
The Mottled Lizard 1962

Nigeria is the largest of African nations and, with a population put at 88.5 million in the census taken in 1992, is the ninth most populous country in the world. This African giant covers 923,768 square kilometres of territory and though some of it presents environmental difficulties they are not of the severity experienced by her Saharan neighbours of Niger and Chad. Her resource base is varied and, importantly, Nigeria possesses large reserves of oil. Yet Nigeria's development path has been far from easy. Her problems and experience exemplify many of the central issues in Third World development and demonstrate vividly the difficulties of translating development models into practice. In particular Nigeria illustrates three facets of development: first that of the spatial dimension and the question of regional development, second that of effective sectoral investments, and third the crucial role of government policies.

Upon independence in 1960, Nigeria was an assemblage of lands and peoples on the coast of West Africa demarcated by the boundaries of colonial parcellation. Her economy and political structure reflected both her pre-colonial and colonial history, histories which have produced patterns significant today. It is important to stress two elementary geographical characteristics of Nigeria. First, Nigeria has a coast and was thus accessible by sea to Europeans from as early as the fifteenth century. Second, gradual but pronounced changes in the environment take place from Nigeria's Guinea coast on the Bight of Benin to her northern limits on the edge of the Sahara. The climate becomes increasingly dry with a more seasonal and uncertain rainfall ranging from over 2500 mm per year in the south to less than 500 mm in the north. Natural vegetation correspondingly changes from high rain forest through the more open woodland of the Guinea Savanna forest to grassland with trees and eventually passes into thorn scrub in the Sahelian zone.

Agricultural potential and practice likewise vary. A series of east–west zones thus present contrasted environments and opportunities. The first Europeans along Nigeria's coast sought slaves and ivory but later traders came to know of, and barter for, the products of the interior. In West Africa there developed two contact zones. One, the older, was along the Saharan edge where the contrasted environments of the desert and the more humid south come together and where a whole string of trading cities grew up. In Nigeria they included Zegzeg (now Zaria), Katsina and, most important, Kano. This same contact zone traded products across the Sahara to the Mediterranean cultural area as well as east–west along the Sahelian edge. In the south the second contact zone was manifest in coastal towns and villages where the products of Europe were traded for the exotic commodities of tropical Africa. Between these two zones internal trade existed to some extent since within Nigeria's climatic range crops were of contrasting types.

The Portuguese made their first contact with the Nigerian coast in 1471 and were followed in turn by the Dutch, French and English. The early and notorious interest was in slaves but this traffic was totally abolished by all European nations, except the Spanish and Portuguese, in 1815. It was trade in ivory and palm-oil which claimed British interest in the Bight of Benin where towns like Calabar, Bonny and Brass grew up and prospered from the trade on this inhospitable coast, much of it backed by the mangrove wastes of the Niger delta. Further west on its lagoonal site, the fishing village of Lagos also participated in trade, first as a slave port but later for the export of palm-oil and other products from the Yoruba country to its north. Since its contact with the interior was not restricted by deltaic mangrove swamps, Lagos came to be the increasing focus of trade.

THE COLONIAL EXPERIENCE

The British interest was essentially mercantile with trade penetration conducted through companies set up for the purpose. The first territorial annexation took place in 1862 when Britain acquired Lagos so as to secure it as a trading base. In 1884 Britain declared the Oil Rivers, that is the southern palm-oil producing country, a protectorate and by the end of the century exerted political control over southern Nigeria. After a series of military forays into the north to forestall the French moving south across the Sahara, the British Government created three protectorates to govern the whole of what is essentially the Nigeria of today. Parts of Cameroon were transferred to Nigeria after the First World War. British colonial rule of a territory called Nigeria (the name was invented by Flora Shaw in an article in *The Times* newspaper in 1897) dates, therefore, from 1900. Lagos was made the capital in 1914. Nigeria's development throughout the twentieth century has been very much influenced by this colonial presence. A colonial power had implanted itself on the coast of West Africa primarily because of trading interests, had demarcated a territory over which it secured political and economic control and was able to promote a colonial form of development by infrastructural investment and the organization of production designed to this end. The structure and shape of Nigeria's economy for the next sixty years was being forged.

The British were concerned with the extraction of raw materials, mainly crops, with an initial concern with palm-oil from the wild trees of the oil-palm bush. The potential of Nigeria for the growth of cotton, groundnuts and cocoa was later developed together with the mining of tin from the deposits on the Jos Plateau. Areas which could produce these commodities received the economic stimulus of the British demand. They were linked to the coast by transport developments, mainly railways, specifically designed to facilitate export, the railway from Lagos northwards being started in 1898 and reaching Zaria by 1912 (Fig. 24A). This rail network has proved to be important in establishing spatial patterns of development.

The companies handling the growing export trade which enjoyed tax exemption until 1939, invested little in Nigeria, their activities being essentially extractive. Nigerian cotton went to Lancashire mills, groundnuts, palm-oil and cocoa to British and European factories and tin to British industry. By 1939 the United Africa Company, the largest involved, was itself handling over 40 per cent of Nigeria's export trade.

The British involvement in Nigeria has, however, a distinctive and significant feature. While tin mining was essentially in British hands, though there were a few small Nigerian operators, and while trade for export was overwhelmingly dominated by British companies, there was no British settlement, no taking over of agricultural land to produce export crops and no plantations. Agricultural production including that of export crops was left entirely in the hands of the village farmers. This meant that the demand stimulus of the export market was felt at the small farmer level. Agricultural systems did not change in any profound way and cash crops were incorporated in bush fallowing, but it did bring numbers of peasants into the cash economy who, though their purchasing power was low, themselves became a market for low-cost manufactured goods. In the Kano close-settled zone these developments stimulated intensification but, for the most part, since population pressure over much of Nigeria was low, the increase in demand created by the export market was met by extending the area under cultivation rather than by the intensification of inputs. Wider areas and more farmers thus became involved. The rail-lines were subsequently augmented by roads motorable in the dry season. These feeder-road networks further diffused production since they could reach and serve the extensive agricultural zones whereas railways were essentially the main arteries of transport linking them to the modern ports of Lagos and Port Harcourt. The former was the more important because it was established earlier, was the capital and terminated the most significant rail link to the cocoa, cotton and groundnut areas. Not only did these transport connections give the two ports growth potential, they also gave locational significance to towns upon them such as Kaduna, Zaria and Kano. Towns not linked, such as Sokoto, Katsina and Maiduguri, stagnated and became Mabogunje's 'decadent cities' though Maiduguri was later stimulated by a rail connection after independence.

Axes of economic development began to be discernible along the transport arteries and were thus skewed or aligned by this particular situation. In it the peasant farmers were the major productive agents. A separate development like that of Kenya, Rhodesia and Zambia with their white farmers did not take place. The cocooned economy of the foreign plantation was avoided and the road

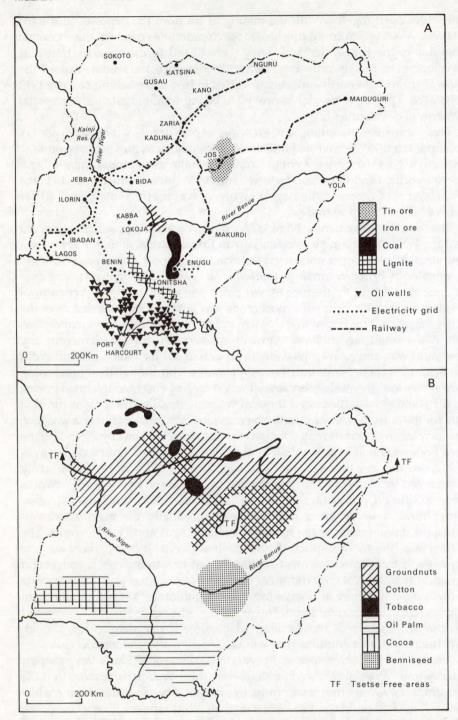

Figure 24 Nigeria: (A) minerals, railways and electricity grid; (B) major cash crops

network and peasant production spread the developmental impact of the colonial cash economy widely. Some farmers, notably the cocoa growers with their lucrative crop, became commercial farmers; most grew cash crops along with their subsistence cultivations, some remained untouched. Prior to the discovery of mineral oil, the small independent farmers of independent Nigeria accounted for 70 per cent of her exports. In the early 1960s it was estimated that some 6 million persons depended upon the sale of groundnuts for their cash income. Cotton production grew rapidly after the Second World War. The collection of raw cotton was organized by a Cotton Marketing Board and the British Cotton Growers' Association which ran the ginneries. This central organization co-ordinated the production of the many farmers, supplied them with high-yielding cotton varieties and gave advice on cultivation techniques. A major industry of minor producers was thus created. Cocoa and rubber production was again associated with small-holdings though there were a few larger plantations of rubber.

The colonial period saw the emergence of spatially specific export crop regions and, of course, of tin mining. The transport system stimulated the growth of towns along its routes. Overall, however, there was little marked differentiation in regional affluence though the South was demonstrating a greater awareness of European methods. Little had been invested which could create development in its full sense, the economy being extractive rather than constructive. Importantly that distinctive feature of the affluent societies of Western Europe, manufacturing industry, was but little developed until the 1950s. Mabogunje makes the point that it was the MacPherson Constitution of 1951 that began the process of self-government which culminated in independence in 1960 (Mabogunje 1978). During this nine-year period the newly-established federal and regional legislative assemblies were able to govern Nigeria effectively and marked its economic transmutation from a colonial to a national economy. It is significant that by almost all economic measures, the economic progress of Nigeria markedly accelerated after 1951. In this period regional differences in economic opportunity began to appear. Manufacturing started to grow and was essentially an urban phenomenon. The transport network, by giving a new centrality to towns on its routes, influenced the spatial pattern of these developments.

There was, however, another spatial manifestation which was a product of Nigeria's history and has proved to be of a very special significance. It has raised dramatically the issue of regional development. The British delineation of that part of West Africa now called Nigeria encompassed a large number of distinctive tribes, each having its own language, its own customs and its own territory (Fig. 25). Upon this tribal variety had been imposed the Islamic conquest of the early-nineteenth century when the Fulani horsemen in a jihad, or holy war, conquered and converted to Islam the greater part of the population of the northern two-thirds of Nigeria. The British contacts throughout the nineteenth century had been with the non-Islamic south which became more fully acquainted with European commercial organization, production methods

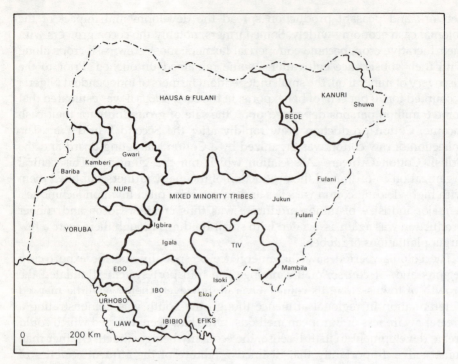

Figure 25 Nigeria: tribal distributions

and transport. With the merchants had come the missionaries who converted the southern tribes to Christianity. Importantly, mission schools were established in many villages and rural towns. English was taught and with it came the European mind and European ideas of society and means of production. It was an education which equipped the young Nigerian more readily to participate in the process of modernization, to become a clerk or an artisan in the European mode. The British did not occupy the North until the twentieth century and there was, by agreement with the emirates, no setting-up of missions and mission schools in Islamic areas. Islamic schools did not, understandably, teach English or equip children for participation in the process of Europeanization. These two factors, the longer duration of the British presence in the South and the exclusion of European forms of education from wide areas of the North, contributed to significant differences in the ability of people to participate in modern economic and technological developments.

On the eve of independence, the South was coming to have the attributes of the 'core' and the North of the 'periphery'. Within this broad division there were exceptions. The major northern towns of Kaduna, Zaria and Kano, all on the railway, and Jos, the tin town, were centres of growth and industrial development. In the South, sizeable areas in the Niger delta and the remaining high forest were areas of relative backwardness. The country was still overwhelmingly agricultural and though farming was far from sophisticated,

Nigeria could feed itself and crops were the major source of export revenue. Energy resources, though largely undeveloped, had considerable potential. Bituminous coal deposits around Enugu possessed reserves of 364 million tonnes, the largest in Africa north of the equator though small by European standards (the Netherlands' reserves total 3764 million tonnes). This coal fuelled Nigeria's railways until they were dieselized and fired the main thermal electricity stations in Lagos, Ibadan, Port Harcourt, Kano and Enugu itself (Simpson 1969, 1992). Large lignite deposits lay untouched to the north of the delta and hydro power was developed along the Jos Plateau's edge for use by the tin industry. Other resources were limited. Oolitic ironstones of the Kabba province were low quality ores. The high forest timber, though valuable, was not abundant while that of the Savannah woodland was of little value other than as woodfuel. The harnessing of these resources began before independence. Investigations into the development of the large hydro-electric potential of the Niger and Kaduna rivers began as early as 1953 though it was the independent Federal Government which passed the approving act in 1962 and work on the Kainji Dam began in 1964. Factories had been established in Kano and Kaduna in the North and in the capital, Lagos. A long and expensive search for oil was rewarded in the middle 1950s by the first commercial production by the Shell-BP company from a field in the delta. A grid network for electricity distribution was beginning in the urbanized south-west while the Kainji Dam Project on the Niger in the empty Middle Belt was of such magnitude that the establishment of a national electricity grid was essential.

The sum total of these developments meant that in the newly independent Nigeria a strong pattern of regional differentiation in terms of manufacturing industry, power supply and utilization, and oil-based developments had emerged and was to influence all subsequent developments. In the period 1958–62 three-quarters of all manufacturing employment occurred in a belt across the most southern part of the country (Fig. 26). Almost all the remainder was associated with the northern towns of Kaduna, Zaria and Kano, all on the main northern railway route. The proposals for electricity transmission developments by the mid-1960s reinforced both the east–west axis and the northern Lagos to Kano axis as foci of industrial and urban development (Fig. 24A). Since these spatial differences were identified with tribal distributions when tribal feelings of identity were strong, the coincidence of tribal distributions with levels of development took on a political significance. The new independent Nigeria of 1960 was faced not only with the development of its natural resources, the husbanding of its work-force and the enlargement of its productive capacity in a manner which would initiate true development, but was confronted with the need to produce a distinctively Nigerian political economy. The hopes and aspirations of diverse peoples whose loyalties were first to tribe and region and only second to Nigeria, had to be encompassed within the new nation. It has proved a difficult task.

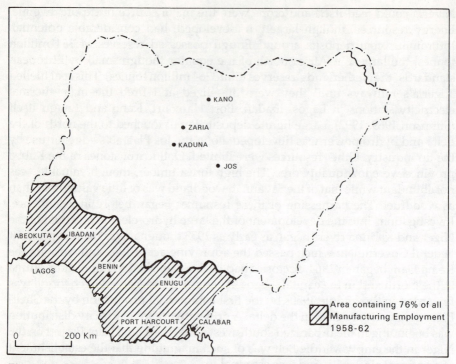

Figure 26 Nigeria: employment in manufacturing 1958–62

INDEPENDENT NIGERIA AND THE PROBLEMS OF POLITICAL ECONOMY

Nigeria inherited and retained the three regional governments of the pre-independence period within a federal constitution. Each region had considerable economic autonomy and sought to promote its own development. Representation in the federal parliament was proportional to population and attempts were made to falsify the population census of 1962 for political advantage and the census was abandoned. In 1963 a second census was carried out. Its results were disputed and subsequent analysis has revealed that it too was manipulated. None the less it remained the only accepted Nigerian census for nearly thirty years and the size of Nigeria's population was extrapolated from it. The World Bank thus quotes a figure of 115.5 million for mid-1990. In 1992, in a census considered accurate by United Nations observers, the population of Nigeria totalled 88.5 million. After a plebiscite, a fourth region, the Mid-West, was created in 1963 to give identity to a group of peoples who felt distinct from the Yorubas, the majority tribe of the Western Region. These events give some indication of the political ferment in the new nation.

An uneasy alliance between the Northern People's Congress (NPC), the political party dominant in the Northern Region, and the National Council of Nigeria and the Cameroons (NCNC), that of the Eastern Region, formed the first federal government. This coalition broke up in 1964 and put the country in a state of political turmoil. It was out of the unstable political alliance of NPC and NCNC that Nigeria's first development plan was born in 1962. As its objective the plan put 'the achievement and maintenance of the highest possible rate of increase in the standard of living' (Federation of Nigeria 1962). The goal was laudable but imprecise. The plan was to run until 1968, funded by a series of sectoral investments over half of which were to be in the public sector. Half the necessary finance was to be obtained abroad but oil production rose appreciably and its revenue gave promise for the future funding of Nigerian development. The plan set out and achieved major infrastructural investments in increasing the carrying capacity of the road network, in the establishment of more schools and four new universities and in the harnessing of hydro-power. Its targets in primary production and manufacturing were not reached.

In the event this first plan, unrelated to regional issues and with all the uncertainties of any initial plan, never ran its full course. The first of a long series of violent political events cut it short. In a coup organized by Ibo army officers, the Federal Prime Minister and those of the Northern and Western Regions were killed, as were many senior Northern army commanders. When the coup was suppressed, General Ironsi, himself an Ibo, was asked to head a military government. It was January 1966 and parliamentary democracy had lasted less than six years. The new regime stressed the need for national unity. The people of northern Nigeria saw the change as a take-over by the Ibos not only of the nation but of their own Northern Region. They had already found themselves as second-class citizens in their own state where jobs requiring the skills and education they did not possess went to those from the South. Southerners were the office workers, the artisans, the engineers and train-drivers; the Northerners were peasant farmers and labourers. Now they saw the Ibos coming to rule over them. In June 1966 the Northerners began to massacre the Ibos in their midst. In July Ironsi was killed and a young Northerner, General Gowan, from a Christian minority tribe and hence regarded as impartial in the North–South clash, became the supreme military governor of Nigeria. In September a second massacre of Ibos occurred in every town throughout northern Nigeria and the Ibos in their thousands fled to their home areas in the south-east. The regional differentiation in development and the political divisions along regional and tribal lines had found expression in a national tragedy. No greater case could be put for the development of a political economy which embraced place and people.

Gowan, acutely aware of the need both for national unity and the regional and tribal expression of identity and participation, proclaimed in May 1967 that Nigeria's four regional states would be replaced by twelve (Fig. 27A). The Ibos saw themselves identified with one small state, the East Central, essentially outside the oil-field areas. Ojukwu, the army officer who had become leader of the Ibos, proclaimed South East Nigeria as the separate nation of Biafra and in the same month the Nigerian civil war began. The war has been explained in terms of the dispute over Nigeria's new oil wealth but to explain Biafra's secession and the war to re-incorporate it in terms of a conflict over oil is facile. It was more deeply rooted in the complexities of Nigeria's history, her polyglot composition, the political manoeuvrings within a new country and the manifest unevenness of opportunity among her people.

THE AFTERMATH OF WAR

The civil war, bitter and destructive, halted Nigerian progress and set back her achievements. In the aftermath of war the Second National Development Plan for 1970–74 was a plan of reconstruction as well as development. The twelve states proclaimed by Gowan in 1967 held, in theory at least, the prospect of more disaggregated regional development. The stress was, however, upon the restoration of national growth. The plan stated 'replacement of assets cannot be undertaken indiscriminately because they have been destroyed by war action. Damages need not be replaced at all or replaced in the form or place of the original asset. The crucial test for replacement is contribution to the growth of the national economy' (Nigerian Federal Republic 1970). The projects which were to be emphasized were those estimated to give 'the highest linkage with growth effects' (Nigerian Federal Republic 1970).

The emphasis in the plan was upon urban-located developments despite the acknowledged problems of rural areas and the increasing drift to the towns.

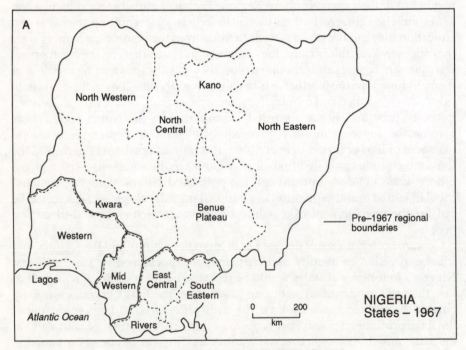

Figure 27   Nigeria states and regions: (A) 1967; (B) 1975; (C) 1993

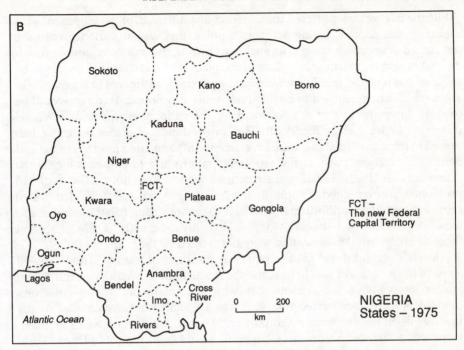

B

FCT –
The new Federal
Capital Territory

NIGERIA
States – 1975

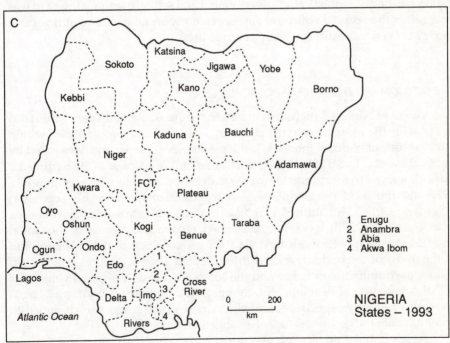

C

1 Enugu
2 Anambra
3 Abia
4 Akwa Ibom

NIGERIA
States – 1993

While the twelve states offered the prospect of a diffusion of development with their regional capitals acting as growth poles, this was at variance with the developed economic axes and as such was pitting the role of regional growth against national growth. According to the plan, regional disparities were to be reduced but it stated that this would not be pursued 'at the cost of stagnation in areas which are presumed to be relatively more developed. To do so would be to slow down the rate of development for the national economy as a whole' (Nigerian Federal Republic 1970). This understandable stress none the less served to reinforce the status quo. Investments implemented tend to persist and become permanent parts of the economic pattern and to emphasize regional differences as the colonial infrastructural legacy had done. The regional problem again emerged. People in the less-developed areas could move to restore the spatial equilibrium but what if there are strong tribal identities and local allegiances which the civil war revealed only too cruelly? These issues and those of urban–rural polarization were not tackled by the plan.

Nigeria's second development plan was born out of trauma. Time allowed some of the wounds of war to heal and the fountain of oil wealth was more than a salve, it gave Nigeria that element missing in so many developing countries, the wealth to finance her economy. It was a wealth further increased by the upsurge in oil prices brought about by OPEC of which Nigeria was a member. Here was a chance to transform Nigeria's status from that of a Third World nation into one of sophisticated advancement and self-sustaining growth. It was time rather than capital which was in short supply and the wise spending of oil money proved more difficult than its acquisition.

THE OIL BOOM AND THE THIRD NATIONAL PLAN

The second of Nigeria's plans had aimed to reduce regional disparities but did not provide the means. The third plan had the means and was designed within a framework of regional intent. Like its predecessors its passage was jolted by political events. The tiger of military coup let out of its cage in 1966 proved a difficult animal to recapture. Gowan was deposed in a bloodless coup in July 1975 and the third development plan was implemented by his successor, Brigadier Murtala Mohammed. He created twenty states to reflect local interests more closely (Fig. 27B). It was proposed that Lagos should cease to be the capital and that a new capital be built at Abuja at the geographical centre of Nigeria and in the midst of the underdeveloped Middle Belt. These measures would help reduce the dominance of Lagos and the south and create a regional vehicle for the dissemination of development, though Kaduna, Port Harcourt and Lagos were also to receive special federal aid. A new constitution was designed to accommodate the new government arrangements.

Nigeria's population increase was put at 2.5 per cent per annum and the new plan's objective was to increase GDP by 9 per cent per annum. Between 1960 and 1970 the growth rate in GDP had been 3.1 per cent per annum. The main thrust of the plan was industrialization in a 'big push' approach but with some balance. Oil revenues were no longer related to the producing states but put under federal control. They were to be used to develop oil-associated industries such as refining and petrochemicals and to establish basic industries such as cement works, fertilizer plant, pulp and paper mills, and iron and steel plant using the lean Kabba ore. This basic industrial development was considered necessary because, hitherto, Nigeria's manufacturing had consisted mainly of the assembling processing of imported commodities for the domestic consumer goods market. The aim was to establish a broad industrial base with vertical and horizontal linkages and convert Nigeria into a secure self-sustaining industrial nation rather than one which, at great cost, was importing the appearance of modernization rather than creating the reality. While oil was exported, refined oil products had·to be imported because Nigeria's oil refinery capacity was insufficient to supply the nation's demand. Cement was imported and so too were fertilizers and iron and steel. Yet agriculture remained organizationally and technically backward though it employed 64 per cent of the country's labour force. It too was allocated investments including the implementation of major irrigation schemes. In other respects farming received but token attention with heavily subsidized fertilizers and insecticides and tractor-hire facilities made available when the rural economy in reality required a radical reshaping. The large and growing population placed demands on educational, medical and other social services. The Third National Plan encompassed all. In terms of its funding and range of objectives, it was the most ambitious plan ever to be launched in an African country and it was to be financed in its entirety by Nigerian resources and that meant oil.

The emphasis on basic infrastructural investments, necessary though they were, was characterized by high ICORs and slow returns so the impact could not be immediate. Meanwhile Nigeria continued to import manufactured consumer goods, components for assembly industries, raw materials and even food. Significantly her exports were of primary products rather than manufactures. The structure of her economy had not fundamentally changed despite the increase in manufacturing. Inflation rates rose from an annual average of 4 per cent during 1960–70 to one of 14.2 per cent in the following decade. It was an inflation created not only by home demand but also imported as the spiralling oil prices of the early 1970s increased the cost of goods from the industrial nations.

Mohammed had come to power amidst chaos, congested ports, corruption and a roaring inflation. The Third National Plan was intended to provide for a better organized and more productive future and, as Mohammed promised, for a return to an elected civilian government. But the tiger struck again and Mohammed was assassinated in 1976. His successor, General Obasanjo had the task of implementing the plan, the vulnerability of which soon became evident when a drop of oil prices in 1977–78 reduced Nigeria's earnings. A balance of payments crisis emerged and Obasanjo began to cut back the plan's investment programme. It was a warning but oil prices resumed their upward movement.

In the election of 1979 a civilian government was returned to power with President Shagari at its head. But the questions still remained. Could investment and entrepreneurial skills create a stable self-sustaining economy before oil revenues ran out? Could the pull and influence of the major industrial centres, of which Lagos was by far the most important, be counterbalanced? Could the new regional capitals and federal capital disseminate development into the untouched stagnant hinterlands? Could these towns be given positive growth functions, functions which can be put to use, or would the capitals become mere bureaucratic replications? Could Nigeria manufacture industrial goods rather than import them in such large quantities and could she export manufactures? Could she reduce her great reliance on oil revenue? Could agriculture and rural life be given the stimulus to an acceptable change? Nigeria had the resources to feed her population and at a higher level of nutrition; could this be done in systems which would allow the farmer a fair return and enable him to participate in the new Nigeria and cease to be a poor follower-on? The second civilian government had as daunting a task as the first. In sum, it was the problem of the efficient use of the great but finite oil wealth.

Under Shagari the building of Abuja was started and the twenty new states began functioning, but the pressures which had led to their existence were still at work. Distinctive groups within each state wanted their own separate territory and representation, believing that other groups dominated them, wanting a greater say in federal decision-making and, importantly, thinking that by having separate status they would gain a bigger share of federal funds. The further subdivision of Nigeria ran the risk, however, that such fragmentation might imperil the unity of the country and a still further replication of buildings and officials would be a woeful drain on resources. The country had proved to be far from healed of its war wounds and its spatial administration was still unsatisfactory.

Equally significant was the failure of Nigerian governments to use oil revenue effectively to restructure the economy. It was true of Gowan's and Mohammed's administrations despite the Third Plan and it was to be true of Shagari's. Oil which had accounted for 3 per cent of Government revenue in 1963 represented 80 per cent in 1982 and 90 per cent of the value of all Nigeria's exports. Shagari inherited the Third Plan and its aims but he also was heir to the financial and technical inefficiencies which had impeded the best laid of plans, the lack of adequate data to monitor allocations and progress, and the continuing inflow of imports while the infrastructural and basic investments took slow effect.

THE FOURTH NATIONAL PLAN AND THE GLUT OF OIL

In 1981 the fourth of Nigeria's national plans was announced. It was based on the assumption that oil would continue to fund the country's economic transformation and, like its predecessor, was ambitious. An annual growth rate of GDP of 7.2 per cent per year was predicted in which manufacturing, communications, electricity supply, water and other public utilities were to be the fastest growing components. Agriculture, having experienced a negative growth for so long, was planned for an increase in output of 4 per cent per year and, after transport, received the second highest allocation of funds in the public sector. In the private sector, manufacturing and construction were the lead components accounting for almost 50 per cent of investment while agriculture was put at 15.6 per cent. The plan did refer to the 'volatility of the international oil market' which might require a reassessment of investment priorities. In the event the slump in oil prices, beginning in 1980 and falling, in real terms, to below those pertaining in 1970, threw Nigeria's fourth plan out of balance giving little prospect for its targets to be met and leaving unobtained fundamental objectives in her structural transformation. It begin to appear unlikely that Abuja would be built on the scale originally envisaged. The high level of governmental infrastructural expenditure could only be sustained by those same high oil prices which had adversely affected so many developing countries. Nigeria began to borrow on the international market and her debt commitments rose significantly. The investment programme was cut back and this resulted in unemployment. Infrastructural programmes which had been completed, such as roads, schools, universities, transmission lines and power stations, all required maintenance and staffing. Major projects which made good, long-term economic sense had to be abandoned or at least indefinitely postponed. This situation can best be exemplified by the aborted project for natural gas. Natural gas is a valuable source of energy but presents problems in its transport. Some 87 per cent of the world's natural gas is consumed in the country of its origin. Nigeria produces natural gas unavoidably with the oil she raises but has an insufficient industrial capacity to make use of it. The gas is flared off and a valuable resource wasted. In addition, substantial gas-only fields exist in the oil field area and remain unexploited because of the lack of a local market and the difficulty of transporting the gas to the industrial markets of Europe and North America. Encouraged by the upturn in oil prices in 1979, Nigeria established a consortium to build a plant for the liquification of natural gas so that it could be transported by refrigerated tanker. The plant was to be the largest in the world and costing US$10 billion required foreign participation. It would have enhanced Nigeria's resource wealth appreciably. The fall in oil prices subsequent to the consortium's establishment led to Nigeria withdrawing its financial support and the dissolution of the consortium in 1982. For Africa's largest country and one of the world's major oil producers the scheme was too costly. Moneys were, however, allocated for a pipeline system to carry previously flared gas to a new gas-fired electricity station in Lagos. Much behind schedule, the pipeline was completed at the end of 1988 (Simpson 1992).

Under Shagari, Nigeria's economy had failed to stabilize and prosper. The economy had not been diversified to any significant degree. Over 90 per cent of foreign exchange was derived from oil whose earnings accounted for 80 per cent of government revenue. The nation was hostage to international fluctuations in oil prices and had created nothing to buffer their impact. Religious and tribal divisions remained in an explosive condition. Elections were due in 1983 but in that year Shagari's government was overthrown, two days before he was due to present his budget, by yet another military coup and General Buhari took control of the nation's destiny.

Shagari's budget had been designed to cut back the profligate spending on imports which was associated with Nigeria's oil wealth and to limit the amount of foreign exchange available for imports. Earnings from oil had dropped dramatically from US$23.4 billion in 1980 to US$9.9 billion in 1983. Foreign debts and the current account deficit escalated. Buhari made further cuts in available foreign exchange. His restrictions were intended to ease the balance of payments situation and to enable export earnings to service Nigeria's borrowings. Since three-quarters of the materials used by Nigeria's industries was imported his actions caused many firms to close, others to work below capacity and unemployment to rise. Drought brought about poor harvests so more food had to be imported. Disillusion followed optimism and Buhari was deposed in 1985 by a fellow officer, General Ibrahim Babangida.

All Nigerian governments have functioned within a political–economic milieu which is very much a product of Nigeria's history and characteristics. It is a situation which relates to the way in which power, patronage, economic decision-taking and non-productive money-making are interwoven. The World Bank speaks of 'directly unproductive profit seeking' and while this can occur in the private sector, it also is associated with governmental policy interventions (World Bank 1987). Political scientists have investigated this situation both in Nigeria and the Côte d'Ivoire – a situation which has a bearing upon economic development (Watts and Basset 1986). When Nigeria was created, the colony encompassed a large number of distinctive peoples with their own methods of government and systems of allegiance. While Britain was the colonial power there was no obvious coherent pyramid of political power among the indigenous peoples. The system of indirect rule favoured by the British ensured that traditional rulers and authorities managed local affairs. The heterogeneous nature of Nigeria meant that upon independence no indigenous power group existed on a national scale. The regional power groups were held together by a system of patronage which disbursed benefits in exchange for allegiance (Watts and Basset 1986). Cohen claims that the award of political office and contracts together with places on state boards became a competitive feature of this patronage (Cohen 1974). At first the Federal Government was little involved as most affairs were managed at regional level with three, and later four, regions covering the whole country. After the establishment of a central military government and with the flood of exchange revenue which resulted from the oil boom of the early 1970s, the Federal Government became much more interventionist and, over time, many economic activities were nationalized and new state ventures initiated. This growth of state activity led to the creation of a vast, loosely controlled bureaucracy, the establishment of scores of management boards and the setting up of many parastatal organizations. Not only did the government become the largest single employer by far, it also became the central pivot of the economy funded by a single resource – oil. Preferment and access to wealth were in the hands of the government. This situation was much exploited and was one which encouraged bribery and corruption. The regulatory nature of Nigerian governments, with their penchant for import quotas, export levies, licences and such, created opportunities for what the economists refer to as 'rent seekers', that is persons who by lobbying and other means gain access to such things as licences and import quotas when these are restricted and are therefore at a premium. Tariff evasion can yield unproductive profits that is a monetary yield when nothing has been produced. Though not peculiar to Nigeria these practices created a group parasitic upon the government's central economic role. The state thus became the means by which capital could be accumulated and thus inhibited the development of organized capitalism (Watts and Basset 1986). This can explain, in part, the lavish expenditure upon non-essential items, the import of consumer goods and the careless approach to the economic viability of major projects. Interventionism on such a scale by the state could well be one of the factors which has retarded Nigeria's economic development.

Babangida recognized that the nation's situation required drastic measures and that these had to be within the framework of a broad policy. He began, significantly, to distance central government from its deep involvement in the economy. The governmental apparatus was itself a massive consumer of revenue with its bureaucracies and state-run schools, universities and hospitals. In addition the state was involved in hotels, breweries, insurance companies, steel works and car assembly plants. Nigerian railways and the Electricity Power Authority were state-owned. Babangida proposed complete denationalization in some instances and partial in other cases. In all state economic activities commercialization had to replace what had been almost the economics of a command economy and commercial accountability was to be the rule. The State Commodity Boards which set the prices for the major cash crops were abolished and this quickly stimulated production in a stagnant sector of the economy. Most importantly, Babangida devalued the *Naira* by 66 per cent and set up a currency auction to maintain a realistic exchange rate. Price controls and import licences were almost completely removed and the import restrictions on many goods previously prohibited entry were withdrawn. High customs duties were maintained on some imports which competed with Nigeria's manufactured products and on some crops (Tallroth 1987). The currency devaluation and uniformly low tariffs gave protection to home producers and made Nigeria exports more price-competitive. A liberalization of the economy and the reduction of interventionism together with steps which reduced the opportunities of those parasitic upon the economy were all steps designed to free Nigeria of its past and to maintain a steady growth sustainable when oil and gas supplies run out in the next century.

One further feature of Babangida's reforms was bold and authoritarian. Political strife had compounded Nigeria's difficulties. Babangida took two actions. First, he enlarged the number of states to thirty-one, including the

Federal Capital Territory at Abuja, to reflect regional distinctions more closely (Fig. 27C). Second, he abolished all political parties and replaced them by two of his own creation in an attempt to divorce politics from tribe and region. Neither party has a distinct political ideology. Elections held in 1992 gave one party a slight majority in both houses of the Assembly and surprisingly voting was not on strictly tribal lines.

A civil war and a succession of coups have distorted plans and predictions but this does not explain by itself Nigeria's failure to achieve the progress of which it is capable. Oil provided the necessary capital; its misuse has been described. The increasing resort to state intervention, with its multitude of regulations and controls together with the maintenance of an unrealistic exchange rate, have undoubtedly done much to cloak inefficiency and impede progress to the extent that public utilities were neglected to the point of collapse and state ventures were commonly economic failures misconceived and mismanaged. Oil wealth was imprudently invested in too many cases and too lavishly spent in others. The establishment of a secure economy which could flourish when oil prices are low and be maintained when oil is gone has not yet been achieved. Babangida's reforms were a step in the right direction. It was important that they were maintained in the new democracy, which the elections of 1992 and 1993 were to have set in place, if international confidence in Nigeria as a secure investment was to have come about. In the event yet another military coup diminished this prospect.

13 Zambia

'Given a period of peaceful development, Zambia has excellent prospects for economic growth, better than those of almost any other country in tropical Africa'

A T Grove
Africa South of the Sahara 1967

Zambia, the former Northern Rhodesia, became independent in 1964. Like Nigeria, she had experienced over half a century of colonial rule but both the acquisition of the territory now Zambia and the nature of her colonial experience were different. Unlike Nigeria, Zambia is a land-locked state in the interior of central Africa and this location meant she remained one of the 'unknown' regions for a longer period. Explored by missionaries, notably Livingstone, there appeared little at first to attract economic exploitation, and Britain was hesitant to embark upon an expensive venture with little prospect of reward. Indeed at one stage it appeared likely that the area would pass into the hands of the Portuguese. In the event it was the striding ambition of Cecil Rhodes to see the British present on African soil from the Cape to the Mediterranean which initiated a complex, and often unsavoury, chain of events which led to much of what is now Zambia becoming part of the British Central African Protectorate (Hall 1965).

Having amassed a fortune in the diamond mines in Kimberley and augmented it by his involvement in the development of the gold deposits on the Witwatersrand, Rhodes, who was Prime Minister of Cape Colony from 1890 to 1896, obtained in 1888 a royal charter for the British South Africa Company which he founded. In effect this chartered company began to administer a sizeable proportion of south central Africa on behalf of the British Government. Rhodes, a complex man, was spurred on by a political ambition and a desire for wealth, and saw his company as an instrument for the development of the riches he thought lay in this lightly populated and little known interior. He was to do this by extending the railway from the Cape Colony into what became Southern Rhodesia, then named Southern Zambezia (Fig. 28A). He hoped to discover other gold deposits and to establish white settlers on suitable agricultural land. In the first he was disappointed; in the second he succeeded. The company was permitted in 1891 to extend its activities north of the Zambezi into what is now Zambia. The frontiers of the territory had been drawn up in Europe as the King of the Belgians received the Congo, Portugal received Angola and Mozambique, and Germany received Tanganika. Rhodes took the railway northwards. In 1902 lead and zinc deposits were discovered at Broken Hill (Kabwe) and the railway

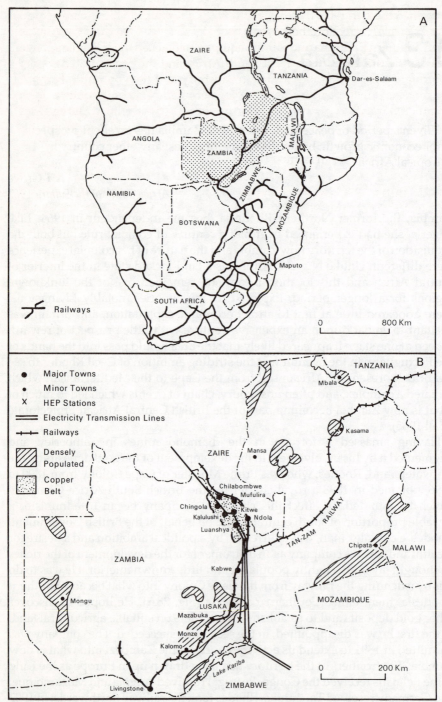

Figure 28 Zambia: (A) location in Africa; (B) economic and population distributions

was taken to them in 1906. By 1909 it had reached Katanga in the Congo pedical where extensive copper ore-bodies had been discovered. Coal from the Wankie mine in Southern Rhodesia and mining equipment reached these Belgian Congo mines via the Rhodes railway from South Africa and the concentrated copper was brought out by the same means to Salisbury and then by rail to Beira. A major modern transport artery bisected Northern Zambezia from south to north and gathered its traffic almost entirely from outside the territory. But the area eventually to become Zambia was now connected to the outside world and this connection was to the south, to the lands south of the Zambezi increasingly being settled by European farmers, traders and mine-owners.

Lean copper oxide deposits had been worked sporadically in Northern Rhodesia in the early-twentieth century but proved uneconomic. In 1923 a major, rich, copper oxide ore-body was discovered at Nchanga near the Congo border. Deeper sulphide ores were identified in 1925 and by 1930 the copper potential of Northern Rhodesia was realized. This led to capital investment in mines and infrastructure on a large scale. Towns were built to house the workers drawn from an otherwise agricultural community. It was a development to have a lasting effect upon the space economy of the country that is now Zambia. The capital invested was foreign, so too were the ownership and management of the mining companies. The earnings of the mines financed the necessary social and physical infrastructure to support them, while those of the miners created a local market for food and consumer goods, but these impacts were highly focused. There was little development of associated industries and the copper, once refined, was exported out of the country to European and North American markets.

Later augmented by road, the railway from Livingstone in the south to the Katanga border was now busy with Rhodesian copper traffic. It became an axis of accessibility and development and was, and is, referred to as 'the line of rail' (Fig. 28B). The greater involvement of European interests from South Africa and Southern Rhodesia encouraged European settlement north of the Zambezi. Using African labour, European-owned farms were established on the better land and with access to the line of rail. The farms were large-scale commercial units producing crops for sale. There were now present not only European science and technology applied to the mining of copper at depths and on a scale unknown to the local population, but also Europeans in agriculture practising an implanted system which was both different and dominant and associated with a wealthier alien group. Digging for copper was one thing, farming African land was another. Since the majority of European settlers had moved in from Southern Rhodesia and South Africa, they brought with them attitudes to agricultural developments and local labour which had grown up to the South. The words of J. F. Hone, written in 1909, illustrate these attitudes.

It is generally recognized by those who have studied the difficulties which arise between black and white races, that the differences in character, temperament, intellect and humanity itself, are so well marked, so clearly defined, that never will it be possible for the two to blend and amalgamate into one . . . It must, however, be recognized that in industry, commerce, manual and domestic labour, white and black races will be indispensable to each other, the former as employers . . . the latter as labourers

(Hone 1909).

The British South Africa Company's charter ended in 1923. In a referendum taken in 1922 the white population of Southern Rhodesia had chosen not to become part of the Union of South Africa and the country was made a self-governing colony. Northern Rhodesia was administered by the Colonial Office in London through a Governor and Legislative Council (Hellen 1968). Between 1955 and 1965 Northern Rhodesia formed part of the Central African Federation of which the other members were Southern Rhodesia and Nyasaland, a liaison which came to an end when the latter country withdrew and became independent Malawi. It was a further episode in Zambia's history which has left its mark on her post-independence development. Not only were Zambian attitudes to Europeans coloured by their experience of European settlement but the status of the country was felt to have been relegated to that of up-country poor relation in the Federation. Under colonial rule the people of Northern Rhodesia were required to pay taxes to help in the administration of their country. This gave them an incentive to enter the cash economy. For farmers in the more remote areas this was difficult to achieve and increasing numbers sought work in the mines, in the towns and on European farms along the line of rail. All these developments heightened the contrast between the towns, importantly those of the copper belt and their economies, and the countryside with its villages and their economies. Development had been highly focused on the mining and export of copper, on the copper towns and the line of rail. Apart from the European commercial farms, the rest of this large country had hardly been touched by the twentieth century save where men had been lost to the mines.

THE BEGINNINGS OF DEVELOPMENT PLANNING

In the year following the collapse of the Confederation, Northern Rhodesia became independent, changed its name to Zambia and declared itself a republic. Within two years of independence the First National Development Plan was published to embrace the years from 1966 to 1970 (Zambia 1966). In his foreword to the plan, President Kaunda clearly states his country's main goals and outlines his assessment of Zambia's colonial legacy.

We in Zambia are fortunate in that the problems of famine and disease are not present in the manner they appear in some over-populated countries. However, this does not mean that the problems of growth are in any way less urgent for us by comparison with the rest of the world, for Zambia, faced by the barrier erected by European domination south of the Zambezi, is in the front-line of independent African countries. So it behoves us to demonstrate, in no uncertain manner, our determination and ability to provide standards of living for our peoples which will bear comparison anywhere in the world. The structure which we inherited from the colonial era resulted in many situations which we have to put right: an education system which was so far below requirements that at the time of independence only a handful of people had anything like sufficient training to enable them to take their place in the service of Government; a transport system which is linked in a situation of subservience to Southern Africa and which has led to all the inconvenience and frustrating difficulties which have since been provoked by the unilateral declaration of independence of Rhodesia; a system whereby all our supplies of petrol and oil come to us by the southern route as did coal for our mining industry; our dependence on the jointly-owned Kariba Dam for our electricity; and trade so organized that we seem irrevocably linked to supply from Southern Africa not only for raw materials but for the ordinary consumer goods of everyday life. So it is that, by opening a frontal attack on these inherited structures – themselves a major factor in holding back the level of economic development in Zambia – the First National Development Plan lays the foundation not only of true economic independence but also of growing prosperity

(Zambia 1966).

This was clearly to be a development plan with a difference. Not only was it to lay foundations for a more productive nation, it was also intended to secure the nation in a strategic sense. It highlights the vulnerability of a land-locked nation dependent for outside access upon lines of communication which traversed, and were therefore in the control of, other nations which, in Kaunda's view, were hostile not only to Zambia but to the peoples of Black Africa. The situation was made more dramatic by the declaration of unilateral independence by Rhodesia which was regarded as a further attempt at white domination of an African people and hence Kaunda's view of Zambia as a nation in the front-line. It is impossible to understand fully Zambia's approach to development unless the political–economic situation surrounding her emergence as an independent nation is appreciated.

President Kaunda chose to make special mention of several of the plan's proposals. First, he referred to the plans to increase both secondary school and university enrolment. Second, he mentioned the intention to complete a new rail and road link to Tanzania which would give access to the port of Dar-es-Salaam (Fig. 28). Third, he described the need to secure Zambia's supplies of energy. He stated that this was to be achieved by constructing an oil pipeline to Dar-es-Salaam, by developing Zambia's coal resources and by constructing a hydro-electric plant on the Kafue river. Both the transport and energy proposals represented large infrastructural investments and the President wrote of using both foreign loans and locally generated capital. Fourth, he made particular reference to the regional differentials and the urban–rural dichotomy in standards of living. The regionalization of economic activity with its social and economic implications is present in Zambia as in Nigeria and likewise it is a consequence of the period of colonial economic development and is manifest in the focus on the 'line of rail'. Kaunda put the regional development problem succinctly, stating:

a constant preoccupation of my Government is the disparity in the standard of living between the rural masses and the comparatively limited urban and industrial sector. Throughout this plan an effort has been made to localize investment and to increase to a maximum, consistent with their capacity of absorption, investment in the less-favoured provinces of the country. This does not mean we have neglected the growing industrial and urban centres of prosperity. A close examination of this plan will show that investment levels in the urban centres still outweigh those in their rural counterparts; for, while we must attempt to redress imbalance, we must at the same time invest in growth and it would be foolish to exclude the areas of prosperity which will themselves contribute directly or indirectly to raising the prosperity of neighbouring provinces.

(Zambia, 1966).

Similar views were expressed in Nigeria's second development plan and point to the need to ensure a maximum return on investments in favoured localities in order to generate capital to fund the disadvantaged.

President Kaunda's identification of the central components in the plan is important. It reveals his view, the view of an able and powerful leader, of the development situation as he sees it, with his emphasis and his interpretation of the mechanics of the plan. The plan itself spells out in great detail its many proposals and the allocation of sectoral investment. It identifies Zambia's resources and refers specifically to seven features: a stable government, a foreign exchange resource in the copper industry, the country's agricultural potential, the potential resources of hydro-electric power, and the basic, though incomplete, transport network. Two human resources are mentioned, first the stable industrial and labour relations and second the people of Zambia themselves, 'desirous for self-improvement and hard work'. The unfavourable legacies to which Kaunda referred are discussed further in the plan. In particular it stresses the way in which the functioning of the Central Africa Federation was to Zambia's disadvantage. It states that the copper revenues were used to fund developments in Southern Rhodesia and that the Federation led not only to a greater extension of a European settlement in Zambia but also to the European community and economy receiving first and favoured treatment with African development regarded as an associated ancillary development. For example, it claims that the quality of education services was higher for the Europeans and that the agricultural extension services provided for European farmers were funded by the Federation whereas the Africans received only locally-funded services. The organization of the railway network was centralized on Southern Rhodesia, as were air services. In the joint venture of harnessing the Zambezi to generate electricity for the Federation, the power station at Kariba had been built on the Southern Rhodesian bank. To it was connected an extensive transmission grid supplying electricity to the Zambian copper belt and to the major urban centres of Southern Rhodesia. During the short life of the Federation this integrated development made sound economic sense. After Zambia's independence and with the unilateral declaration of independence in Southern Rhodesia under a white-dominated government, the location of the Kariba power station posed a potential threat to the power supply for Zambia's copper mines (Williams 1977). Power supplies were not interrupted but the need to secure them was regarded as a priority in the plan. A further legacy of the Federation was to be found in Zambia's limited manufacturing sector; most of the plants were branch plants of Southern Rhodesian establishments rather than independent or parent firms. The European farmers, some 700 in number, accounted for a £7.7 million production of cash crops, chiefly tobacco and maize, and of cattle, while 450,000 African farmers sold produce worth only £3.2 million. The European output was a positive contribution to Zambia's economy and was a reflection of the scale, modernity and organization of their farms and of their access to transport via the line of rail. It was, however, a situation which highlighted the low productivity of African farmers and the plan included measures designed to involve them more fully in the cash economy and to spread this involvement more widely throughout Zambia.

All these considerations of resources, legacy and regionalism were incorporated into the plan. Schemes to provide alternative connections between Zambia and the coast, to diversify the industrial sector and modernize the agriculture, to secure a wider and more varied raw material base and to develop energy supplies, were all presented in detail. It was felt that in its land-locked position Zambia must develop as far as possible a self-reliance, and establish its own basic industries such as iron and steel, cement and fertilizer production upon which other developments could be founded. It was a philosophy of import-substitution within broad-front planning with the establishment of vertical and horizontal linkages. It was a first plan and like those of most other countries, it was ambitious and, perhaps, over optimistic. In its four year span it set itself the objective of increasing employment by 100,000 and of raising per capita output by over 50 per cent which, taking into account the increase in population, would require an increase in the GDP of 11 per cent per annum over the plan period. Two-thirds of investment was to be in the public sector which ranged across the whole field of economic, social and administrative activities and included very substantial infrastructural investments. Copper mining was left essentially funded by private finance.

In agriculture the intention was to reduce imports of food and move into a situation where agricultural products could be exported. A detailed specification of the extent to which self-sufficiency would be reached by the terminal stage of the plan in 1970 was constructed, with a further five years allowed for the attainment of full self-sufficiency. In the rural and agricultural sector investment was to be made in roads, water supply and land conservation with a focus on land of high potential. Resettlement schemes were planned to move people from the overpopulated line of rail area into tsetse consolidation areas where the fly, which inhibits both settlement and cattle-keeping over wide areas of Zambia, had been eliminated. The resettlement of 12,000 families was envisaged. Additionally the establishment of co-operative farms, the provision of credit and the introduction of subsidies and grants were developments which it was hoped would both increase agricultural production and revitalize rural areas. It is significant that these agrarian proposals involved a considerable degree of state participation. There were to be state co-operatives, state dairy farms and beef ranches, and state farms concentrating on crops with an export potential such as tobacco, citrus fruits, coffee, tea and sugar. Tractor mechanization schemes likewise signalled a centralized state involvement. In concert these proposals were expected to raise total agricultural production from K15 million in 1964 to K35 million in 1970. In addition to manufacturing and agricultural proposals, the plan paid some attention to tourism as a means of diversifying the resource base and earning exchange revenue essentially by building more hotel accommodation. The prospects for tourism were, however, limited. Though there are game parks with the attraction of African wild animals, the fauna is not as rich and varied as in Kenya and Tanzania. Over much of Zambia the natural landscape lacks the beauty of neighbouring Malawi. The major scenic attraction, and it is indeed one of the greatest natural spectacles in the world, is that of the Victoria Falls on the southern border with Zimbabwe. Here a new hotel was built to augment the old rest-house. The major activity by far in terms of the generation of revenue was, of course, copper mining. From this was to come much of the money to finance other developments. In this venture the Government did not wish to interfere save to ensure that it was able fully to monitor the marketing and pricing policies of the mining companies. A special expert governmental unit was established for this purpose.

In the event many of the targets of this first development plan were not achieved and its terminal period was extended for a further year. Importantly it had effected little in the diminution of the imbalance between the central corridor, from the copper belt through the capital, Lusaka, to Livingstone in the south, and the rest of the country. Indeed, in a mere four years any shift could only have been small but the imbalance had actually increased. Of the total national investment, 69 per cent had been allocated to the central area because, as President Kaunda had stated, it was there that investment was most likely to be productive. In practice, however, this area had absorbed 82 per cent of investment with only 18 per cent entering developments spread throughout the rest of the country. In 1969 this central corridor contained 98 per cent of all manufacturing establishments. The legacy of colonial development was still deeply etched upon the economic geography of Zambia.

The regional imbalance remained, and still remains, to be resolved. Here, as in Nigeria, there is the regional issue of the built-in momentum of a more developed axis with, on the one hand, its economic advantages and investment rationale and on the other, the problems of disparate development leaving problem regions with tribal affinities. Greater infrastructural investment in these areas might link them into the developing economy but would inevitably give a slower return and in consequence retard the development of overall national progress in a directly economic sense.

During the plan's life, wage employment increased by 77,000 – somewhat short of the 100,000 planned – but investment as a proportion of GDP increased and reached 28 per cent in 1970. The average annual growth in GDP was, at 10.6 per cent, very close to the planned rise but it was an average derived from marked oscillations which were essentially related to copper prices. So important was copper in the Zambian economy that Zambia was, and still is, almost a hostage to world copper prices. Like Nigeria, Zambia's main source of exchange revenue was related to a single and dominant mineral resource whose value was determined outside the country.

The poorest performance was on the agricultural front. Targets for crops were rarely met, output was often very low and income disparities between town and country increased. Only 79 per cent of the planned investment in agriculture was achieved. Farmers did not appear to respond, the co-operatives were unsuccessful, agencies such as the Marketing Boards performed badly, investment schemes were not fully implemented. In all explanations of this unsatisfactory situation the reason given has been the lack of trained manpower. A reading of the assessment of performance between 1966 and 1971 which is made in the Second National Plan, reveals the failure in agriculture and, indeed, in forestry. Energy, initiative and drive do not appear to have been stimulated by the Government's efforts among the 70 per cent of Zambia's population who are agriculturalists.

In manufacturing, excluding copper mining and refining, although employment targets were not met, a number of significant new industrial units were established but the majority were in the copper belt of the north where a brick and tile factory, a lorry assembly plant, a tyre factory and a cement works were built. A brewery in Lusaka, the Zambian Sugar Corporation refinery, and the two large establishments at Kafue, a textile works and a fertilizer plant, were the other major installations overwhelmingly reinforcing the line of rail axis.

The achievement in infrastructural investment was considerable. Here were enterprises laying foundations for future development and increasing the country's economic security. The construction of the 600 MW hydro plant at Kafue was begun, with the first stage completed in 1972, and in 1969 the decision was taken to build a power station on the Zambian side of the Kariba dam (Williams 1977). The railway from Kapiri Mposhi in Zambia to Dar-es-Salaam was begun, financed by an interest-free loan from the People's Republic of China and built by 10,000 Chinese and 30,000 Zambian and Tanzanian labourers. It was completed in 1975. The oil pipeline through Tanzania was operative by 1968.

THE SECOND NATIONAL PLAN AND SUBSEQUENT DEVELOPMENTS

With the experience derived from the first plan its successor came to terms with economic reality and, in particular, attempted to address the problems created by the unreliability of copper earnings. By making cautious estimates of copper revenue and allowing for the inflation of the prices of imports, the plan assessed the likely limits on investment. The growth in population of 2.9 per cent per annum set the lower limits on developmental achievement if it were to solve the employment problem which, with the movement in to the line-of-rail axis with its towns, was becoming serious. The plan stated, 'a simultaneous development of industry and agriculture is a basic pre-requisite for achieving optimal results in employment and a rise in living standards' (Zambia 1972).

In its broad objectives the plan gave agricultural production the highest priority again, expressing the hope that it would achieve self-sufficiency, reduce imports, provide exports and produce raw materials for industry. The aim was to concentrate on the traditional farming community and convert it into one of commercial farming. Again the state intent was to focus on areas of greatest potential. Manufacturing industry was to be further diversified and in the case of the giant copper industry, its processing functions were to be enlarged so as to increase value-added revenue. The so-called Tan-Zam railway was to be completed and feeder roads built to open up the countryside while new administrative and economic organizations were specifically instituted to stimulate rural development. The plan was thus much as before but characterized by a greater realism. Throughout its life this plan's intent was beset by the problems south of the Zambezi and increasing hostility between Zambia and the new Rhodesia. Zambia became a base for thousands of freedom-fighters whom she had to support and her routeways to the south became increasingly vulnerable.

By the termination of the second planning period, much outlined in the first plan had still to be attained and the following plan carrying the period up until 1981, had to contend with the same issues of regional imbalance, of an inadequate and little changed agriculture, of not only a dependency upon copper revenue but of an inability to invest it effectively to stimulate a true development process. The inadequacies of rural opportunities and the illusion of the city continued to promote an influx into the towns notably the capital, Lusaka, which is now surrounded with wide areas of squatter settlement.

The plan continued the programme of rural road-building, agricultural marketing, credit and advisory work but in addition identified two particular initiatives. First, there were to be set up specific agricultural projects which included tea plantations, a sugar scheme, a wheat project with Canadian aid and an irrigation project with German support. These were large-scale state-owned commercial schemes rather than developments involving the traditional farmer and in a sense could be regarded as State replications of the large-scale European-operated holdings. Secondly, a rural reconstruction programme was inaugurated. This aimed to give compulsory training in new agricultural techniques to a captive work-force, namely the National Service Units of the Zambian army in which 42,000 could be trained each year. Both these initiatives represented a further direct involvement by Central Government in the economy but were not the only instances of the process. All land in Zambia was nationalized so that all farmers became tenants of the Government. This included the large commercial farms operated by Europeans, a group which produced 70 per cent of all the beef, fruit, vegetables, milk and poultry sold in Zambia, as well as all the seed for maize improvement. In industry, the state-owned Industrial Development Corporation, INDECO, now dominated. The state also participated to a much greater extent in commerce and mining. The Zambia Industrial and Mining Corporation, ZIMCO, is the main holding company of many industries including all insurance companies, all building societies and all mining activities, with the running of copper mines passing to total state control in 1982. Newspapers and all cinemas have been nationalized and there are national energy, transport and hotels corporations. This movement to the overwhelming state dominance of the economy could be explained as a product of the war-footing upon which Zambia was placed during the long years from Rhodesia's unilateral declaration of independence until the emergence of Zimbabwe but at best this can only be a partial explanation since state intervention and dominance persisted after the Rhodesian situation was resolved. It must be seen as the outcome of President Kaunda's political philosophy. The nation was a one-party state and the Government, party and economy became increasingly one. It may have been seen as a means to solving Zambia's considerable development problems but it was a policy open to all the abuses of privilege, patronage and preferment which were discussed in the Nigerian context. It is a policy which also can be questioned on other grounds since it is one which puts so great a burden upon limited administrative, economic and technical skills within government agencies and, importantly, it tends to stifle, and indeed prohibit, individual initiative.

Faced with increasing energy costs and higher prices of imported manufactures following the OPEC instituted increase in oil prices, Zambia's economic problems have been exacerbated and foreign indebtedness has grown. In this situation, overseas aid has risen dramatically in importance. Between 1975 and 1980 it increased by 400 per cent (Wood and Smith 1984). A large proportion of this aid has gone to the funding of agricultural projects and has been augmented by overseas technical assistance. As Wood and Smith indicate, much of the location and direction of these aid projects is in the hands of the donors because Zambia is still deficient in appropriate provincial manpower and has no provincial planning centres. Although an Integrated Rural Development Programme (IRDP) has been established, the donor countries tend to allocate their aid according to their perception of Zambia's needs. While the Swedes and Dutch act in accord with the IRDP and locate their projects so as to spread developments into the less developed areas and so reduce the rural–urban differentials, both the World Bank and EC authorities advocate a concentration upon areas of highest and quickest returns with the goal as the maximization of national output (Wood and Smith 1984). The latter approach is that adopted by Zambia in the early years of planning. The two approaches present the dilemma facing all regional development issues. They occur in countries as small as Fiji, as large as Nigeria and indeed, as will be discussed, underlie development issues in China, the world's largest nation. In Zambia the inadequacy of indigenous professional manpower to implement the IRDP and the opposing views taken by donor nations, mean that an integrated, well-considered plan for rural development has yet to emerge. It is an issue central to the development dilemma and will be returned to in the concluding chapter of this book.

The Zambian economy has continued to decline. Copper prices still dominated and dropped sharply during the period 1975–80 and production fell. The resulting deterioration in terms of trade was such that four times as much copper was required to be exported to import the same volume of goods as it had in the early 1970s (Clark and Allison 1989). When copper prices rose towards the end of the 1980s, Zambian mines were unable to respond and increase output because of their long neglect. The state control of crop prices led to difficulties as food supplies became short. When the price of maize was doubled, riots in the towns followed. Prices were reduced and farmers already hit by high inflation suffered. By 1991 maize had become scarce yet farmers were paid very low rates for their crop by a state which controlled the economy and heavily subsidized prices to the townspeople. Agricultural returns became so low that even more farmers left the countryside for the towns. Unrest led to political pressures for Kaunda to call an election and, fearful that food shortages would cause him to lose the election, Kaunda maintained his price controls. The World Bank and International Monetary Fund which had made the free-market determination of prices a condition of their help withheld their negotiations. In November 1991 Kaunda lost the election and President Chiluba came to power with promises of paying farmers more for their crops, of retaining food subsidies for only the poorest, of privatization and of a diversification of the economy away from copper. The nation had come to the edge of bankruptcy with a huge foreign debt and budget deficit. The average annual growth rate in GNP per capita between 1960 and 1990 had been a negative 1.9 per cent compared with a 2.9 per cent increase on average for other low-income countries. Zambians were poorer than they were before Kaunda came to power; his command economy had failed. Chiluba's promised reforms met the approval of the international community and foreigners have begun to invest. By 1993 Chiluba was selling off state-owned enterprises, had cut down the cost of his inherited government bureaucracy, controlled spending by ministries and removed foreign exchange controls. Zambia currency's value was to be determined by market forces. This liberalization of the economy has meant that foreign investment in Zambia is rapidly increasing while foreign aid has been resumed on a large scale.

Zambia has still to energize her agriculture and create a meaningful rural economy. It can be argued that unless she can do this she cannot sustain an efficient industrial sector. The new injection of foreign capital and management skills will do much to help, provided President Chiluba maintains an open market economy. The disaster wrought by a state-controlled command economy and the remedies implemented by Chiluba strongly parallel the situation in Nigeria; it is a pattern of democratic liberalization hopefully to be repeated in other African nations as one-party states and autocratic leaders are increasingly called into question. The solutions are not easy and throughout most of Zambia's independent history have been made more difficult by events beyond her borders. As yet her development process is still in its initial stages.

14 Sri Lanka

'. . . it is possible that – precisely because of popular awareness of economic problems – exaggerated or unreasonably hopeful expectations could be placed in the efficacy of planning in general, as well as in a particular planning exercise – such as the Five Year Plan – as the panacea for Sri Lanka's prevailing ills'

L A Wickremeratne
Sri Lanka (K M De Silva ed) 1977

Lanka, the Sinhalese name for the island so long referred to as Ceylon, the Taprobane of Ptolemy and 'the utmost Indian isle' of Milton, is a country of plantations and indigenous agriculture, yet a country which finds itself importing food. Being an island it was able spectacularly to eliminate the malarial mosquito which, in considerable measure, led to a massive increase in its population. It is a former colony. It is a country, therefore, with many of the ingredients of the Developing World and for this reason its development path is worthy of examination. But there are other features which make this Asian nation a significant case study of the realities of development planning in practice. One is the lasting imprint of its early history upon its human dimension. Another is the persistence of a parliamentary democracy which in so many former colonies has not survived the rigours of independent politics.

Three-quarters of Sri Lanka lies less than 150 metres above sea level. Its northern half is lowland broken up by north–south ridges separating the drainage basins. The whole of this lowland surface is covered by a mantle of deeply weathered rock called *kabouk* which conceals much of the underlying solid rock so that its nature is imperfectly known. It would appear to be largely of ancient metamorphosed material and of low mineral-endowment. There is neither coal nor oil and little indication of any metalliferous or non-metalliferous deposits of consequence with but one exception. Rich deposits of gem-bearing gravels yielding rubies, sapphires, garnets, amethysts and opals have been worked over the centuries in the hills of the south-west around Ratnapura. Out of this lowland of hills and valleys rises abruptly a central core of mountains with plateaux of 900 to 1400 metres and peaks rising from 2000 to 2400 metres. This area accounts for less than 10 per cent of Sri Lanka's area but it is an important tenth since it carries the tea plantations. The increase in altitude reduces the temperatures from the 27°C which prevails throughout most of the year on the coast to a range of between 15 and 20°C in the mountains, and favours not only the growth of tea but also some temperate crops. Sri Lanka experiences both a south-west and a north-east monsoon with the former being the more significant of the two, bringing heavy rains to the mountains and the south-west where everywhere precipitation exceeds 1300 mm and wide areas experience more than 2500 mm. To the north and east of the central highlands the south-west monsoon yields less than 500 mm and most of the rain comes from the weaker north-east monsoon from November to February. This area is referred to as the Dry Zone because of its high evapotranspiration losses and the greater variability of the rains even though its rainfall exceeds 1300 mm. The Dry Zone extends over two-thirds of Sri Lanka and is a significant physical component in development planning. The wet south-west and the central highlands are green throughout the year and, as the focus of commercial cropping, have the greatest density of road and rail networks, and support the bulk of the population (Figs 29 and 30).

Upon independence Sri Lanka was a country with a marked internal contrast in environment and level of economic development, and with a resource base which was almost entirely agricultural. There was another significant division and that was in ethnic composition. It too has had a bearing upon social and economic progress.

THE PEOPLES OF SRI LANKA

The aboriginal people of Sri Lanka, the Veddahs, may still survive in small numbers. Originally collectors and hunters, they pre-date the other peoples of the island. In 1770 the Abbé Raquel described them thus: 'The Bedas who were settled in the northern part of Ceylon . . . go almost naked and upon the whole their manners and government are the same with that of the Highlanders of Scotland.' By 1945 they had been reduced to a few thousand in the eastern Dry Zone. The great majority of Sri Lanka's population are of Aryan and Dravidian stock from the Indian sub-continent. It is necessary to review their early history since the consequences in population distribution and attitudes are so strongly present today. Prior to the fifth century BC there were Dravidians in Sri Lanka identical with those of southern India. Towards the end of that century peoples of the Aryan stock characteristic of northern India entered Sri Lanka, intermarried with the Dravidians and became the forebears of the ethnic group described as Sinhalese. This race occupied the northern part of the island. They were a civilized, cultured people of high material and intellectual attainment, skilled in architecture, irrigation and agriculture. From the fifth to the third centuries BC a sophisticated civilization flourished in what is now the Dry Zone based on an intricate irrigation system. The Wet Zone with its rain forest was regarded as inhospitable peripheral territory. Buddhism was introduced around 300 BC and adopted by the Sinhalese, 90 per cent of whom are Buddhists to this day. Shortly after the advent of Buddhism, Dravidians from southern India began to invade the island. They were referred to as Tamils because of their language and were Hindus of the Saivite cult. Religious, linguistic and cultural differences kept the two races apart and armed conflict followed. The civilization of the Sinhalese withstood these early assaults but from 300 AD to

1100 AD the history of Sri Lanka is one of repeated Tamil invasions. By the twelfth century AD the Sinhalese kingdoms were broken, their irrigation works abandoned and the Sinhalese retreated to the hills and forests of the south-west. Jungle grew over the ruined cities of both Anaradhapura and Polannaruwa destroyed in Tamil invasions, water stagnated in abandoned tanks and canals, and malaria became rife. The Dry Zone which had once cradled the elegant Sinhalese kingdoms became known as *Wanni* – the wasteland. In the fifteenth century a strong Sinhalese king defeated the Tamils of the north and united Sri Lanka. In 1505 the Portuguese landed at the site of the present capital Columbo and the period of colonial interest and interference had begun.

THE COLONIAL PERIOD

The Portuguese interest was in spices, especially cinnamon, since Sri Lanka produced most of the world's supply. They were followed by the Dutch with their East India Company which had a similar interest in the spice trade. Having allied themselves with the Sinhalese, the Dutch fought the Portuguese and had displaced them by 1658. They introduced a legal system, organized the spice trade and built a series of forts to protect the ports; that at Galle survives almost intact. These early colonial exploits were essentially coastal and extractive. Their impact on Lanka its peoples and economy was slight. The British were to play a role of much greater and lasting consequence. They had two interests. First, the British East India Company, like the Portuguese and Dutch, had an interest in the spice trade but, secondly, the British Government coveted Trincomalee Bay as a valuable naval base on the Bay of Bengal. The conflict in Europe between the Dutch and British in the eighteenth century gave Britain the excuse to fight and displace the Dutch in Ceylon. In 1796 the Dutch were ejected and the British established themselves securely in the coastal lowlands, only to find themselves opposed by the King of Kandy in the central hills. He was defeated and deposed in 1815 though revolts continued into the mid-century. The British occupation was no longer simply a matter of spice trading ports. A governor was installed and a road network commenced. Between 1820 and 1834 Governor Barnes built some 650 kilometres of road, producing a network which not only allowed the inland areas and central highlands to be controlled and subdued but also, importantly, to be opened up to economic developments in a manner previously not possible. Colonialism as an economic system was being implemented.

Cinnamon, at first the only important export commodity, was cultivated on British-managed estates in the hot, wet south-west. The British monopoly was destroyed by competition from cinnamon exports originating in China, India and Java, and the investors looked for other crops. Now accessible because of Barnes's roads, the central highlands provided both climatic and edaphic conditions suitable for coffee. Plantations grew apace until 1865 when the market collapsed. The surviving plantations emerged with better organization and techniques so that by 1878, 101,000 hectares were under coffee, a railway had reached into the interior mountains and Ceylon had 4800 kilometres of road. A colonial infrastructure was developing but unlike the situation in Nigeria it was characterized by a plantation economy in which the British were owners and managers. Ceylon became in international terms a one-crop export country. A rust disease began to attack the coffee plants in 1868 and by 1880 all plantations were infected and coffee was abandoned. The colony's chief resource of revenue had gone; coffee had provided the money for roads, railways, harbours and for the repair of old irrigation works. With coffee had also gone the livelihood of thousands of plantation workers. The solution was found in tea plantations. Previously less profitable than coffee, tea was planted extensively from 1884 and it was found that Ceylon's environment was ideal. It became a highly organized and scientifically based agricultural system with an associated curing industry exporting almost all its production to the United Kingdom and its Dominions and to the USA. Tea emerged as the most important single commercial activity in the country and by 1945 with 227,000 hectares under tea, about 17 per cent of the cultivated area, Ceylon produced 30 per cent of the world's tea. A highland wet zone crop, with 90 per cent grown at altitudes of over 600 metres, production became increasingly based on large estates as plantation ownership passed into the hands of the big tea companies which also operated the tea factories. Labour costs represented two-thirds of production costs. Both the Sinhalese and the Ceylon Tamils, who were peasant peoples working their own land, were uninterested in employment as hired plantation labourers so the labour force was recruited almost entirely from southern India. The Indian Tamils were regarded as quite distinct from the long-established Ceylon Tamils of the colony. Not all tea production was in European hands. Some estates were owned by Sinhalese and 12 per cent of tea was produced on small holdings but a classic plantation economy with foreign-owned and managed estates dominated Ceylon. Profits went to Britain though some revenue was retained to support the colonial administration of Ceylon and maintain the physical and social infrastructure.

The growth of tea was augmented by other agricultural commercial ventures (Fig. 29). Rubber was introduced, and as the market for it grew in the twentieth century, extensive areas in the lowlands of the wet south-west were planted with rubber, the area totalling 265,000 hectares by 1950. This lowland crop, unlike tea, came to occupy land used in shifting cultivation and again, unlike tea, was much more a crop of the Sinhalese with 50 per cent owned and managed by them. Cocoa was a further, but minor, export crop with similar environmental requirements to rubber but important only in that it offered a further diversification of cash cropping. Coconuts were more important in terms of area and Ceylonese involvement, with less than 25 per cent being grown on European estates. Furthermore, they constituted an important subsistence crop providing food, oils, timber, fibre and alcohol with two-thirds of the product consumed in Ceylon. The cash economy for agricultural export was therefore one in which the Sinhalese and Ceylon Tamils were little involved either as owners or workers. The tea plantations were worked by temporary, imported

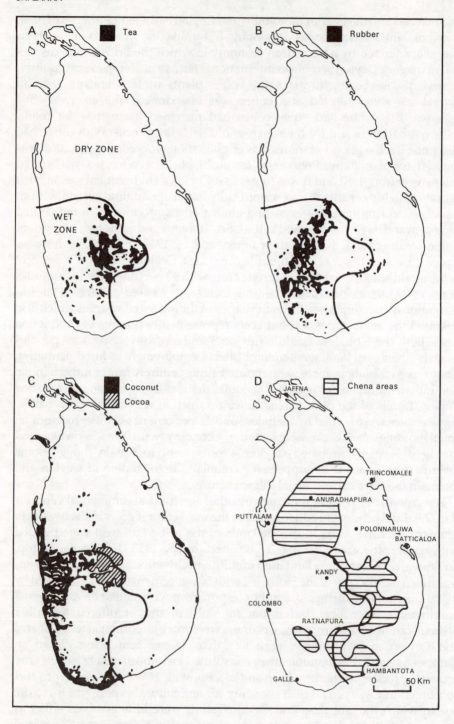

hired labour. In other export crops the Ceylonese were essentially part-cash and part-subsistence farmers.

As an Asian hot and wet country, Ceylon is a land where rice is the staple of the diet. Paddy, or wet rice, is characteristic of the south-west where two crops per year can be grown. A crop of the lowland and alluvial valley bottoms, it is also grown in impressive irrigation terraces on the western hill slopes. In the Dry Zone paddy was grown in large compact blocks of irrigated land where the ancient tanks or reservoirs had been restored and in a myriad of tiny plots near small more recent tanks. Malaria was rife and the population small, and Dry Zone yields were characteristically low but throughout Ceylon inadequate manuring, poor seed, an absence of transplanting and a fragmentation of holdings were the hallmarks of a poor and undeveloped peasant agriculture. Ceylon imported half its rice needs. In addition to the permanent farming of paddy rice, throughout the Dry Zone and in the forested areas of the Wet Zone, shifting cultivation, *chena*, was practised (Fig. 29). This extensive form of farming could persist because of the low population densities in the Dry Zone and made use of land which could not otherwise be used since it could not be irrigated with existing tank systems. The crops grown were hill rice, maize, millets and a range of minor crops without irrigation. As in other countries, the Government looked at the possibilities of improving *chena* systems by, for example, introducing rotation and by incorporating cash crops. More importantly, however, the Colonial Government in the 1930s implemented plans to regularize the settlement, the so-called colonization of these Dry Zone areas, giving tenure to settlers and financial grants in return for a requirement to cultivate the land continuously (Farmer 1950, 1957).

In an overwhelmingly agricultural colonial economy, manufacturing was hardly developed. The country had no significant raw materials other than the products of agriculture so its industrial plant consisted of tea-curing factories, rubber mills processing latex, and coconut-oil mills. Domestic consumer goods such as cloth and clothing were manufactured on a very small scale, as were foodstuffs and beverages. The countryside was a place of small handicraft industries. In 1950 manufacturing accounted for 4 per cent of the Gross Domestic Product.

DEMOCRATIC INDEPENDENCE AND DEVELOPMENT PLANNING

When colonial rule came to an end in 1948, Ceylon possessed an economy the die for which had been cast in the colonial period. It lacked balance and diversification. Earning exchange revenue from a few primary products, it was at risk from world price fluctuations. Its people were poor and their agriculture far from productive. Manufacturing was almost non-existent. A population, regarded as alien, of nearly one million Indians working the tea plantations had

Figure 29 Sri Lanka distribution of: (A) tea plantations; (B) rubber; (C) coconut groves; (D) *chena* agriculture

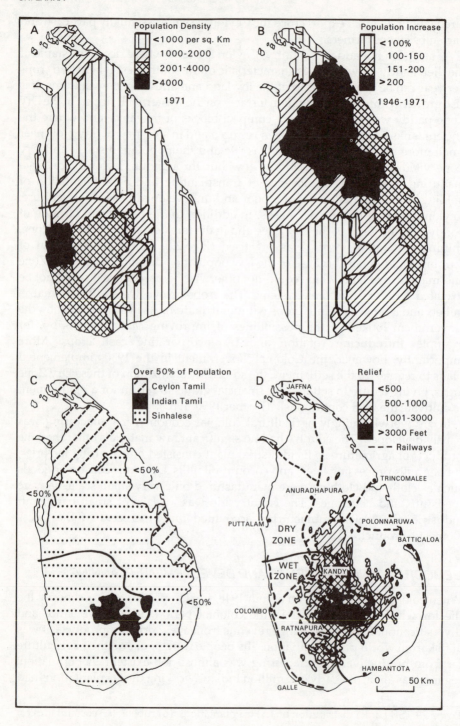

been brought in by the British. These were features which could be seen but the explosion of population had only just begun and was hardly appreciated let alone built into projections. The imprint, far older than British colonialism, of duality of race, religion and language was still clear; it too was to have a bearing on the shape of Ceylon's political economy.

The party which formed the first government was the moderate United National Party (UNP). It had at first little idea of structural planning. The view that there should be state control of basic industries was mooted, reflecting, perhaps, feelings that after the experience of colonialism it was necessary to be in control of the nation's destiny and that private enterprise was associated with foreign domination. The colonization of the Dry Zone initiated by the British was examined and plans were made for new major irrigation projects to further this aim, particularly since malaria was in the process of eradication there. Like so many new governments of the period, that of Ceylon had little previous experience to draw upon in identifying the problems and potentials of development or of techniques in resolving the one and developing the other. In 1952 the International Bank for Reconstruction and Development, the World Bank, sent a mission to Ceylon which made an analysis of the economic situation on behalf of the Government (World Bank 1953). In 1954, drawing upon the findings of this report, the Government issued a six-year programme of investment (Ceylon Planning Secretariat 1955).

The World Bank's mission stated that the resource of agriculture must form the basis of development. It argued that Ceylon had the capital, derived from buoyant exports and external assets which had accrued during the Second World War and the Korean War, to fund a major development programme. Investment could be made in hydro-electric irrigation projects on major rivers associated with peasant colonization (Farmer 1957). The World Bank viewed manufacturing as a means of diversifying the economy, of providing employment and of replacing imports by home-manufactures but stressed that industrialization itself would not be able to solve the country's problems. The resource base was too weak and development was unlikely to take place at a rate which would provide sufficient employment. The Bank recommended that private enterprise be given every incentive to develop the manufacturing sector. Taking up these points, the United National Party in its Six Year Programme laid emphasis upon agriculture and advocated a higher input into the land and improved methods of cultivation in the case of the export crops of tea, rubber and coconuts. It thus advocated a greater production without an extension of the cultivated area. In most cases, it was advice little needed and in the case of tea, the industry was already highly efficient and organized. In the case of food production, the programme aimed at reducing the large imports by increasing yields of rice from the paddy fields of the Wet Zone and by extending the area under paddy rice in the Dry Zone. This was to be achieved by restoring old

Figure 30 Sri Lanka: (A) population density 1971; (B) population increase 1946–71; (C) ethnic groups; (D) relief and railways

irrigation tank systems and, as the World Bank had recommended, by constructing entirely new large-scale irrigation schemes. It had been estimated that of the 243,000 hectares in the Dry Zone which were capable of being irrigated, half was not yet irrigated. The largest single scheme to emerge from this programme was the Gal Oya multi-purpose scheme though its beginnings pre-dated both the World Bank's report and the Six Year Programme. Designed to control floods on the Kala Oya, energize a 10 MW power station and irrigate 48,565 hectares, it involved a massive investment. It was calculated that it would be farmed by 125,000–150,000 persons who together with supporting services might total a population of 250,000. To settle such numbers was a big achievement but it is put in a truer development perspective when placed against a population expansion of 200,000 per year. In the case of manufacturing, the programme outlined plans of the development of small-scale private enterprises to operate import-substitute industries and larger scale schemes for the production of tyres and fertilizers. It was not envisaged that the state would operate large-scale industries at least as a permanent feature of the economy.

This first planning programme was imprecise and uncertain. It was formulated at a time when the prices for Ceylon's export products were high. When these dropped markedly in 1956 they reduced the income available for development and exposed the vulnerability of Ceylon's economy, dependent as it was upon the export of a few primary commodities. It was an experience shared by many developing countries. The situation was made more difficult by the increasing evidence of population expansion. An ever-growing mismatch between employment demand and employment opportunities was one manifestation. A second was the growing gap between food production and food need. The Government held rice prices low by subsidy but this itself drained away government resources. The population explosion hit hard at a goal, which, to their great credit, has been held dear by all the governments of independent Sri Lanka no matter their political complexion. To the fore has always been put the provision of health and education facilities. This has undoubtedly contributed to the subsequent reduction in birth rates which make Sri Lanka anomalous in its income group (Chapter 4) but in its initiation its increasing costs, so directly related to population size, became a growing call upon central finances (Wickremeratne 1977). A leisurely patrician attitude could no longer be taken to national economic problems; they had become urgent and immediate.

In the second general election of 1952 the United National Party had been returned to power. In that of 1956 it was defeated and the Sri Lanka Freedom Party (SLFP), a left-of-centre party, came to power in a coalition which included a Marxist faction. Mr Bandaranaika, its leader, had courted the Sinhalese rural vote by stressing that the Sinhalese language and the Buddhist religion were the strength of the nation of Ceylon. The first UNP Government had enacted legislation determining citizenship which effectively removed it from almost all the Indian Tamils. As B. H. Farmer has described, throughout the post-independence period both Indian and Ceylon Tamils had experienced political and psychological pressure from the Sinhalese majority. The colonization of the Dry Zone had settled Sinhalese in areas which had previously been dominantly Tamil (Farmer 1963b). Matters were brought to a head when in April 1956 the newly-elected government introduced legislation which declared Sinhala to be the only national language. Not only did the Tamils see this as an attack on their culture and religion but also an exclusion from the economy if fluency in Sinhala was necessary. Rioting with a considerable loss of life occurred. A second violent outbreak took place in 1958 and a state of emergency declared. The ancient racial division was becoming manifest.

In 1959 the SLFP leader was assassinated and in the ensuing elections his widow Mrs Bandaranaika took over its leadership and led the party to a second victory in 1960. The Bandaranaika governments had come to power at a time of economic crisis and had as their goals the creation of a socialist society. A planning council had been set up in 1956 and a Ten Year Plan was published in 1959. It involved a massive investment by the state with an increasing control of basic resources. Government expenditure was to be cut over a wide range and taxation was to be increased. The aim was to increase production, become more self-reliant and reduce unemployment.

The Ten Year Plan was faced with one problem which no amount of political rhetoric could disperse. Ceylon's rapidly growing population was increasing the size of the development task as each year went by. The plan stated:

a sizeable proportion of the resources for investment have to be devoted to meeting the social needs of the rising population in the way of school buildings, houses, sanitation systems and so on. Investment in these fields competes with investments in directly productive spheres and result in an allocation of resources unsuited to the needs of maximum growth. Such a maldistribution of resources would itself have a cumulative depressing effect on the level of future investment. Ceylon is clearly facing problems of this sort. Not only does the consumption need of a growing population tend to depress the rate of investment but the social services, education, health and housing themselves make increasing claims on investment resources

(Ceylon Planning Secretariat 1959).

The Government must have been sorely tempted to reduce this social provision and indeed contemplated it.

Over its ten year span 41 per cent of the total investment was to be devoted to the manufacturing sector, compared with only 22.9 per cent to agriculture. It was a radical change from the less precisely defined UNP views with their agricultural emphasis. Sri Lanka was to be put on a new course. The plan was as costly as it was ambitious. It was hoped that foreign investment would facilitate this structural transformation of the economy into one with a much stronger manufacturing sector. The economic situation continued to deteriorate and in 1962 a Short Term Implementation Scheme was introduced. It had the same goals but was less ambitious and characterized by a realism sharpened by the lack of funds.

It could be questioned whether it was wise to put this elaborate industrial cart, constructed by the state, before an undernourished and infirm agricultural horse. Sri Lanka is but one of many countries to do so. The arguments underlying the proposed industrialization seemed reasonable: the vulnerability on the primary export market had been demonstrated; there was an obvious need to diversify the economy; there was a need to reduce the costs of imports which could be made at home; crucially there was an urgent need to provide employment. Industrialization seemed the answer but the real question was one of timing and balance. Despite the vagaries of the export market, the earnings of tea, rubber and coconut products were essential to any investment programme and Mrs Bandaranaika's government was aware of this. It was, however, critical of the large-scale colonization schemes of its predecessor with their high ICORs and high costs of social provision. It felt that, instead, more attention should be focused upon the individual peasant farmer since much lower ICORs would pertain and results would be quicker. The industrial emphasis was to remain because of all its perceived virtues and its likewise quick returns. This experiment of Sri Lanka's first socialist government failed. As the situation deteriorated, massive import restrictions were imposed as import costs went up and as, in the face of a rising population, per capita GNP declined. Between 1960 and 1965, despite the emphasis given to it, the contribution of maufacturing to GNP rose by only 0.9 per cent. In the general election of 1965 the Government was defeated.

The party to return to power was the centrist UNP which established a Ministry of Planning and Economic Affairs but was wary, in the light of its predecessor's experience, of over-ambitious plans and long-term planning proposals. Indeed, the new government was coming to power in conditions of economic crisis some seventeen years after independence when Ceylon's development plans appeared to have achieved little and her enlarged population was enjoying no improvement in her economic and social conditions. It was felt appropriate to institute a recovery programme to establish a securer basis upon which longer term programmes could be founded. This programme set itself a target of an annual growth in the economy of 5 per cent which would encompass the expansion in population and allow a slight increase in the per capita GNP. In establishing foundations it laid emphasis upon the agricultural sector in an attempt to improve the condition of the mass of the peasantry and to make more secure what was seen as the central strength of the economy, the export crops. Measures were taken to improve the output from the traditional paddy farms and irrigation schemes constructed to extend the area devoted to paddy rice. In addition, encouragement was given to a diversification in food crops grown, such as potatoes and onions and other vegetable and fruit crops, not only for consumption on the farm but as cash crops for the urban markets. In the central hills and highlands, slopes on non-tea land were terraced to produce food crops favoured by the more temperate conditions. These steps served to reduce Ceylon's dependence upon food imports and together with the securing of the export crops were designed as a necessary prerequisite for subsequent industrial developments. The plan incorporated industrial funding but to an extent only two-thirds of that devoted to agriculture. The agricultural effort was repaid by increases in paddy rice production which rose by 16.5 per cent between 1965 and 1970 and by the successful establishment of intensive vegetable farming. Growth in per capita GNP more than offset the population increase and allowed a sizeable proportion of the population to experience an increase in real income. This progress was set back by events external to Ceylon. Prices for her exports of tea and rubber fell and, as the decade ended, the price of imported goods increased, a trend to be further accelerated in the early 1970s as oil prices escalated. Ceylon had no overseas assets with which to pay for her imports and had to resort to borrowing.

Both the SLFP and the UNP had followed policies which stressed Sinhalese paramountcy rather than a national unity and the feeling of isolation remained among the Tamils. It was not, however, this issue which brought down the UNP Government in the election of 1970 but the Government's decision to reduce the ration of subsidized rice (Jeyaratnan Wilson 1977). Expenditure on social services had expanded to such an extent that it accounted for 35 per cent of government expenditure by 1963. The state had taken over the running of all schools in 1960 and added to its financial commitments. Since rice subsidies were the biggest call upon resources the Government, in the face of a deteriorating economy, cut the subsidy and lost the election. The new government, once again led by the SLFP with Marxist support and calling itself the United Front Party, found itself in a further insurrection unrelated to the Tamil disputes. An extreme left wing group, the Janatha Vimukthi Peramuna, or People's Liberation Front, drawing on the discontent of the educated unemployed, often of middle-class groups, attempted to install by force what they regarded as a true socialist society. Eventually crushed by the army, the insurrection damaged the prospects for economic development and delayed the development initiatives of the UFP.

A Five Year Plan was produced at the end of 1971 and set out the proposals for a socialist society and the involvement of the people in planning, and the country's name was changed to Sri Lanka. In reality the plan was little different from that of the displaced UNP party. Again there were proposals to increase the output of the peasant farmer, extend the amount of irrigated land and to widen the agricultural export-base by extending it to pineapples, passion fruit and other crops including the revival of cocoa exports. As before, the goal was to reduce export vulnerability and make Sri Lanka more self-sufficient in rice. Importantly, the Government turned its attention to land holding. As an ideological tenet it intended both to redistribute land-ownership more widely and to extend the degree of state ownership and control. Evidence existed which indicated that land tenure systems in Sri Lanka inhibited the increase in the productivity of peasant holdings of foods for home consumption. Tenurial forms, though several, were essentially variations in forms of share-cropping referred to in Sri Lanka as the *ande* system. The landowners provided seed, draught animals, usually water-buffalo, and other inputs; the tenant provided the labour and turned over between 50 and 75 per cent of his harvest to the owner. Many landowners had moved out of agriculture into other occupations but had retained their land as a source both of income and security. A symbiotic traditional relationship between landlord and tenant was common in many rural communities. Nor was it all bad. The Paddy Lands Act of the socialist government in 1958 had addressed the issue and converted shares into rents. It was unsuccessful. Rents were equated with half-shares of cropping and what had been a far from rigid relationship involving friendship and often kinship became a commercial transaction.

In its Land Reform Act of 1972 the Government did little to change the status of the peasant working the paddy land with its share-cropping systems since the Act did not apply to holdings of less than 10 hectares. Most paddy holdings were smaller. Private individuals were not allowed to own more than 20 hectares and any excess was to be redistributed. The holdings of public companies were exempt. In the event, under this Act the state acquired waste and uncultivated land and land growing tea, rubber and coconuts. Of the order of 72 per cent of the tea area and 34 per cent of rubber was removed from private ownership. It was a drastic change but it did not lead to land redistribution since the Government was aware of the dangers of fragmenting the large plantations; instead they became state plantations. Thus very little land was redistributed. It went to co-operatives, district authorities and the state, while a further act set production norms on private land with the threat of confiscation if they were not met.

In its industrial section the plan laid emphasis on the provision of employment with a stress on labour-intensive methods to secure this goal. It was a policy with much to recommend it in the field of small-scale import substitution industries, but could not be implemented in industries such as steel, chemicals and cement if modern large-scale methods were to be used. The policy was that such industries would be state-owned and operated. In all these policies the issue of unemployment was ever present. Its political ideology put the Government in difficulties. On the one hand it believed the state should own and administer as wide a section of the economy as possible. On the other, it was aware of its own inefficiencies and inabilities to do this effectively and that it therefore required the energy and initiative of private enterprise. Its policies had done much to stifle this enterprise. The plans of the United Front Government, whether wise or unwise, were not able to prove themselves. Buffeted by the rising prices of all imports subsequent to the OPEC rise in oil prices, Sri Lanka's economy became further burdened with indebtedness. With but one or two exceptions, Sri Lanka had experienced an adverse balance of payments every year since 1957 and in some years before that. Import costs had gone up as prices had inflated. Import levels had increased both to service Sri Lanka's development proposals and supply her increasing population. Her educated population had developed an increasing awareness of higher levels of material well-being and its level of expectations increased accordingly. The UFP did not survive the 1977 election nor indeed subsequent ones as the middle-of-the-road UNP party led by K. R. Jayewardene proved itself more acceptable to the electorate.

Under Jayewardene, the Government has moved away from the restrictive, command economy of its socialist predecessor and endeavoured both to open up the country and broaden the basis of its economy. This had been attempted earlier but there were now three further dimensions of economic consequence. First, tourism had been encouraged on a scale far greater than ever before, and with Sri Lanka able to offer much of attraction with its climate, dramatic scenery, ancient cities and modern temples, it became a development of success earning extra exchange revenue. Second, there has developed a considerable temporary migration of Sri Lanka's workers to the Middle East and Singapore, chiefly into the industrial and construction sectors in countries where wage-levels are much higher than in Sri Lanka. The remittances home were, and still are, economically significant. In 1970 they totalled US$ 3 million and by 1990 had reached US$ 401 million, representing 8.2 per cent of GNP. The overseas movement of labour eases to some extent the pressure on employment. Third, and perhaps most significantly, the government of Mr Jayewardene established an Industrial Export Zone north of Colombo. Foreign capital and expertise uses Sri Lanka's labour, producing goods not for the home market but for export. The zone is a significant source of industrial employment and has been successful to date.

Throughout her post-colonial history Sri Lanka has maintained a parliamentary democracy. It has survived considerable oscillations in ideology, economic policy and economic depression. It has given a voice to the people of the nation and undoubtedly allowed the release of political tensions but its very oscillations have made the development path a far from straight one. So climacteric have been political events that it has been questioned in Sri Lanka itself as to whether democracy could survive. Resort has been made on several occasions to rule by decree in periods of disturbance; the events which began in 1983 have been, however, on a scale and of a degree of violence for which the word disturbance is inadequate.

Never far below the surface, the dissatisfaction of the Tamil minority with their position of inferiority in a supposedly united nation burst out into armed conflict as a guerrilla group styled the Tamil Tigers committed a number of terrorist acts. It will be recalled that both major political parties in Sri Lanka had used nationalism as a political lever in elections. Nationalism meant the characteristics associated with Sinhalese culture, language and Buddhism. The repatriation to India of the Indian Tamils of the tea areas because they were seen as allies of the Ceylon Tamils was begun in the 1960s. As a solution to the discrimination against them, the Ceylon Tamils had sought autonomy for the Tamil-dominated areas of the north and east within a federal constitution as long ago as 1956 but this had been rejected by the Sinhalese majority. As the prejudice against them persisted, even the moderate Tamil United Liberation Front began to demand a separate state for Tamils. The Tamil Tigers carried this demand with them into armed conflict calling for a *Tamil Eelam, Eelam* meaning 'the precious land'. In the backlash to the first terrorist acts of the Tigers, Tamil property and Tamils themselves were attacked by Sinhalese throughout the country. Some Tamil entrepreneurs left their businesses and fled the country, others left the Sinhalese dominated areas for the north. Further killings and the counter-insurgency activities of the Sri Lanka army further polarized the country. The emerging economic initiatives to bring some prosperity to a poor country were blunted as tourists fought shy of the island's violence and as the work-force and infrastructure was disrupted. Mr Jayewardene's government was confounded by this violent uprising. Tamils were in the cabinet, held key jobs and were playing on important role in the country's economic recovery. The long years of the discrimination the Tamils had felt, of not being full and equal citizens of Sri Lanka had, however, not been appreciated. India, which has so much influence in Sri Lanka, was asked to help with her troops to quell the insurrection; it was a failure and Indian troops were withdrawn at the request of Mr Premadasa, who succeeded Mr Jayewardene as President in 1989. The economy had begun to recover as tourists returned at least to the coast resorts of the south-east but Mr Premadasa still faced a problem unresolved as the Tigers, though much subdued, were still undefeated. In 1992 a Tamil member of the Sri Lankan cabinet again proposed autonomy for a Tamil region but was strongly opposed by the Sinhalese majority. None the less discussions continued with both the Government and the opposition agreeing to consider the proposal, provided that the northern and the eastern Tamil areas did not merge into a single state which the Tamil Tigers demand. The economy was disrupted by the severe drought of 1992 which caused crop failures and affected manufacturing industries also. The International Monetary Fund withheld half of the loan to Sri Lanka because the nation's budget deficit was too high. Mr Premadasa's unpopularity grew and an attempt was made to impeach him; in 1993 he was assassinated. The insurrection continued.

The civil war in Sri Lanka, for that is what it has become, has done much to hinder the island's hesitant economic development. It is a conflict which could have been avoided if, in the 1950s, Tamil requests for autonomy within a federal system had been met. As it is, Sri Lanka is currently spending around 6 per cent of its GNP on the war and half a million of the population have left the country as refugees from the war. Others abroad are the overseas workers referred to who remit earnings to their home country; their number is now put at 600,000 and net workers' remittances in Sri Lanka are the highest of all low-income countries apart from those of the much larger nations, Bangladesh, India, Pakistan and Egypt. Many professionally qualified citizens have left Sri Lanka and this represents, of course, a loss of much-needed talent.

On the one hand the ethnic divide and its associated armed conflicts have undoubtedly retarded Sri Lanka's progress and it is not possible to predict confidently a solution to the problem being attained. On the other hand Sri Lanka's economic policies, which until the Jayewardene era were so full of directives, regulations, quotas and red tape, have themselves thwarted enterprise and initiative. Socialism and state controls served Sri Lanka no better than they did Nigeria and Zambia. Jayewardene and Premadasa nurtured an economic revival by freeing trade of constricting regulations, allowing privatization and creating supportive financial services; foreign investments were allowed in rather than discouraged; tourism was seen as a valuable development of resources. A resolution of the Tamil problem could free the energies to take advantage of this new liberal economic environment. Tamil Eelam may be the only acceptable price. The World Bank's comparison between Sri Lanka and Malaysia succinctly compares the consequences of their differing policies (World Bank 1991). Both countries are former British colonies which developed primary exports from tree crops; both initiated and maintained democratic forms of government, and both saw to it that their populations were well-educated and healthy. Each nation was markedly divided ethnically with the majority group economically underprivileged. Both subsidized rice farming and both developed state enterprises in the agricultural, manufacturing and service sectors. Sinhalese had priority access to positions in government employment as did Malays in Malaysia. The similarities are striking yet over a period of thirty years Malaysia increased its per capita GNP much faster than Sri Lanka. Two main factors explain this contrast. First, Malaysia contained and resolved her ethnic differences while Sri Lanka at great cost has not. The World Bank estimates that apart from causing tens of thousands of deaths, the conflict in Sri Lanka has resulted in a loss of revenue due to disruption of economic activities which together with the cost of destroyed infrastructure totals nearly 66 per cent of GDP. The second factor relates directly to policies pursued. Malaysia's policies were much more liberal and sensitive to changes in economic conditions. Apart from high tariffs designed to protect a few specific commodities, Malaysia's trade was open. Private enterprise was not fettered by licences, currency controls or threats of nationalization. In contrast, the Sri Lankan policies described were severely regulatory, suspicious of foreign involvement and dominated by state controls. State involvement was such that it appeared to be fostering the ethnic divisions rather than giving equal representation to all views. After 1977 Jayewardene's reforms did much to redress the balance but over thirty years had been lost and the Tamil problem still eats away at the nation's very substance.

15 China

'The mansions of princes and nobles all have new lords
And another generation wears the caps and robes of office'

Li Po (699–762 AD)
Autumn Meditation

The development path chosen by China has a significance beyond that of any other country in the Developing World. Her size alone, with 1133 million representing almost 22 per cent of the total world population, contributes to that distinction. Of even greater significance is the uniqueness of her approach to development problems. It is an approach which had sought to transform not only her economy but the whole structure of her society. From a country which has suffered the occupation of an aggressive and destructive foreign power, Japan, and had been torn by a bitter civil war, from a country characterized by corruption, chaos and poverty, China has converted herself into an orderly, organized nation of pride and purpose. Her development path has not been a straight one, nor has her progress been smooth and uninterrupted. China remains a poor country, yet her achievements have been great. It could be argued that they are without equal in the Developing World. But it has been a success dearly bought. The price has been the near total loss of freedom by the people of China for three decades.

In the complex interaction of China's planning strategies, one aspect stands out – it is the attention paid to the countryside with its farms, villages and country towns. It has been a concern which has not only embraced agricultural production but the whole rural economy and with it the issue of the industrial–agricultural relationship, the urban–rural dichotomy and the rural to urban drift. It is precisely in this ensemble of issues that least success has been achieved in most developing countries. The failure to maintain adequate agricultural productivity, to give a self-sustaining vitality to rural areas, to stem the flood of cityward migrants, has underlain the problems of so many developing countries. While the Chinese have held to this concern, their approach to it has evolved from experimentation. Alternative strategies have been tried and the final solution, successful though it has been in terms of its own objectives, is proving itself to be but a stage in an unending process of social and economic evolution.

Central to all developments, from the founding of the People's Republic in 1949 until his death in 1976, was Mao Tse Tung. The reputation of Mao and the esteem in which he was held stood above the internal intrigues and strife of the new republic. A dominant personality and a great thinker, in Mao were combined the roles of spiritual leader and political philosopher. Though he created a totalitarian state he felt strongly that the state must not be an entity separated from the people. He saw an increase in national production as a central task but not as one designed solely to create revenue for the state. In part this was, no doubt, a reaction against the policies whereby the state throughout China's long imperial history had kept alive its political and administrative functions and huge bureaucracy by heavy taxation. If people were the state then, in Mao's view, the benefits of increased production should go to them. Equity was another, and obvious, goal of his new People's Republic. Sustained by a Marxist–Leninist ideology, he none the less considered entrepreneurship to be important but the entrepreneurial roles were to be taken by groups rather than individuals. He believed that China could achieve the material levels of well-being experienced by the Industrialized World and that this could be attained quickly. Within his ideological pronouncements were often strong pragmatic undertones.

The new China, heir to one of the richest and oldest cultures in the world, was as complex and paradoxical a society as any which had preceded it. In it were contained both spiritual and ideological components; it aimed for equity yet it was characterized by a party elite and personality leadership; it had grown out of revolution but required stability; it despised the bureaucracy of the Mandarins yet created one of its own; it required expertise yet denied the existence of the expert; it clung to the anchorage of an alien ideology from which its indigenous pragmatism was continually trying to free it. The enormous and forbidding reality with which it was faced was that of providing over 500 million people with food, shelter, health, employment and, most importantly, hope. Politics were indeed tempered by pragmatism.

In facing this challenge one of the most striking features of China's approach has been her recognition of the importance of psychological conditioning. By a combination of devices both loud and brash, muted and subtle, the thoughts and energies of the Chinese people have been harnessed to the radical transformation of China's society and economy. The manifestations of religious beliefs, competitors with the new ideology for hearts and minds, have been suppressed, and temples turned into museums, factories and grain stores. The reception of external radio transmissions was prevented and the availability of foreign literature restricted. Travel outside China became impossible. Levels of expectations could thus be confined to circumstances within China and the population kept ignorant of alternative systems and their levels of attainment. The bad conditions of pre-revolutionary China were kept alive in the minds of the young who had never known them, by displays of photographs, models and artefacts in museums on housing estates even twenty-five years after the Revolution. It was seen as essential that the young should have some measure of the progress which had been made if the revolutionary fervour was to be kept alive. Commercial advertisements were totally absent; in their place were quotations from Marx, Lenin and Engels, exhortations to work and sustain the Revolution and, everywhere, the picture of Chairman Mao. The great aesthetic legacies of China's past, among the finest artistic achievements in the world, were not rejected but took their place in the new order. They gave to it a sense of belonging and continuity and nourished a feeling of national pride in the great civilization to which the people of China are heirs. This astute awareness of the need to control minds if events were to be controlled, produced a disciplined, patriotic, enthusiastic and largely compliant populace.

THE EMERGENCE OF MODELS OF DEVELOPMENT

In 1948 China was a wide expanse of agricultural peasantry. The distribution of the population was in large measure topographic and climatic, densely packed in river valleys and on the coastal plains, thinly spaced in the mountains of this mountainous country and sparse to almost non-existent in the high arid interior of central Asia. Modern industrial installations were located in a few large, and mainly coastal, cities. Many had been established in the 'treaty ports' set up by European powers in the nineteenth century. Private land-ownership, with landlord and tenant systems, characterized a countryside in which craft industries were scattered throughout villages and the rural towns. A complex system of indigenous marketing at local and regional scale had developed over the centuries to handle internal trade. China's foreign trade had never been very large but that which had grown up in the nineteenth century was very much in foreign hands. The Chinese Communist Party realized that only a massive reorganization of the whole country could transform it into the socialist society of the People's Republic. It began by transferring to the state all means of production. In the industrial sector, the state requisition of the large, relatively modern manufacturing units seemed obvious and simple. Agriculture and the countryside with its millions of small landowners and peasants presented an entirely different and more difficult problem. Clearly both major sectors were essential to China's development aspirations, both needed to be encompassed within the socialist beliefs of state ownership and management and the goal of equity, and both needed revitalization and development if the dreams of enough for all were to be realized. The problem was urgent since China's population was growing apace and the momentum of the revolution could be lost.

The only model of development which was at hand in a Marxist state in 1949 was that of the Soviet Union. Ideologically the two nations appeared to have much in common. The Soviet Union had not entered the war against Japan but had been an ally of those who had. Mao's rejection of the old society and his endeavour to create a socialist state in China led him to base his initial experiment in development upon the Soviet model. In the USSR the economy had been developed on the basis of a concentration on heavy industry on the one hand and the collectivization of agriculture on the other. A policy of rapid incremental growth especially in capital-intensive industry had taken precedence over agricultural investment. Mao's China followed suit.

The first Chinese planning period from 1953 to 1957 attempted to organize the structures necessary for subsequent economic development and the inevitable socialist bureaucracies were constructed. The Soviet balance between agriculture and industry was, it could be argued, less suitable for China's socialism and its advocacy of heavy-industry investment questioned, but Soviet policies were adopted and heavy industrial plant from Russia was transported to China and set up with the help of Soviet technicians. These large units were established in the major cities and functioned at a scale which their scale-economies required. They thus reinforced the pre-existing concentration of manufacturing and they made large demands on available capital.

In assuming control of the land, the Government encouraged the formation of mutual aid teams which were groupings of six to ten families to work the land jointly. By 1952 the process had incorporated approximately 40 per cent of rural households. It was the initial and elementary way of accommodating the transfer from private to public land ownership but the teams were essentially private rather than state enterprises. Out of these simple groupings co-operatives were established. These in turn were reassembled into large more elaborate co-operatives where the individual was subordinate to collective action and where the co-operative produced to meet targets set by the state. By 1957, it has been estimated, 95 per cent of Chinese peasants were working on these large co-operatives. This sequence of events, briefly stated, marked the transition from private ownership, individual enterprise and hired labour in a market economy to a collective organization in a command economy. These agricultural developments were, in part, a corollary of those in manufacturing. The state controlled all forms of agricultural production. Since manufacturing was to receive the major emphasis in investment because it could more effectively use it, the reorganization of agriculture had to be carried out in such a way as to minimize its capital requirements. Since the state controlled the cost of all inputs and the price of products, it could manipulate the agricultural economy so as to extract a 'surplus', a revenue, from it which could then be reinvested wherever the state chose and not necessarily in agriculture.

The confiscation of land and the dispossession of landlords, the rupture of market relationships and the reorganization of agrarian labour and production, all made for deliberate and drastic changes in social and economic systems which had evolved over the centuries. The past was to be broken, agrarian conservatism shaken and, with the aid of the cadres, salaried party officials drafted into the countryside, the world of the Chinese peasant opened to new developments. This initial reorganization of agriculture did increase production and at first led to a rise in the income of agricultural workers. This latter improvement was, however, the product of the more equitable distribution of income rather than as a result of increased output. As the state continued to manipulate the agricultural economy, increasingly its surplus was siphoned off to fund industry and the prospects of greater affluence for the peasants receded. The huge central bureaucracy attempting to manage the co-operatives on a national scale was far too big, too remote, too inexpert in agronomics to be able to articulate policies sensitive to the wide range of environmental conditions which are present in China. The contact with the land was too insulated from reality. It was a manifestation of the now familiar problem of over-centralized state control. These agrarian developments did achieve a widespread parity in agrarian incomes and they did not result in agricultural disaster such as that which the Soviet Union had experienced in its enforced collectivization in the late 1920s but they were not a satisfactory solution to China's agricultural problems.

The large industrial units proved more easy to co-ordinate and manage, and industrial production of basic products rose appreciably. During this first Five Year Plan, with manufacturing growing more rapidly than agriculture and with industrial investment being city-based, the contrast between countryside and town was being heightened as growing industrialization was accompanied by growing urbanization. The familiar pattern of over-industrial emphasis, of core and periphery, of backward agriculture and forward industry, of change in the town and stagnation in the countryside, was beginning to appear in China. This was taking place despite the agricultural reorganization. As China's population continued to increase rapidly and become more urban-based so the pressure of demand on food production grew. Agriculture was not producing enough raw materials for industry and retarded its growth; food supplies for the towns began to fall. The mismatch between the two sectors became apparent and the Soviet model was called into question. As Mao Tse Tung stated in 1958, 'Since we didn't understand these things and had absolutely no experience, all we could do in our ignorance was to import foreign methods' (Schram 1974). The Chinese began to re-examine their development problems themselves.

THE DEVELOPMENT OF THE PEOPLES' COMMUNES

Mao introduced the idea of the Peoples' Communes in 1958. They represented a significant shift away from the Soviet view of state-managed agriculture and were a distinctively Chinese creation. Larger than the co-operatives which they replaced, they covered a much wider range of activities which they managed themselves as discrete administrative and economic entities. Mao claimed 'the agricultural policy of the Soviet Union has always been wrong in that it drains the pond to catch the fish and is divorced from the masses' (Gray 1969). In stating this he rejected the pre-existing Chinese collective system with its inbuilt disincentives to enterprise and productivity. By the autumn of 1958, 26,000 communes had been formed out of 740,000 co-operatives. It was a development which marked also a move away from the heavy industry-led development of Soviet strategy.

The commune covered a large area of land and its boundaries were generally those of pre-revolutionary administrative areas. Within it there was occasionally an old market town and a number of villages. The size of communes has varied with time and place. In 1962 the 26,000 communes were subdivided to make 74,000. The more fertile areas have the smaller communes. In 1977 the range in size quoted by Sartej Aziz in a sample of 42 communes was from 800 to 4000 hectares though some are known to be 8000 hectares in extent. Populations ranged from as low as 1000 to 80,000, with the majority in the range 10,000 to 40,000 (Aziz 1978).

Each farming family lived in a village, often in its own house. Each had a small plot of land for the family's use growing crops for themselves or for sale but with restrictions on the sale of staple food grains. The members of the family worked for a wage in a production team with the wages based on sex, age and amount of work. The wages formed by far the greater part of incomes. The production team was the working unit. It managed its own budget and administered its own work programme but all its work tasks and production targets were set by a decision body at a higher level in the commune of which it was a part. The commune provided the necessary working equipment. Production teams were organized into production brigades composed of a dozen or so teams, with a similar number of brigades forming the commune. Norms of working and production and the marketing of crops were determined at commune level by the Workers Revolutionary Committee, the management committee, and were passed down via the brigade to the production teams. Thus at each level of activity the participants were involved and each level was integrated with the next above it in the hierarchy. Not only were agricultural functions thus planned and managed but also social services. For example, primary schools and first-aid stations were operated at brigade level while secondary schools and hospitals were operated at commune level. One of the most notable deficiencies in the developing countries has been the lack of provision of adequate health services for their populations and particularly their rural populations. Commonly doctors and hospitals are urban-focused and in many cases they are not free. The poor and the rural areas remain beyond the reach of medical care. In China the training of 'bare-foot' doctors, essentially medical orderlies operating at brigade level, and the provision of fully-trained physicians in commune hospitals, has made medical care available to everyone. With the characteristic appreciation of the psychological factor, both traditional herbal medicines and modern drugs are utilized. In a hierarchical arrangement of facilities the Chinese peasant can receive treatment for minor ailments in his local clinic or undergo major surgery in the nearest large town. To accommodate a thousand million people in such a health service is an achievement of which any nation would be proud.

The commune was, itself, part of the organization of the state and as such had cadres upon the committee. It received high-level decisions from regional and central government and passed them down to its constituent parts. In return it could forward ideas, initiatives and responses to the higher levels of government. It was a system which thus gave an intimacy of contact and allowed individual communes to develop their own initiatives but it co-ordinated this into a national agrarian effort. The commune system thus grew into a remarkably successful way of managing the world's largest population in a system which was highly structured and yet allowed individual or group ingenuity to nourish it. It is, perhaps, the most original and significant of development initiatives in the Developing World, harnessing the abilities, resourcefulness and energies of whole communities in endeavours which they could see as of direct benefit to themselves and not solely for some distant government. It gave to the Chinese peasant the opportunity for pride in his achievement, a justifiable pride which is apparent to any visitor to China.

THE GREAT LEAP FORWARD

Initially the commune system was introduced as part of the strategy to increase agricultural output through a labour-intensive effort achieved by injecting the whole community with an urge and zeal for greater efforts and ultra-efficiency in what were essentially traditional methods of husbandry. A simple concept but an enormous psychological task, it was destined to fade in time. The strategy recognized the possibility of running the economy in two discrete sections, each with a distinctive mode of operation, technology and development-objective. On the one hand the modernization of large-scale industry could take place using large investments of capital in modern plant with the resulting production being re-invested in further production and, to some extent, in the production of exports. Its activities were to be largely unrelated to the 80 per cent of Chinese population who lived in the countryside. On the other hand, the communal system of agricultural production using little capital, deriving little from the industrial sector but gaining its energy from millions of peasants, could feed the country and supply its agricultural raw materials. Mao spoke of 'walking on two legs'. He was referring to the need for balancing the two sectors of the economy and the possibility of utilizing, at one and the same time, two levels of technology. It was, perhaps, an unconscious adoption of Nurkse's views on the substitution of labour for capital but it involved much more (Nurkse 1953). This phase in China's development was promoted as 'The Great Leap Forward'. The development of appropriate technologies and the emergence of industry within the countryside were also associated with the ripening of the commune concept. They came to have a special significance when the impetus of the Great Leap was fading and when the issue of population became pressingly evident.

THE DIMENSION OF POPULATION

There was clearly a limit to what could be done in the agricultural sector by exhortation to greater effort and by ideological fervour and zeal applied to a labour-intensive input without concomitant technological advances. Output fell off and by 1962, coupled with bad harvests, the situation was such that resort had to be made to food imports. The Great Leap Forward had not secured the basis of the nation's food supply nor agricultural raw materials for industry. The situation described was present also in India, the dimension of population was a major factor.

In 1955 China's population was estimated by the Chinese as 615 million; in 1977 it was 934 million. Investment in China during the Great Leap Forward was at such a low level that full, effective use could not be made of the nation's labour force. China had, and still has, a population surplus to the means of fully using it. Labour was substituted for capital but it is an input to which the law of diminishing returns applies rigorously unless other steps are taken to increase per capita output. An over-large population can retard development. In the early 1950s China had adopted a policy of family planning but by the time of the Great Leap Forward this had been dropped. Mao claimed in a classical Marxist manner, that the answer to the problem of population was production and that, therefore, there was no population problem; the problem was one of production. These two items are, of course, interrelated in all development issues. There is a problem created by a rapidly expanding population and this can be compounded by large, absolute resultant numbers. Both features applied in China's case, as in India. It was not a problem which could be dispelled by rhetoric. China came to realize that production was not enough and that some control of population growth was necessary. Family planning became official policy again in the early 1960s since when, though it has been subject to modifications, it has never been dropped (Banister 1987; Yuan Tien 1989, 1991). Propaganda and persuasion have raised the age of marriage; young mothers are urged to the point of coercion to restrict the size of their families; from time to time punitive measures on the one hand and privileges on the other have been used to reduce the birth rate. Birth control advice and appliances are readily available, as are abortions. China has, as a result, achieved the demographic transition in thirty years. Even so its population between 1989 and 1990, when the most recent census was taken, increased by over 21 million. Any further reduction in the rate of population increase is likely to be a product of the growing affluence of the Chinese people and as this differs regionally so too will its demographic effects.

INDUSTRY IN THE COUNTRYSIDE: DISPERSED DEVELOPMENTS

By 1962 the population issue was clearly recognized and with the stagnation of agricultural production a new phase of planning was introduced which laid stress upon the need to increase agricultural output and to relate, in some measure, industrial developments to it. But during the Great Leap Forward developments were also taking place in the communes which enhanced their contribution to the process of development and helped cushion the impact of population growth. These developments were a part of Mao's wider concept of the commune and were spread by the advocacy of Premier Chou En Lai who toured throughout the country. The concepts involved self-reliance and self-sufficiency and the creation of innovations from within Chinese society and by its people. Clearly labour input by itself was insufficient; it required technical support. Mao saw mechanization as more meaningful if it grew up within the community which was to use it than if it were brought in as an alien thing, little understood. The centralized tractor-stations of the Soviet Union seemed an imposed factory system unsuitable for the rural society of China. Chemical fertilizers and machines were needed but these could be produced in small factories in the countryside within the communes, run by the communes and for the communes. Rural workers would be involved in their design and production and by doing so would understand how machines work and how to repair and modify them. It was a development which could draw upon the inventiveness of the people and was a further example of Mao's shrewd awareness of the role of psychology. The significance of this view is apparent to all who have witnessed in so many developing countries, imported machines and vehicles lying idle because they have not been understood by their users who, as a result, were unable to repair them.

The creation of small industries in rural areas, which began in the Great Leap Forward and has continued since, had other consequences. By the early 1970s the increase in population meant that between 10 and 15 million Chinese were entering the job market each year. The cities could absorb less than a million of this increase so any drift to them would only lead to the problems experienced in many developing countries. Chinese policy has been to restrict that drift and until recent years neither a free movement of population was allowed nor any freedom in the choice of employment. Industrial developments in the countryside could provide employment, increase industrial productivity and give technical support to agriculture. Mao's communes came to incorporate these features and an integrated development of the rural economy characterized by a disaggregated pattern of production began to emerge.

Carl Riskin has pointed out that while local rural industry was not promoted in China's first development plan, by the end of the period the need was appreciated and designs for small-scale plants to be set up in rural areas, worked by rural people and utilizing local raw materials, were being produced (Riskin 1979). This development became a special feature in the Great Leap Forward and many thousands of small manufacturing units were set up at brigade and commune level. By the mid-1960s, 60,000 were operating at the Hsien or county level and 200,000 in the communes. They produced iron and some steel, agricultural machinery, the walking or two-wheel tractor, chemical fertilizers, some non-ferrous metals, coal, bricks and engineering production materials. In addition and very importantly, electricity was produced in these rural areas for their own consumption. Civil engineering works were completed almost entirely by manual labour as dams were built for irrigation and hydro-electrical power, canals dug and roads constructed. The rural infrastructure to support both agriculture and the new small industrial enterprises was thus consolidated. Many things went wrong and many mistakes were made. Resources were not deployed efficiently, products were sub-standard and craft production suffered, but these problems were tackled and the industrial plants which survived became more efficient and more attuned to the reality of local economic needs and resources. Furthermore, the peasant had gained experience from the involvement. As may be imagined, the central bureaucracy with its nationalization mentality saw this proliferation of small producing units as chaotic, unorganized, untidy, unplanned and unmanageable. It did not appear to be part of the grand design for the new China and yet it was. Those who understood saw it as utilizing local resources often too small to be worthwhile to any large industrial enterprise but with the virtue of proximity; they saw it as developing local initiative and creating self-reliance, and as participation in true development rather than external imposed growth.

The virtues of a dispersed rural development became more obvious as it became more effective. It was a feature which survived the otherwise disastrous Cultural Revolution which put itself above economic reality and humanitarian concern and in doing so brought about its own eclipse. In so many respects the Cultural Revolution retarded China's development quite apart from the human suffering and thousands of deaths which accompanied it. Its ideological vanguard, the Red Guards, in searching for revisionist tendencies so utilized transport in their countryside movements that the transport of essential raw materials could not take place. Factories were closed or run only part-time, disputes and struggle further interrupted production and national output fell. Ironically one manifestation reinforced the development of the disaggregated rural economy. The leaders of the Cultural Revolution not only stemmed the movement to the cities but reversed it by their policy of rustication, the Hsia Fang or back-to-the-countryside policy. It is estimated that between 1968 and 1975, 15 million were transferred from the cities to the rural areas.

By the mid-1960s the virtues of local production and self-reliance had been appreciated and were preached. Rural raw materials supplied rural factories which made products needed in the countryside. Cement for water-management works, small hydro-electric generators, diesel engines, water pumps, insecticide sprayers, chemical fertilizers are examples of the involvement. By 1971 local rural fertilizer factories accounted for 60 per cent of China's total production. The product, ammonium bicarbonate, was typical of the approach. As Riskin points out, ammonium bicarbonate is a poor fertilizer being very low in nitrogen and losing much of its ammonia if stored or transported. It is not an apt product, therefore, for large modern fertilizer plant and indeed is not used outside China. To the Chinese, however, it had great virtues. It involved a simple technology, could use local deposits of coal, lignite, methane or other sources of hydrogen, and could produce a solid fertilizer for local use so that neither transport nor long storage were necessary. It was produced locally, from local materials for local use cheaply and with a technology and labour force the countryside could muster (Riskin 1979). Local iron and steel furnaces, using methods and working on a scale more reminiscent of early eighteenth-century England, were in 1971 producing 20 per cent of China's pig-iron. While the early furnaces in the 1950s had been less than successful, their more efficient successors came into their own as they became vertically integrated into engineering concerns producing agricultural machinery. Many rural factories concerned with processing agricultural products on a small scale were likewise dispersed throughout the countryside. The electrification of rural areas which has been mentioned, taking inanimate power into places which had never before experienced it, was brought about by local initiatives and did not come from the tapping of a huge national grid. Some came from small local thermal stations but most from tiny hydro-installations,

35,000 of which accounted in 1972 for 16 per cent of the total hydro-power capacity installed in China.

By the early 1970s an integrated rural economy had emerged in the Chinese countryside. Small-scale, locally-based industrial units integrated horizontally and vertically and operating for the most part at Hsien or county level, had become part of the commune concept. As such these units could seek expert advice, and indeed obtain more sophisticated equipment, from a higher level in China's hierarchies while at the same time they could relate to local needs. It was not a development evenly distributed throughout the whole of China. The economically backward areas developed more slowly.

As may be imagined, though many were ingenious, these self-developed technologies were unsophisticated. They represented the so-called intermediate technologies which have attracted much attention. Their significance lies, however, not in their being an intermediate technology but in being an appropriate technology and this is the lesson they have for the rest of the Developing World. They represented a crucial stage in development, a stage omitted by almost all other Third World nations, a stage which has so appropriately bridged the development gap between the agricultural peasant and the skilled industrial worker and brought the countryside with all its vast population into the development process. It has been a truly remarkable development.

Two cautionary and explanatory points need to be made. First, the scales and technologies of industrial and power production used were not, in absolute and clinically abstract terms, economically efficient. They did not make the most efficient use of resources in the sense of minimizing factor costs. It was therefore essential that this dispersed local production should not compete with modern large-scale and efficient industrial units for resources of capital, raw materials or energy. In China it did not. It made use of resources which would otherwise not have been deployed. It was not, nor ever had been, the intention of the Chinese Government to neglect the development of large manufacturing plant using the most modern methods. This development could, and did, run parallel to that in the countryside. The 'walking on two legs' doctrine of Mao implied that the two should operate in harness. It is this combination in which the countryside is not abandoned to neglect and stagnation which makes the Chinese experiment so remarkable and so significant a pointer for other countries. It was a strategy which made use of a whole range of resources including those too small and, in the case of labour, too untrained, for modern large-scale technology. Scattered low-grade mineral resources could be drawn on and used, power sites capable of generating only small amounts of electricity could be given a local significance. People were given the chance to use their initiative and the opportunity for everyone to participate in the development process was provided. The second point to note is that the industrial developments in the countryside did not play a major role in providing alternative employment to farming. The new rural industries, unsophisticated though they were, none the less were more efficient in the use of labour than the craft industries or farm and family methods of material and food processing. They released labour to agriculture rather than drawing labour from it. Even at peak harvest times the rural industries processing agricultural products used only around 5 per cent of the rural labour force (Riskin 1979). The new local industries did play an exceedingly important role in the agricultural sector. Their activities and products not only increased agricultural production but also enabled a greater input of labour to be used more effectively and so counteracted the diminishing returns on labour input. The rustic industrial development played a part in diminishing the size of China's 'surplus' population.

There were other benefits resulting from the development of the rural economy. One was to minimize the movement of raw materials and products in a country which, for its size, was, and is, poorly endowed with transport facilities. As Leung has commented, China's transport policies since the revolution have been much concerned with connecting the peripheral, border areas with the main focus of population and industry on the coastal plains (Leung 1979, 1980; Leung and Comtois 1983). The main trunk roads were constructed by the state while the local road system was in the hands of communes as a part of the policy of self-reliance. Much development has taken place, though the Cultural Revolution effectively put a stop to road improvements for four or five years, but the carrying capacity of many roads and the absence of bridges still inhibits large-scale, long-distance movement. In such circumstances, localized developments eased the pressure on the road system. These advantages, the minimizing of transport costs which compensated for sub-optimal production costs and the retardation of rural–urban drift were, therefore, further attributes of the development of a dispersed, but integrated, rural economy. It should, however, be appreciated that it was not conceived of as a static achievement. As industries in the countryside grew and matured they could move out of the stage of intermediate technology into another more appropriate technology, larger in scale, more demanding of capital and much more sophisticated in technique. As they developed they would become a more directly integrated part of the national economy requiring national levels of research and development and major national resources. The new developments would more efficiently meet national needs. It would therefore be a development in which the distinction between the rural self-help economy and the modern world progressively disappeared. This remarkable Chinese manifestation must therefore be thought of as a stage in the development of the world's largest nation. It is one which has meant that the countryside has become something more than a mere adjunct to the town, that towns have not been the only centres of change and modernization, that the contrast between core and periphery has been diminished, that a form of integrated development of rural areas and societies has come into being. For the Developing World, essentially a rural world, it has been a development of great significance and achievement. Since it involved men and not the gods it has been a flawed development but it has laid foundations for future progress and the building of the new China. As such it must be seen.

After the death of Chairman Mao in 1976 and the subsequent and rapid eclipse of the 'Gang of Four' who had led the Cultural Revolution, China entered an entirely new phase of development. It was not simply a cry of 'The king is dead long live the king'. The China which has emerged is a country considerably different in its approach both to development problems and to the outside world. The achievements of Mao's China had been great. It was still a poor country but one which had made provision for the essentials of life for all. Few other countries in the Developing World could claim as much and in none had the size of the task been greater. Some would suggest that progress would have been quicker and more substantial by taking another route but there is no means of testing the proposition. What is certain is that the road has been a difficult one and has involved much suffering (Chang 1991).

A NEW PHASE IN DEVELOPMENT: CHINA AFTER MAO TSE TUNG

The new regime in China made a close and searching scrutiny of what had gone before and identified its many imperfections. China was opened up to the outside world and its inhabitants given a much greater freedom; it was this freedom which was the striking contrast with the China of Chairman Mao. Mao had built foundations but they were for economic purposes; he revealed no intention of liberalizing the totalitarian nature of his communist party. Likewise the new freedom of Deng Xiaoping was not to embrace democracy. Deng's reform of the economy and the opening up of China to trade and foreign investment were small beginnings but they have proved to be of profound consequence. Self-reliance is diminishing, the use of intermediate technology is giving way to techniques associated with 'modernization', the declared goal of Deng Xiaoping, and bureaucratic interference is beginning to be replaced by modern management. The government of China is concerned that it may not be able to control the genie it has let out of the bottle. A market economy, private enterprise, the concepts of capitalism are uneasy bedfellows of a command economy with its state-owned industries and artificially fixed prices. A free society is much more difficult to manage and direct, and freedom once savoured is a never-forgotten taste. This is the situation which the government and the peoples of China are now facing. The young people of China misread the signs and equated economic revolution with political revolution but democracy was not on offer; the massacre on Tiananmen Square in June 1989 was the tragic result. For some two years after the massacre the rest of the world stood at a diplomatic distance from China as its masters appeared to reinforce strict political control over the people. Commercial and economic intercourse did not, however, cease. The economic changes, once begun, developed a momentum which is propelling China's economic transformation at an increasing rate. The changes began first in the countryside and these will now be examined.

THE RURAL SECTOR

It was in the rural sector that Mao's China had made such a distinctive contribution to development ideas and it is there that recent changes have been considerable. The communes with their rigid three-tier structure have been progressively dismantled. The system had been fatally damaged under the radical and extreme Marxist ideology of the Cultural Revolution and become more bureaucratic and inflexible. As the Grays reported, there emerged after Mao not one but several new systems with every indication that experimentation would produce a new development model (Gray and Gray 1983). The production of cotton, sugar cane, the edible oils, tea and China's most important meat, pork, fell to a considerable extent as the policies of the ultra-left faction during the Cultural Revolution stressed the need to concentrate on grains. Deng's government saw this action as stifling rather than stimulating agrarian production. Even when the Red Guards enforced rustication and put a million or so city dwellers on the land, production had fallen. Clearly food grains were vital to the country's food supply but the new government felt this could be secured by incentives rather than enforcement. The restrictions which had pervaded the rural scene had clearly blunted endeavour. Families had been prevented from growing crops for local sale in addition to the prescribed cereal targets. During the Cultural Revolution a party member described to the author these individual plots for private production as undesirable residues of capitalism. The Cultural Revolution had starkly demonstrated the inadequacies of a political direction of farming matters. The commune had played an important part in the development of China's rural economy. It had absorbed rural labour, encouraged local initiative in the use of appropriate technology and it had managed local social services. The commune had given cohesion to a society previously in chaos but it was a self-limiting system blighted by political control and the dead hand of bureaucracy. At all levels management had been inefficient as political cadres had been placed in positions of power to execute decisions beyond their competence. Commonly cadres had little idea of how to formulate and implement development plans, kept no accounts whereby programmes could be monitored and controlled, and had an insufficient scientific grounding in agriculture (Gray and Gray 1983). Brilliant in its concept, the commune system required for its successful operation a level of competence, a mixture of skill and an amount of capital investment rarely available to it. Often a large proportion of a commune's revenue was absorbed in its administration rather than deployed in production. Peasant workers in the production teams saw, in consequence, a low return on their efforts and their incentive to work was progressively diminished. In the command economy of China in which prices and market conditions were fixed by the state in a state monopoly, the energy of input was again made remote from the benefit of return. The state had set itself a task of management which it was unable to accomplish. It too had begun to drain the pond to catch the fish. Despite three decades and more of political indoctrination, the incentive of service to a remote state had not replaced that of personal self-interest and the removal of opportunities for self-improvement, other than that of political advancement, achieved a deadening effect.

The situation described was recognized by the government of Deng Xiaoping which has progressively returned initiative and responsibility to the individual peasant farmer. Personal gain has been allowed to energize China's agriculture. At first communes remained intact but contracts replaced labour tasks on the part of the individual and quota production by the tiers of the commune. The new arrangements were described as 'production responsibility systems'. In areas with different environments and hence low levels of production, the standardized administration costs had in the past taken almost all the revenue leaving little for the commune workers. Under the new arrangements in such areas all the means of production were handed back to individual farmers who worked what became rented farms. Increasingly the structure of the communes has been dismantled. The family unit has become the basis of millions of rented farms working a designated parcel of land on the territory of the former commune. There still exist controls and requirements such as an obligation to supply grain to the state in some cases, but to all intents and purposes China's agriculture is in the hands of farming families which benefit directly from their own efforts (Leeming 1985b, 1989; Saith 1987). Of great significance has been the comprehensive removal of price control over a wide range of products which allows market forces to operate. One result has been that in areas surrounding the big cities, the market so presented has allowed farmers to prosper by specializing in supplies of fruit and vegetables. The former system of government-forced procurement of agricultural products has been replaced by a negotiated contract system between farmers and government agencies. These changes have resulted in a very substantial increase in agricultural production and a corresponding rise in farm incomes. China is a vast country of varied terrains and climate and not all areas have shared in the new wealth, but the more fertile and accessible areas present a contrast with the situation twenty years ago. City workers have always been better paid that the country dwellers in communist China but the difference is now being reduced partly because of the liberalization of the agriculture but also as a result of a further rural development.

The commune system was a lavish user of manpower, indeed this could be argued as one of its merits in the populous new People's Republic. There is, however, a limit to labour inputs on the land and China's population though growing at a low rate is so large that millions join the work-force each year. It has been estimated that in rural areas there are some 150 million people in excess of current agricultural requirements. Since the cities, which themselves have unemployment problems, cannot absorb the surplus, alternative sources of employment are required in the countryside. The development under Mao of an integrated rural economy based on the commune and with small manufacturing units dispersed throughout the countryside has proved to be of benefit. Despite the unsophisticated nature of rural industry it had introduced to the farming communities the concept and the practice of non-agricultural employment. The liberalization of the rural economy which Deng Xiaoping initiated took place in an environment supportive of a new and modern industrialization.

THE MANUFACTURING SECTOR

The large-scale production of raw materials and basic manufactures has always been in the ownership and control of the government of the People's Republic. This situation is a legacy of the Soviet-inspired early years of the economy of communist China but is a tenet of socialist ideology. In 1979 the Chinese Government began to examine the possibility of modernizing heavy industry to achieve greater productivity and profitability. The enterprising manager working within the highly-structured and vertically-integrated manufacturing sector was frustrated at all points by the superimposition of state plans and controls. He was unable to evaluate the cost-effectiveness of any particular process since both prices and costs were artificially controlled, nor was he able to attempt any market analysis. From 1949 to 1978 heavy industries had been sustained by large state injections of capital to the detriment of other parts of the economy yet most plants remained inefficient, with a quarter in 1978 making a loss (Leeming 1985a). The self-sufficiency approach practised in many installations involved large opportunity costs and hindered the emergence of specialization. The whole of state industry was permeated by the philosophies of the Maoist period and change was incompatible with these beliefs which were still held by Chinese leaders. The system, which may have made sense in the early days of the republic when development was isolated from the rest of the world, was as in agriculture, now impeding progress. Deng Xiaoping began to introduce change. No thought was given to denationalization. Most changes related to management procedures, where responsibility schemes were introduced, to the financing of developments and to the abstraction of revenue from the industries in the form of taxes rather than the retention of profits by the state. At first state prices rather than market-determined prices were maintained. Further reforms in 1983 and 1987 were introduced and increasingly costs and prices came to reflect the evaluations by the market rather than those set in state plans. The price control of raw material inputs has been removed in most cases and, importantly, economic power has been decentralized from Beijing to the provinces with the result that inter-provincial competition has arisen (The Economist 1992). The state-owned industries have still not freed themselves from the cloying bureaucracy of the state. Their further reform runs so counter to the tenets of communism that the state monoliths are changing the most slowly of all Chinese economic institutions. Changes are, however, taking place which are subtly altering the state industries by infiltration rather than frontal assault; in this the opening up of China to international involvement is playing a part.

THE GROWTH OF INTERNATIONAL INVOLVEMENT

In both agriculture and industry, in town and country, management initiative was needed. To be effective this initiative required training in and experience of modern management techniques and production methods. The advanced technology which China wishes to use, whether it was designed and developed in China or made more quickly available by importation, needs a scientific work-force of technicians. These requirements in turn meant that education had to be more selective in its intake and more specialized in its focus. Both these features had been anathema in the Cultural Revolution; furthermore they are developments which require an international dimension. Isolation, so long characteristic of the People's Republic and at its most absolute during the Cultural Revolution when all scientific intercourse with other nations ceased, is incompatible with the professed goals of modernization.

After Mao's death the new government took positive steps to develop international contacts which involved not only trade, but also foreign investments on Chinese soil. The internal changes relating to the organization of labour, the development of special skills, the emergence of the money incentive, the growth of managerial responsibility, have all brought China closer to an outside world which, with the collapse of the Soviet Union, is essentially non-socialist. The changes, though they still have a long way to go to transform China completely, have been dramatic both in character and scale as production and trade have grown rapidly. The great Chinese industrial and trading cities of the nineteenth century in which European nations invested and which were inter-linked by the Chinese rail network, once again have become recipients of foreign capital. By 1985 the coastal cities of Dalian and Tianjin in the north, Shanghai in the centre and Guangzhou (Canton) in the south had been declared open cities for foreign investments. At first direct foreign investment was not as common as licensing agreements, sub-contracts and joint enterprises but this has developed so that foreign investors are buying shares in Chinese companies and investing directly. In 1992 the estimate for direct foreign investment in China was running at between four and five times the rate of that in Japan.

THE SPECIAL ECONOMIC ZONES AND RURAL INDUSTRIALIZATION

The earliest and most overt manifestation of China's changing relationship with the world of international capital was the establishment of four Special Economic Zones (SEZ) on the coast of southern China. The first, established in 1976, was Shenzen across the border from Hong Kong; a second in Guangdung Province (Zhuhai) was set up near Macao. The Shantou and Xiemen zones were sited in the Fujian province on the coast opposite Taiwan. In these zones few restrictions have been placed on foreign capital which to an overwhelming extent has come in from Hong Kong. Independent capitalist enterprises have taken root as well as joint ventures with Chinese state organizations. The output is primarily for export either as finished goods or as components to be assembled elsewhere. For the investor the benefit lies in cheap land and cheap labour; both are very cheap when compared with Hong Kong, Singapore, Taiwan and South Korea. There is also the attraction of breaking into China's growing, and potentially vast, domestic market. China can benefit not only from taxes and other fees but importantly from participation in modern commercial and production enterprises. Whether the SEZ facilitate the transfer of the technological and business skills has been queried. They certainly provide needed employment but they could draw in too many of the skilled and enterprising from surrounding areas in the same way in which urban opportunities commonly do. If, however, they achieve a spread effect by the circulation of labour and management through them their creation can have positive and beneficial effects. Without this circulation they could remain insulated enclaves detached from the main body of the Chinese economy. In fifteen years of flourishing growth these misgivings have proved to be unfounded. The SEZ at Shenzen exemplifies their success; what was once a small rural town is now a commercial and industrial city of some one million people, with factories, high-rise apartment blocks, hotels and retail centres. The initiative, energy and capital has come in the main from Hong Kong Chinese entrepreneurs who not only speak the language, Cantonese, but being Chinese themselves are fully acquainted with Chinese methods of business negotiation and have demonstrated their abilities by their successes in Hong Kong. The SEZ and the general opening up of China to outside contacts has removed the protective shield of isolation from her pure socialist ideals. Apart from any matter of labour circulation, in the SEZ the market place and its success is there to behold and its methods and attitudes are seeping into the fabric of Chinese society over wide areas.

The Hong Kong Chinese are not alone in their investments in China for other nations have joined the flood of investment; Taiwanese via Hong Kong, Singaporeans, South Koreans (now that China has recognized the sovereignty of the Korean Republic), Japanese and Americans are all investing in China. A remarkable transformation is taking place. Between 1980 and 1990, China's GDP grew by an average of 9.5 per cent per year; her growth in manufacturing averaged 14.4 per cent. Chinese exports, growing at a rate of 4.8 per cent per year between 1965 and 1980, rose to an average of 11 per cent between 1980 and 1990 as the decentralization of trading regulations and a currency devaluation to realistic levels took effect. China has begun to benefit from her comparative advantage of low labour costs in a wide range of labour-intensive industries; her GDP is the tenth largest in the world.

In this economic transformation non-state industrial enterprises have played a major part and though the state firms have not remained static they have changed but slowly; their role and contribution to the economy will diminish as the new non-state firms grow more rapidly around them. The attitudes and structures of the commune have, as has been indicated, helped to foster the new industrial growth much of which has taken place in the countryside. It is estimated that 28 million were employed in 1.5 million rural non-state industrial enterprises in 1978; by 1991 the number of such firms had risen to 19 million and employed a total of 96 million people. These rural establishments account for nearly 40 per cent of employment in manufacturing, over 25 per cent of industrial output and almost a quarter of exports (The Economist 1992). Rural industrialization has spread out far beyond the SEZ but is largely associated with the broad coastal provinces from Guandong in the south through Fujian, Zhejiang and Tiangsu to Shandong in the north. Imbedded in this zone are major cities, themselves large industrial centres. The rural enterprises known as 'township and village' enterprises though not owned by the state are not necessarily private companies; some are private but most are owned and operated by local government bodies – some as small as that of the village, others at county level. The commune inheritance is still discernible. Such enterprises function by identifying products and markets in which they can successfully compete both domestically and internationally. Many have, of course, failed but the survivors have prospered. Their industrial output is shared equally between light manufacturers and heavy industrial products. Partnerships with foreign management and the use of foreign investment are often sought. Like the now dismantled communes, these rural industries are a peculiarly Chinese development. Prosperity is spreading throughout the countryside of the coastal belt.

China has become a place which attracts foreign investments, investments which have spurred on her economic growth and modernization. There could be no greater contrast with the days of the Great Leap Forward and self-sufficiency. China's slow, sequential development involving all aspects of society had produced foundations upon which to build within the restrictive confines of her totalitarian society. The rapid changes in which China is now involved will require careful management but they are changes which cannot be reversed. Rich in raw materials and energy resources, with the world's largest labour force which has given every indication of adaptability to modern practice, China's recent economic growth and industrialization has been at a pace and on a scale that has few parallels in history. Poverty is still widespread, unemployment is high, but prosperity is spreading throughout the coastal provinces and the major industrial cities. Free-enterprise development is permeating the now less-than-complete totalitarian barriers of the Marxist state. The surviving state enterprises are unlikely to be dismantled but will simply be eclipsed by the economic success of others. Political ideology may become less relevant to the people of China but China will no doubt produce a uniquely Chinese solution to the challenges of her further development. The transformation quietly begun behind closed frontiers has put China on the path to becoming the major economic power of the twenty-first century. The political and economic consequences of this achievement will have a profound significance for the rest of the world.

16 India

'There just does not seem to be any one panacea for India's vast socio-economic problems; the most to be hoped for is to be able to suggest means to ameliorate major areas of difficulty'

T Scarlett Epstein
South India, Yesterday, Today and Tomorrow 1973

By virtue of her size, her history, and the diversity of her rich and complex cultural heritage, India selects herself as a case study of development issues. The focus of this exemplification will be upon the dimension of population but not exclusively so. Their Asian proximity and size makes comparisons between India and China inevitable. Their paths to economic development and social transformation are contrasted both in the manner of their execution and in their success even though their problems have been similar.

The second largest nation with a population of 866.5 million in 1991, India is also one of the poorest. In 1991 her GNP per capita stood at US$330 putting India in the bottom half of the World Bank's category of low-income countries. Switzerland, the richest nation, recorded US$ 33,610. India's population grew at an average annual rate of 2.3 per cent from 1965 to 1980 and by 2.1 per cent per year between 1980 and 1991. Growth in per capita GDP averaged 1.9 per cent from 1965 to 1990. This slow rate of improvement would have been poor enough had it been equitably distributed among India's population but it was not. No less than 65 per cent of the total of family incomes accrues to the top 20 per cent of families. It has been estimated that 420 million people lived below the poverty line of US$370 per year in 1985. The World Bank considers that appropriate action by India's government could reduce this figure to some 250 million, the present size of the United States' population, by the end of the century. This is one measure of India's development problems and failures. It is easy to be critical but a population already some one-third as large again as that of the whole continent of Africa and growing by the size of Australia each year presents severe economic and logistic problems. China faced a similar situation.

Most of India's large population, some two-thirds, are agriculturalists, around 15 per cent work in industry and the rest are in the service sector. By the basic parameters quoted, India is a large, poor, agricultural country though these generalizations conceal both substantial industrial developments and more affluent sections of the population. Within any reasonably conceived demographic trends, one dimension is certain, namely that India's population will continue to grow considerably beyond its present size. In a hypothetical but reasoned projection the World Bank estimates that India's population will reach

1006 million by the end of the century and will eventually peak at a stationary population of around 1800 million. While in terms of policy and plans this ultimate figure is of little significance, since it is so far in the future, present plans and objectives have to be conceived in the context of population increases which are large and inevitable. India is at a stage in the demographic transition which yields large increases. Death rates have fallen steeply in the last thirty years and reached 11 per thousand in 1991. Crude birth rates are also falling but at 30 per thousand need to fall much further. The Chinese birth rate has been reduced to 22 per thousand.

INDIA'S DEVELOPMENT PLANNING

Economic development involves raising the productivity of a country by more fully and efficiently utilizing its resources so that this eventually leads to a transformation in the economy with an associated and large increase in production. In the case of India this would involve transforming the economy from that of a mainly pre-industrial society, prior to independence, into an industrialized society in which further growth was self-sustaining, and from a situation of many localized economies into one of macro-regional economies within that of the nation. It could be argued that this increased material wealth would allow for social developments as this wealth was distributed, as it funded health, education and other social services, and as it raised the opportunities of a fuller life for all.

The independent India of 1947 set about planning for precisely such a social and economic development though the tragic upheavals of communal rioting and massacres which marked the separation of Hindu and Muslim communities set back the institution of formal plans. The first development plan appeared in 1951. It encapsulated, as have all subsequent plans, four major goals, by the attainment of which it has been felt, a true and continuing development can be engendered. These objectives are: first, a high rate of national growth to create material wealth for distribution and productive reinvestment; second, to achieve over as wide a front as possible national self-sufficiency and self-reliance so as to be free of any foreign dependency; third, to provide employment for all; and fourth, to reduce economic inequalities in the population, all being measures towards the attainment of economic security, a fuller life and human dignity by India's millions. Many development ideals would wish as much. The attainment of these goals has been marked by difficulties, uncertainties, by qualified successes and unqualified failures.

During the long years of the British Raj, India's resources had been developed for export, and commodities such as tea and jute were shipped to the European market. But India was a large country with a population, even in the nineteenth century, much larger than that of any other British colonial territory. It represented a large internal market in a country rich in craft skills and traditional marketing systems. It was Indian initiative, and not British, which led to the establishment of the first large-scale cotton textile mills in India. The great Indian industrial entrepreneur Janshedjee Tata set up India's first modern iron and steel plant at Jamshedpur in 1911. Both the textile and steel industries catered for the home market and were, in effect, import-substitution activities. The country remained, however, overwhelmingly agricultural, with three-quarters of its work-force on the land and a manufacturing base of craft and cottage industries rather than one of blast furnaces and steam-driven looms and this situation pertained after independence. Furthermore, the nature of Indian society, the beliefs, customs and attitudes of the vast majority of its population, were unchanged. The caste system, associated as it is not only with status, but the kind of work in which the members of a caste can participate, was strong and led to a social and economic inflexibility less conducive to change than in a less-codified society. Both before and after independence India was a country of contrasting variety. Large cities stood in a teeming countryside of peasant farmers. Industrial plant using sophisticated science-based techniques were at work alongside ancient crafts drawing upon inherited skills. Despite the changes of the last four decades much of this contrast remains and so too do old values and beliefs. This mix makes it difficult to generalize about the Indian situation. Although there are modern industrial centres, the existence of the greater part of the population is little changed. It is rural, poor and under-nourished. The urban centres too are thronged with the poor after thirty years and more of economic forecasts and plans.

PLANS FOR DEVELOPMENT

In striving for self-reliance India has followed the Soviet model of industrialization. She placed, quite deliberately, an emphasis upon the basic capital goods industries and on the range of industries producing intermediate goods which represent a processing of the output of the basic industries and which in turn become the components and raw materials of end-product industries. Structures for vertical linkage were thus established. The process began in the First National Plan, covering 1951 to 1956. This plan was the least definitive and detailed of all India's plans, almost as though it was to serve as a breathing space before coming to firm decisions. Its balanced projections were overthrown by the harvest failure of 1955–56 which raised food prices, led to food imports and upset the balance of payments. The insecurity of the agricultural base was thus early made apparent. In the second plan, with much more ambitious schemes for development, came definite proposals and here the Soviet model, emphasizing investments in basic heavy industries, began to emerge more strongly with state involvement more conspicuous. The emphasis on laying foundations for self-sufficiency meant that investment goods received priority over consumer goods and so the consumption of the general population was purposefully held in check. In the Draft Outline of the Fourth Five Year Plan this industrial policy is described:

A special feature of industrial development, especially since the commencement of the Second Plan in 1956–57 has been the growth of capacities in steel, aluminium, engineering, chemicals, fertilizers and petroleum products.

Apart from these, large investments have been made in industries producing heavy electrical equipment, heavy foundry forge, heavy engineering machinery, heavy plates and vessels etc – all of which will become available in increasing quantities from now on

(Indian Planning Commission 1966).

Within the industrial sector, investment allocations to these industries were 70 per cent in the Second Plan and 80 per cent in the Third Plan. In the Second Plan, overall there was a shift away from agricultural investment. The Indian economists Bhagwati and Desai stated that this shift was vigorously debated at the time and in the event has proved to have retarded the country's economic development (Bhagwati and Desai 1970). The heavy industrial sector is not labour-intensive and the Second Plan looked to labour-intensive cottage industries both to meet consumer-goods needs and to provide employment. Neither it nor the subsequent Third Plan specifically addressed the issue of unemployment which was rising rapidly with the growth in population. The Second Plan's ambitious growth targets proved unrealistic and were not attained since the resources to carry out the proposals did not exist. Poor harvests again underlined the fragility of an economy which could not secure the food supply for its growing population. Food imports contributed to an adverse balance of trade and to foreign exchange problems. The planning period, beginning so optimistically with large projected investments, saw its close characterized by low levels of investment.

The Third Plan gave greater emphasis upon agricultural developments. It too was buffeted by unforeseen events. The wars with China on the Himalayan frontier, and with Pakistan in 1965, both diverted resources. The harvest failure of 1965–66, due to drought, was on a major scale requiring international aid. The balance of payments situation deteriorated still further and the rupee was devalued. A further bad harvest put millions of Indians at risk and they were only saved by international efforts. The Achilles heel of India's situation was once again dramatically revealed. All her investments, many not directed to agriculture, were not reducing this vulnerability to harvest failure which, when it occurred, not only faced millions with death and starvation but made nonsense of the careful calculations of India's development plans. Resources were not being harnessed to secure, even at a basic level, a large proportion of a population growing by the size of Australia each year. The chaos of the mid-1960s delayed the inauguration of the Fourth Plan until 1969. It was little different from its predecessors though it did enunciate a concern with social change and the more equitable distribution of wealth.

R. H. Cassen, in his penetrating analysis of the travails of India's economic development, attributes the unsatisfactory situation which prevailed throughout the 1950s, 60s and 70s and which was characterized by slow industrial development and even slower agricultural process, to a number of interrelated factors (Cassen 1978). First, he argues that the emphasis upon heavy industry in the early plans produced an unbalanced industrial structure. To utilize heavy industrial products there must obviously be a demand for them on a scale to allow them to operate efficiently. This demand could only be generated if there were investments – sustained and continuing – in other sections of industry not only of the intermediary goods type but also the tertiary industries. The heavy industries producing boilers, generating equipment and machines for making tools, required lighter manufacturing activities to use them yet these other industries were deliberately subdued in early plans and so vertical integration was truncated. Secondly, the level of productivity in agriculture, since it produced not only food but industrial raw materials such as cotton, jute, sugar and oil seeds, had an important bearing upon the production levels of a range of significant industries. When agricultural output fell, not only had food to be imported but also raw materials for industry or industrial capacity had to be left under-utilized and capital diverted from productive investment to an emergency deployment to stave off starvation. The goal of self-sufficiency meant that no manufactured consumer goods were imported so the main imported commodities were raw materials, oil, fertilizers and food. Cassen argues that the cost of these imports and the difficulties of ensuring a secure and regular supply, were increased by the bureaucratic inefficiencies of exchange and import controls. To these problems Cassen adds those of technical failures in electricity-transmission systems, of inefficiencies in the transport network so that goods arrived late and damaged, and the unreliability of power supplies because of hydro-electric shortfalls during drought. This dismal ensemble of lack of demand, foreign exchange shortages, technical unreliability and administrative inefficiency, together with labour problems in part associated with India's vibrant political scene, makes it small wonder that India's industrial expansion has been slow. Indeed, it is an even greater wonder that so much has been achieved. The lack of a balanced, broad-front and integrated approach which accords true importance to agriculture would seem to have limited India's development prospects.

The Fifth Plan, designed to cover the period from 1976 to 1981, played down the goal of self-sufficiency. Oil-field exploration and development within India had yielded sufficient production to ease the economic difficulties created by oil imports. Remittances from Indians overseas again made a minor but significant contribution. But in many ways, the fifth of India's plans was not a major diversion from the path marked by its predecessors and like them it did not address in any direct way the major problem, the size and growth of her population. As Cassen points out, some thirty years of investing in heavy, capital-intensive industry does not make provision for the employment of the many millions entering the labour market.

THE CASSEN ANALYSIS

The social and economic development of India since its independence has, despite many achievements, been diminished by the interrelated issues of population and agricultural production. They are not issues peculiar to India; indeed it can, and will, be contended that they are the most significant among the development issues facing developing countries. India has not achieved her third and fourth development aims of providing employment and reducing economic inequalities among her population.

R. H. Cassen has examined the interplay of population dynamics and economic development in India and this analysis will now be considered (Cassen 1978). The demographic transition referred to in Chapter 3 leads, in its second phase, to a marked increase in population as death rates fall while birth rates remain high. India's position is shown in Figs. 9A and 11B. Not only does the population increase but so too does the dependency ratio as more children survive and as old people live longer and so form a large proportion of the population which consumes but does not produce. Consumption thus goes up but not production, and investments decrease as savings are diverted to support dependants. It takes place not only at family level but also at national level. A larger share of national investment has to be directed to providing social and health services and educational facilities for this enlarging dependent sector rather than be placed in productive investment. These processes can retard economic growth, though it could be that their effects are of too small a dimension in the overall economy to be of any significance. It could also be that, for some reason, population growth had the effect of stimulating the economy. Cassen refers to the work of Coale and Hoover who examined precisely this last issue in the Indian context and modelled alternative fertility situations (Coale and Hoover 1958). One postulated alternative compared a situation in which fertility rates fell by 50 per cent over a period of fifteen years with one in which they remained static. The analysis suggested that with the falling fertility rates GNP would be 17 per cent higher than in the alternative with static rates. In another projection, the number of consumers after fifty-five years of high fertility rates was calculated at twice that where fertility rates were lower while the output per consumer had halved leaving the GNP the same. Even when altering the model to allow for higher rates of productivity of capital and labour so that GNP grew faster with a bigger population, it was never enough to compensate for the growth in population. The evidence of Coale and Hoover, and indeed of other similar analyses, is that population growth does not stimulate an economy sufficiently to accommodate the growth.

Cassen argues that the Coale–Hoover calculation does depend upon the effect which the dependency burden has upon savings and investments. He points out that a large proportion of India's population simply do not have any savings to divert away from investment in order to support their children. It is only the better off who have savings to invest and they are the group characterized by smaller families and who are not contributing to high fertility rates. India's poorer families simply consume less per head as their families grow. There can be no diversion of savings which are non-existent. Secondly, Cassen comments that social investments in health and education are not wholly unproductive. In time, as was noted in Chapter 8, these can contribute to increases in production when circumstances are favourable. In India's case they formed a relatively small item in total national investment so again it can be argued that increases in them are not of major significance. So while these twin ideas of the increasing dependency ratio drawing upon both private and public resources and diverting them from productive investment do play a part, Cassen's opinion is that things are not of the magnitude to change economic performance dramatically. Instead he argues that underlying India's problems is what he describes as the changing composition of the resource costs of producing basic consumption goods. By this he means changes in the relative significance of the resources which are used to produce a basic consumption good. The way in which the failures of the agricultural sector have affected the whole economy has been described. It is the agricultural sector to which Cassen is referring.

From independence into the early 1960s agricultural output was increased by both bringing more land into cultivation and by increasing, to some extent, output per hectare. Both activities demanded greater inputs of labour and were in large measure due to increases in labour intensity though use was made of fertilizers, better seeds, an extension of irrigation and improved techniques. The Green Revolution of the 1960s and early 1970s increased output considerably and was almost entirely due to yield rather than area increases. As was discussed in Chapter 5, the increased yields were consequent upon higher yielding seed varieties, the greater use of fertilizers and pump irrigation with pump engines for tube-wells. This agrarian development did absorb more labour but much more significantly it required capital. To be fully effective, fertilizer, irrigation equipment, pumps and bore-holes were essential in many areas. This involved investment in fertilizer factories and engineering plant. It involved the import of chemical raw materials to make the fertilizers, and indeed fertilizers themselves, and the purchase of foreign equipment. All these imports utilized exchange revenue. These developments saw agriculture becoming more dependent, more of a drain upon capital resources.

The sequence of events described represents a transformation of the input needs of agriculture. Initially output is increased by a greater input of labour, derived from the growing population, applied to the finite resource of land. This is subject to diminishing returns unless the effectiveness of the labour input and the responsive quality of the land base can be improved. To achieve this, technical innovations requiring more capital are needed. Cassen argues that this stage of a change-over in the kind of inputs required was reached in India in the 1960s. Investment which could have been productively invested elsewhere went into the agricultural sector and this was made necessary because the population was growing so rapidly and so much in absolute numbers. A slower growing population would have made this level of investment in agriculture less necessary and would have delayed the change-over from the labour-based to a technological-based input for a little longer. In economic terms the marginal cost of increasing food output had gone up. It is of course possible to postulate a situation where the ability to produce lifts the production curve above that of population (Chapter 8). The more rapid the increase in population in absolute and relative terms, the more difficult is the task of raising that curve. Cassen thus ties agricultural performance to population growth and demonstrates that the growing population, by placing increasing demands on agriculture, is causing investments to be diverted. It is not that investments should not be made in agriculture, indeed the evidence is, and this will be debated later, that insufficient investment and attention is paid to the rural sector, but that the primary, physiological needs of a vastly growing population prevent adequate attention to other developments. It is not therefore simply a matter of 'unproductive' investment going into schools, hospitals and social provisions on a larger scale than would be necessary with a smaller population, but of a restriction of the range of 'productive' investments by the need to cover the basic essential of food production. A Malthusian element is thus present in the Indian situation. The pressure on food supplies has meant pressure on the land which in turn has led to soil erosion and river siltation with an increased risk of flooding so that the very resource base itself is being diminished. There is a mismatch between India's population and the resources she has harnessed to sustain it. As Sartaj Aziz has pointed out, the minimum size of land holding which is economically viable in India is between 3 and 5 hectares. If all the available land were redistributed on this basis, with owners who can afford a pair of bullocks to work the land with a few basic implements, there would still be 20 million rural families in India with no land whatsoever (Aziz 1978).

In India population growth and size appear to be retarding social and economic progress. In western Europe population increases during the demographic transition were never of the order experienced by India and furthermore took place at a time when manufacturing industry was much more labour intensive. India has been striving to reduce the size of families by widespread and rigorous family planning campaigns. Crude birth rates are still high and suggest that the social and economic conditions conducive to smaller family size are still not being experienced by a large section of the population. She has not been able effectively to utilize her growing population, and hence growing labour force, in furthering the productive capacity of the nation. There is chronic overmanning as jobs are spread thinly among increasing numbers so that both income and living levels remain low. The millions on the land cannot move into other sectors where diminishing returns are less evident so that despite developments in India's manufacturing sector, the proportion of the working population employed in agriculture has diminished very little since independence.

Even the relationship between education, values, economic priorities and family size, which were discussed in Chapters 3 and 4, appear to be weak in India. Nor has education markedly influenced skills and productivity.

THE BEGINNING OF ECONOMIC CHANGES

For the reasons cited by Cassen, India's population, with its large size and growth and which needs to be fed and employed, appears to have retarded the country's economic development. The population problem is, however, but part of the explanation; the remainder is to be sought in the manner in which India has addressed her problems. China faced with a similar problem has, in large measure, overcome it within a totalitarian regime. India's democratic society has, again in large measure, failed. The fault lies not in freedom and democracy but in the economic policies followed by all her governments since 1947. China's People's Republic, born out of revolution, was a Marxist society adapted to the history, culture and condition of the Chinese people. It was a peculiarly Chinese manifestation. Independent India with its polyglot peoples, ethnic, caste and religious diversity required a society and economy peculiarly Indian in that it embraced this diversity and created cohesion.

The origins of India's development attitudes are to be sought in the immediate post-independence period. It will be recalled that among the broad aims of Jawarharlal Nehru's first government were two aims of particular significance: the attainment of national self-sufficiency freeing the country from dependence on foreigners and, second, employment for all. Nehru and his colleagues, equating world capitalism with colonialism, rejected the idea of participation in a free international market and turned inwards and adopted a socialist command economy with the Soviet Union as its model. The adoption was but partial, unlike the radical change in China. Private enterprise remained at various levels but the state was to command major means of production, to own and operate state industries and to determine in some detail many aspects of economic activity. The Soviet Union had appeared a well-ordered nation of great economic and undoubted military strength with all achieved by a state-controlled industrialization. China had followed the same policy of industrialization. In the Soviet Union agriculture had been collectivized. China attempted a similar policy but rejected it for one of its own design which harnessed a rural work-force to good effect and eventually created an integrated rural economy. India, obsessed with its industrial involvement, did neither and neglected the rural economy so central to its well-being and so related to its population problem and mass poverty.

Most external analysts of India's economic situation, and some internal, have pointed to the neglect of the rural economy. The Indian economist Isher Judge Aluwalia stated that low rural incomes need to be raised if a sufficiently large domestic market was to allow the growth of manufacturing (Aluwalia 1985). It is a supportive statement but it is from the industrial view point. The Food and Agricultural Organisation of the United Nations, FAO, points out that agriculture in India accounts for one-third of GDP and two-thirds of employment yet agricultural output needs to be raised by 25 per cent before the end of the century if the population is to be fed at present levels. The FAO claims that India's high levels of industrial protection have drawn resources away from agriculture and increased agricultural costs of production (FAO 1991). Raj Krishna refers to grain surpluses because of the lack of purchasing power of the poor (Krishna 1980). The landless of the rural areas, the jobless of the cities of India may be too poor to buy the food available.

Agriculture and the rural community requires a special focus of development initiatives in India. It is easy to recommend but difficult to implement in any circumstance. In India much effort has gone into programmes such as those characterizing the Green Revolution where, as was discussed in Chapter 5, the need for investment meant that the richer were differentiated from the poorer in their degree of participation in, and benefit from, the programme. Even in this effort India's achievement compares unfavourably with other Asian nations involved in the Green Revolution. Can a momentum for the improvement of agricultural production, rural-based industry and services, and rural infrastructures, be established and maintained in such a fashion as to benefit most and not a few? Can the quality of rural life be so improved that the socio-economic mechanisms which seem to lead to family planning come into play and begin to stem the flood of rural births and allow the emergence of wider horizons of a better life? A substantial increase in food output is necessary and will require technical support when beyond its initial stages. Associated with this would be the essential requirement to spread participation and benefits equitably so that increases in per capita incomes were not skewed. Secondly, in the non-agricultural sector an industrialization policy less focused on basic heavy industries and using appropriate labour-intensive techniques, would afford more employment, create demand for basic and intermediate products and so use their capacity more fully. It could not only substitute for imports but importantly enter the export market. The task is forbidding with 170 million or so new extra jobs being required by the end of the century. Additionally, manufacturing employment could be more rurally diffused in small towns to help provide employment for the rural landless and limit the further subdivision of holdings.

Without providing employment and alleviating poverty, India's movement through the demographic transition will be all the slower. It is a condition which increases the need for carefully organized family-planning programmes to be pursued rigorously but humanely, but it also requires the creation of circumstances in which smaller families appear desirable. A way forward may be a greater focus upon policies designed to ensure a more secure and productive agricultural base, to create a more cohesive rural structure in which manufacturing as well as farming is incorporated, and which places the enlargement of employment opportunity as the pre-eminent goal.

The economy was stifled by bureaucratic intervention, was starved of the benefits of foreign technical innovations while the inefficiency of state-controlled industries was maintained by government subsidies. Import-substitution bred protectionism and nurtured inefficiency in captive markets. Trade was inhibited by quotas, tariffs, import and export permits and by long delays. Self-reliance was exacting a huge price. The desire to secure full employment, laudable though it was, was achieved by preventing failing factories from closing rather than by creating conditions for the birth and growth of new, efficient and profitable ones. Afraid to allow foreigners, particularly the ill-regarded multinationals, to participate in the economy, the governments of India assumed responsibilities for the management of the economy far beyond their capabilities. The machine so created has slowly been grinding to a halt. The situation discussed has developed over many years as politicians have acted out their confrontations as though players upon a stage.

The fifth development plan showed little change from the predecessors in its reluctance to approach the reality of India's problems. The sixth, for the period 1980 to 1985, associated with a surge in the growth of the GDP and had, however, given the first indications of tentative steps to liberalize the Indian economy under the direction of Mrs Indira Gandhi. Her son, who succeeded her, carried this policy further so that the seventh plan, 1985–90, incorporated some relaxation of state controls. Importantly this plan set out to encourage a greater role for private enterprise and to allow in more foreign investment. India was beginning to put in place economic reforms of the kind initiated in China some ten years earlier.

After Rajiv Gandhi's assassination and with India in political and sectarian turmoil, the maintenance of the new liberalizing policy looked questionable. The man who has pressed further with a significant reforms policy is Prime Minister Narasimha Rao. His major economic changes are designed to strip away the mantle of restrictions. Industrial licensing has been all but completely scrapped, multinational companies have been admitted, the rupee has been significantly devalued, foreign exchange controls liberated and tariffs cut. The huge subsidies supporting state-owned companies have formed a major part of the cut in public spending. An inward-looking India characterized by a degree of state regulation rarely found elsewhere was being converted into a outward-looking country with an increasingly international stance. It is a beginning; much remains to be done. The attitudes of forty years are hard to change. The Iron Curtain has been drawn back in Europe and revealed the failure of the socialist command economies of the Soviet Bloc. The Soviet Union itself, India's economic model, is no more. The changing world has given India's politicians food for thought.

Inefficient state-industries still exist, as they do in China, some restrictions persist and India has still to get her new development policy under way. Already she has attracted direct foreign investments and companies such as General Motors, Ford, BMW, Suzuki, General Electronic, IBM and Coca Cola have established themselves. The ability to draw in overseas investment is a significant marker of India's new direction. It remains to be seen whether these first steps in freeing India from her self-made shackles can release the undoubted creative and entrepreneurial energies of the India people. The pattern which has emerged in China is taking shape in India and within a free democracy.

17 Brazil

'Why did the United States industrialize in the nineteenth century, keeping pace with the nations of Europe, while Brazil in the nineteenth century evolved in the direction of transforming itself into a large underdeveloped region?'

Celso Furtado
The Economic Growth of Brazil 1959

'Brazil ranks among the few nations of the world which have potential for the achievement of major power status within the next century'

J Saunders
Modern Brazil 1971

One of the nations of Latin America whose colonial history began over four centuries ago, Brazil's independence was achieved in 1822 long before many of the former colonies of the Developing World had become colonies. The chronological sequence of her first settlement by Europeans, her colonial development and her final emergence as an independent sovereign state is not greatly different from that of the United States but, as Furtado suggests, while the United States followed the industrial road of Europe, Brazil did not (Furtado 1959). Though one of its more affluent members, with a per capita GNP which places her securely in the upper middle-income group of the World Bank's classification, Brazil is still today in the Developing World. This position in the continuum of the development spectrum selects Brazil as a case study but she possesses other characteristics which call for special comment.

Brazil is one of the giant nations of the world. Her area is exceeded only by that of the USSR, Canada, China and the USA. Her present population ranks her as the sixth most populous nation in the world. Secondly, it will be recalled Brazil exhibits a highly skewed distribution of income (Chapter 4). The top 20 per cent of family incomes groups accounts for two-thirds of the national total of family incomes. This feature is itself a product of a third characteristic, the existence of a strongly marked dual-economy. It is this which allocates Brazil to her middle-income classification since the productivity of one sector is diluted by that of the other. It is a feature which has resulted both from Brazil's history and the development path she has followed. Her transitional position between the Developing World and that of the industrial market economies in no small measure is the product of this characteristic. It is thus possible to visit Rio de Janeiro and form the impression that Brazil is a modern affluent nation little different from the United States. Even its backstreets and shanty town areas, the *favela*, give no clear indication of Brazil's less developed status; New York and

Chicago have their not dissimilar areas of deprivation. To visit the dry but populated lands of Brazil's north-east is, however, to see a different world and an economy which puts the stamp of the Developing World upon Brazil. Those features are all associated with that spatial characteristic of Brazil which results, in part, from her great size. After nearly five centuries of European involvement, the settlement of Brazil is still essentially peripheral. Most of the population remains concentrated along the Atlantic coast. Three hundred kilometres inland the population density falls sharply; beyond 650 kilometres the terrain is barely populated (Fig. 31B). This means that the greater part of the nation is devoid of towns, and huge areas, notably in Amazonas, contain no people. Brazil's present position is suggestive of that in the United States one hundred and fifty years ago when her population was eastern and coastal in its disposition and the 'West' was a vast and almost untouched interior. This issue of the interior adds a distinctive spatial dimension to Brazil's development problem. It becomes one not only of transforming an economy but also of involving and integrating a large and little-touched area. The former colony has become a colonizer and the geographical periphery an economic core. The precise form of Brazil's development of its interior has raised international interest and debate since it contains the largest area of tropical rain forest in the world. It is argued that the preservation of this forest is essential first as a reservoir of a wide range of floral and faunal species necessary for global species variety and secondly because of its role in reducing global warming.

THE IBERIAN LEGACY

The Portuguese first made contact with Brazil in the opening years of the sixteenth century and a scattering of tiny coastal settlements resulted. In 1530 the Portuguese Crown began to formalize the organization of settlements under the *capitanias* system. Large land grants were made which, in effect, established a landed aristocracy in Brazil. In the first emergence of a plantation system, sugar and tobacco were early among crops grown for export. Initially the labour input was supplied by Portuguese peasants and South American Indians but by the middle of the sixteenth century African slaves were being shipped into Brazil in increasing numbers to work the land. By the seventeenth century Brazil, though it was then neither a defined territory nor an economic entity, had become the world's major source of cane sugar with Bahia as the most important producing area. Further inland, in what was then the 'interior', similarly large land grants were formed into cattle ranches with hides as the main export. Brazilian sugar output subsequently declined in the face of competition from more efficiently run plantations which British and Dutch interests established in the Caribbean. In 1650 the non-Indian population of Brazil was put at 74,000 Europeans and 110,000 African slaves (Vianna 1961–62).

In 1694 the discovery of gold in what is now the province of Minas Gerais helped to compensate for the decline in sugar output. It also led to a shift in economic emphasis from the declining sugar areas of the north-east to the

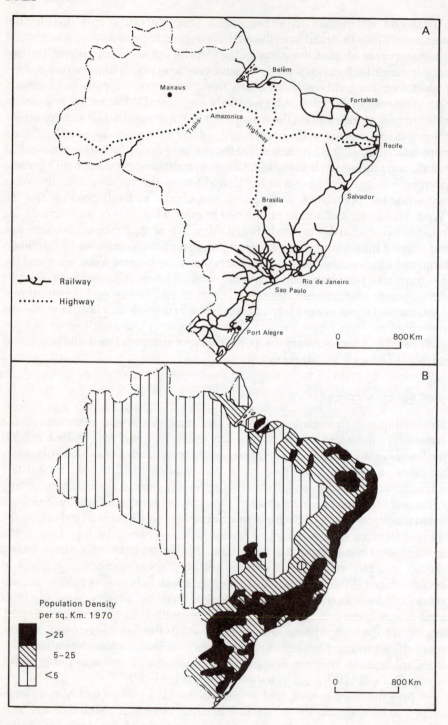

A

Manaus

Belém

Fortaleza

Trans Amazonica Highway

Recife

Brasilia

Salvador

Railway

Highway

Rio de Janeiro

Sao Paulo

Port Alegre

0   800 Km

B

Population Density
per sq. Km. 1970

>25

5-25

<5

0   800 Km

central and southern coastal provinces. Sugar planters and their slaves abandoned their plantations and settled as prospectors in the gold areas. In 1729 the attraction of Minas Gerais was enhanced by the discovery of diamonds and these together with the gold, sugar and tobacco came to characterize the colony's exports together with cotton, rice, cocoa and some coffee. The colonial capital was transferred to Rio de Janeiro in 1763. By the eighteenth century Britain was in the throes of the industrial revolution which in Portugal had barely begun. Anxious to protect her home manufactures from possible competition from her own colony, Portugal, in 1785, passed a law prohibiting the establishment of manufacturing in Brazil designed to serve the Portuguese market (Cardozo 1969).

IMPERIAL BRAZIL 1833–89

To escape from the invading armies of Napoleon, the King of Portugal and his whole court fled from Portugal in 1808 and set up a government in exile in Brazil, giving the trans-Atlantic colony an additional tier of administration to support. This presence, which lasted until 1822, served to centralize the administration of Brazil and give it an individual identity. When the King returned to Portugal, his son whom he had left behind declared Brazil to be independent and set himself up as Emperor. The political ties with Portugal were effectively severed but Portuguese influence was pervasive and lasting.

Both land and labour were cheap in Brazil, the one because it was abundant and the other because it was slave labour brought in, as required, from Africa. Where these two factors of production were of such low cost it was almost inevitable that they were used extravagantly. Since the owners of these two factors were made wealthy by the sheer size of their possessions, there was little incentive for intensity in their use or their careful husbandry. The Brazilian economy was essentially agrarian but it had two parts. One was concerned with producing for export focusing on the crops which have been mentioned but with sugar, still chiefly in the north-east province, and coffee, initially in the province of Rio de Janeiro, as the outstanding items (Galloway 1968). Both crops were produced on large estates worked by slaves. The other sector was very much larger and involved the greater part of the population. Its concern was with feeding Brazilian bellies. In part it was made up of subsistence farmers working small-holdings, the *minifundios*, and in part by farmers producing crops for sale in Brazil. This latter group was characterized by large land-owning units, the *latifundios*, but not by slaves or exclusively by large producing units. Most farmers were tenant farmers of one sort or another. Maize, manioc and beans were typical products. Wage rates for labourers were low because labour was in abundance. Share-croppers were poor and in no position to improve their lot.

Figure 31   Brazil: (A) railways and Trans-Amazonica and Brasilia–Belem highways; (B) population density

On the frontier of settlement in the interior, systems of shifting cultivation were common and, of course, did nothing to improve the land. This sector, made up of subsistence and cash-croppers for the domestic market, accounted for by far the greatest part of Brazil's agricultural economy and it remained the most important throughout the nineteenth century. By its very nature in terms of land tenure, technology, systems of husbandry and in the absence of market stimulus, it remained backward and non-progressive.

The Government, the merchants and the more enterprising landowners concerned themselves with the commercial export sector. It yielded foreign exchange revenue and was more readily taxable but it too was technically backward, leisurely and inefficient. While sugar continued to decline in importance, that of coffee increased. By 1850 the main focus of coffee growing was in the Paraibo Valley between Rio and Sao Paulo, with easy access to the coast for export, and Brazil had become the world's leading exporter of coffee. Increasingly the slave labour which once had been applied to sugar was diverted to coffee estates. It has been estimated that between 1800 and 1852, when the British navy put a stop to the shipment of slaves into Brazil, over 1.5 million slaves had been brought in from Africa (Curtin 1969; Bethell 1970; Leff 1982). In 1888 the number still in a condition of slavery in Brazil was half a million. In that year, slavery in Brazil was abolished and in the political turmoil which followed, the Emperor abdicated; in 1889 Brazil was declared a republic.

THE REPUBLIC OF BRAZIL

During the Empire the population had grown from 4.6 million to 13.9 million largely by natural increase but also augmented by the shipments of slaves from Africa and immigrants from European countries. The Brazil of the new republic, in the last decade of the nineteenth century, was little different in terms of economic development from fifty years earlier. Exports were still made up of primary products of farm, forest and mine. Much of the economy and most of the population were concerned with the domestic market and subsistence, and most of the country remained unsettled. The territory of the new republic was still a vast and largely empty expanse devoid of an integrating network of modern forms of transport. In 1884 there were only 6300 kilometres of railway in the whole country and most was in the vicinity of Sao Paulo and Rio and associated with the coffee exporting area. The road network was similarly rudimentary. The physical means to integrate Brazil's vast spaces into a cohesive whole did not exist. Importantly also, as Leff points out, these transport inadequacies had the effect of raising the costs of products arising in the interior, and not simply the deep interior, and in consequence restricted export opportunities to the accessible and essentially coastal areas (Leff 1982).

Cheap labour had been a characteristic of the Empire. Its costs were kept low by the system of land tenure and the use of slaves. When slavery was abolished both the landowners and the republic's government, which was drawn from their ranks, made every effort to maintain the supply of labour and keep its costs low.

The new republic began to subsidize the passages of migrants from Europe whose numbers greatly increased. This new labour was used mainly on the land and in the mines but there was no significant redistribution of land holdings and they remained as wage labourers (Hutchinson 1961). Increasingly the growth points in the economy came to be associated with the south-east states, leaving the sugar and ranching areas of the north-east province even more deprived and backward.

Up until the First World War Brazil's economy remained dominated by agriculture both in terms of the proportion employed in farming and in its contribution to the Gross National Product. Manufacturing did emerge and grow but only slowly at first and it too was characterized by technical backwardness and a slowness in adopting innovations. The industrial labour force like that in agriculture was, for the most part, poorly paid and badly educated. Little government investment had been applied to the social infrastructure and illiteracy was widespread. It will be recalled that Portugal had restricted manufacturing in Brazil but this is not a convincing explanation of the slow growth after independence. Neither is Brazil's lack of large coal deposits significant, though coal was important in the age of steam and coke iron making. Other coal-less countries industrialized more quickly. Industry which did develop was directed towards import-substitution and most of it was concerned with processing agriculture products so that cotton textiles and food products were the main manufactures. They were made behind protective tariff barriers for the home market and Brazil's exports remained dominated by primary products.

Much of the lethargy in the Brazilian economy of this period can be related to her agriculture. It had developed in a situation of abundant low-cost land and labour. The subsidization of immigration when slavery came to an end and lack of any reforms of land-holding ensured that this situation remained. As a result the majority of the population was poor, landless and wage labouring and this condition persisted throughout the nineteenth century. Brazil's internal market was thus too small, because it was so poor, to generate a large demand for consumer goods and the market stimulus of manufacturing remained weak. This impoverished mass was unable to accumulate savings for investment; nor did it represent a major taxable resource from which the Government could extract revenue for investment. The Government derived revenue from export taxes and it taxed land but the abundance of land and its low value, together with the political influence of the large landowners, kept taxation levels low. The agricultural export sector was not a sufficiently large component of the national economy to fuel the engine of economic development and generate enough investment capital. In the years 1911–13 agricultural exports represented only 16 per cent of the GDP. Had a much greater part of the land and the agricultural population been energized into effective production, the impetus to move forward would have been greater. As it was the inertia of the dead weight of an unreformed agriculture and rural poverty held back economic progress. Infrastructural developments which the Government should have financed were slow in coming because of its meagre resources. Transport in particular was neglected even allowing for the difficulties of terrain and the distances involved in Brazil. In consequence many resources could not be adequately harnessed. It was not until the government of the republic began to borrow on the overseas money market in the last decade of the nineteenth century that railway building on a large scale began. Railway construction did stimulate settlement and promote other initiatives but the focus was mainly in the province of the south-east, leaving the north-east region once more disadvantaged (Fig. 31). Towards the end of the century, Brazil began to import industrial capital goods. Her small market hedged around by tariff barriers was not a great attraction to external suppliers of consumer goods. Apologists for Brazil's development cannot therefore attribute blame to this form of mercantile imperialism; indeed the country's balance of payments remained consistently healthy.

Brazil enacted no legislation to facilitate land settlements and achieve a more equitable distribution of land holding and resources comparable with the Homestead Act and other similar land acts which in the United States in the nineteenth century achieved these ends. She adopted a policy enforcing low returns to labour. Her policy led to the fossilization of much of Brazil's agriculture and created a burden which all the development initiatives of the nineteenth century could not lift. Increased borrowing from overseas to fund investments for development did occur before 1914 and foreign industrialists did begin to invest in Brazil and, as a result, manufacturing began to grow apace. The late nineteenth-century initiatives in railway building, continuing into the twentieth century, undoubtedly further facilitated these developments. They survived the collapse of the rubber export trade which along with coffee had been a leading export commodity. As British and Dutch rubber plantations came into production and destroyed the market for wild rubber from the Amazon, Brazil's rubber exports fell from holding 90 per cent of the world market in 1910 to 10 per cent in 1920 (Bergsman 1970). Furthermore, the cutting off of industrial imports into Brazil during the First World War still further stimulated Brazil's manufacturing sector which had grown up to serve the home market in an essentially import-substituting role. Between 1914 and 1922 industrial output rose by 150 per cent (Furtado 1976). None the less, in 1920 no less than 68 per cent of Brazil's labour force was still employed in agriculture. Coffee remained a successful export crop and the industrialization thrust weakened. It had been associated with the three main components of food products, textiles and iron goods, together with the construction industry. Between 1922 and 1929 industrial output grew hardly at all. In 1929 came the world economic slump bringing coffee exports down with it. The contribution of the industrial sector to the GDP in that year was 11.7 per cent (Furtado 1976). Brazil's political scene was thrown into ferment and in a political coup the presidential link with the vested interests of the landowners of the coffee region was broken. The new president, Getulio Vargas, saw Brazil's future as one best served by a more broadly-based economy and, for the first time, the government of Brazil began actively to stimulate industrial developments.

The fall in the world coffee prices was dramatic and its effects far-reaching. It reduced Brazil's ability to service her debts and led to a devaluation of her currency. In turn these events made imports into Brazil more costly and at the same time rendered coffee a less attractive investment proposition. The price of coffee was subsidized so that its production did not collapse and the home market remained more buoyant than it would otherwise have been. Brazil's import-substituting industries, because imports were made more expensive by the devaluation of the currency, could expand rather than contract. Investments were diverted from coffee planting into manufacturing industries. The economic collapse of the industrialized world made used industrial plant cheap and available and this was brought into Brazil to help manufacture investment goods and intermediate products too expensive to buy in. Encouraged by President Vargas, the process of industrialization was resumed.

AN IMPORT-SUBSTITUTION POLICY OF INDUSTRIALIZATION

It has been remarked earlier that import-substitution has been a commonly adopted approach to industrialization. It was the road Brazil had embarked upon in the nineteenth century and it was this road which Vargas and his successors were to follow for over thirty years. In Brazil's case it was not, however, concerned solely with the production of consumer goods for the retail market; it went to the foundations of industrialism. Under a cover of protectionism, all sectors of the manufacturing spectrum were involved. Energy resources were developed – notably that of hydro-electricity but also that of the meagre non-coking coal deposits in the state of Santa Caterina. Oil exploration was begun and production, though in very small quantities, had commenced by 1940. The rich iron ore deposits near Belo Horizonte in Minas Gerais, together with Brazil's manganese ores, were further developed and bauxite mining and aluminium production began on an increasing scale. A whole series of regional cement plants were set up and transport facilities, particularly roads, were improved. While some of these activities did yield exports, their *raison d'être* was to secure the bases of industries for the home markets. Until 1946 almost all Brazil's increasing output of iron was produced almost entirely with the use of charcoal. In that year a new large iron and steel plant at Volta Redonda, which itself enlarged Brazil's iron and steel making capacity by 25 per cent, was brought into operation using specially processed Brazilian coal in its blast furnaces. By 1964 there were ten large steel-producing companies in the Itabira–Belo Horizonte area. This laying of foundations allowed the establishment of vertically and horizontally-linked industries. Steel rolling mills, engineering plant, aluminium fabricating units and eventually, in the late 1950s, motor vehicle plant, all came into operation. Chemical production and the long-established manufacture of textiles were increased.

Such was the progress of this industrial expansion that home production came to meet almost all the needs of the domestic market across the whole spectrum of industry. In 1949, 65 per cent of capital goods had been imported; by 1959 it was 33 per cent and by 1964, 10 per cent. Intermediate goods were likewise produced almost entirely in Brazil and her reliance on imports fell from 27 per cent in 1949 to 7 per cent in 1964. The import of consumer durables fell dramatically from 65 per cent in 1949 to 10 per cent as early as 1955, falling further to 2 per cent in 1964. Non-durable consumer goods had long been produced almost entirely in Brazil but even in this sector imports were halved (Bergsman 1970). This process of rapid industrialization involved both the state and foreign firms. Strict protective measures effectively excluded the import of goods manufactured abroad. The import of motor vehicles, for example, was prohibited but foreign motor vehicle manufacturers were encouraged to set up plant in Brazil and produce within her borders. The funding of this expansion came from taxes on the export of primary products, notably those from agriculture, from borrowing on the international money market and from the investments in Brazil of foreigners. The three components referred to in Chapter 9 as sources of development revenue were all utilized in Brazil but agricultural exports were the most significant source.

Vargas was finally replaced as President in 1954. His successors maintained the policy of industrialization via import-substitution which was so wide-reaching that it was tantamount to a policy of self-sufficiency but one which accepted and used foreign capital and expertise. Manufacturing was essentially a matter of replacing imports so that even as late as 1960, 97 per cent of Brazil's merchandise exports were made up of primary commodities. By extending industrial developments into all strata of industry, initially with the use of revenue derived from agriculture and hence dependent upon agriculture, by the mid-1960s a broadly based and much more fully integrated industrial structure had been created. This structure was much less dependent upon agricultural exports, but in creating it, Brazil had become much more dependent upon international capital.

Import-substitution has its own self-limiting features. It caters for the home market and in doing so is limited by the size of that home market. It is necessary, therefore, for that market to grow by population increase or by becoming more affluent or both. This issue of an expanding work-force creating itself an expanding market was discussed in Chapter 8. The process of industrial integration which characterized Brazilian development itself, of course, helped to ensure a market for intermediate and other goods which are not consumer end-products. This is in contrast to the situation in India where the focus on the basic industries to the exclusion of industries which processed their products, effectively curtailed the industrial market and led to unused capacity.

Brazil's market difficulties lay in the distinctive nature of her internal market. In the mid-1960s industrial output in Brazil began to fall, partly due, no doubt, to the socio-economic disturbance associated with the period just before and after the military coup of 1964 which overthrew the presidency. However, as Furtado points out, while industrial production declined, a larger proportion of its output was exported, indicating that it was the home market which placed limits on output (Furtado 1976). The character of Brazil's internal market is a product of her long history of abundant labour. This keeps wages low in the industrial sector and confines demand to a limited range of consumer goods. The agricultural population is itself poor with limited disposable incomes, and movement into industrial employment does not raise incomes significantly. As industry has become more sophisticated it has tended to become more capital-intensive so that its level of output is less tied to labour input. In a situation of import-substitution, these features effectively retard the growth of the domestic market and a mismatch occurs between productive capability and demand. The more affluent sector of society (the top 10 per cent of families by income level account for 46.2 per cent of the total of family incomes in Brazil) represents a significant and different market which grows as greater incomes, the product of commercial and technological advances, accrue to this sector. Its tastes are more specialized and for the individual groups of goods within its demand spectrum the demand is small and inhibits economies of scale in their production. Unused capacity in Brazil has developed in this situation which restricts the level of returns on investment and holds back industrial dynamism (Furtado 1976). The nature of Brazil's domestic market has a significance which is detrimental to economic development based on import-substitution. Its second feature is as much of social as of economic significance. It is the persistence of economic inequalities on a large scale. On the one hand it was produced by the systems of land-holding and agriculture going back to the days of the Portuguese conquest and the use of slaves, and on the other it was allowed to develop in the manufacturing sector in the nineteenth century and has persisted into the second half of the twentieth.

The lowest 30 per cent of household incomes groups account for only 8.1 per cent of total household incomes. In rural areas subsidies and tax structures favour large farms and the poor find it almost impossible to buy land and become independent farmers. On the large farms state-subsidized mechanization has reduced the need for labour, both skilled and unskilled (World Bank 1990). Social injustice and a limitation of opportunities for economic expansion are features of this situation in both agriculture and industry. In this respect it would not be inaccurate to say that Brazil is growing but not developing.

MANUFACTURING FOR EXPORT

In the late 1960s the Government attempted to reduce surplus industrial capacity by enlarging the market for Brazil's manufactured goods in two ways. First, following Brazil's long-established practice, the Government stimulated those industries which served the small affluent sector of Brazil's domestic market. Second, and much more significantly, the Government adopted a policy of promoting the export of manufactures. It was hoped that the market enlarged by these endeavours would not only absorb surplus capacity, but also permit economies of scale to be achieved while at the same time allowing a degree of specialization in those items where Brazil might have comparative advantage.

In 1960 manufactured goods had accounted for only 3 per cent of Brazil's exports. By 1977 the proportion had grown to 26 per cent and reached 56 per cent in 1991. Some 55 per cent of these manufactured goods went to other developing countries. Brazil has moved strongly into the industrial-export field. In doing so she has broken free of the limitations which an import-substitution policy imposes even in a country as large as Brazil.

It can be argued that Brazil retarded the structural transformation of her economy and her economic development by unduly delaying entry into the international market for industrial exports in those commodities where she was competitive. It can also be argued that the maintenance of a low-wage economy was not only socially suppressive but likewise held back the industrial dynamism which Brazil undoubtedly possesses. None the less, in the period between 1960 and 1980 manufacturing in Brazil grew rapidly, with annual industrial growth averaging 10 per cent while that of the GNP was 7 per cent. Average per capita incomes had increased greatly in real terms from around US $600 in 1950 to over US $2000 by 1981, with both levels measured at 1980 values. In 1991 the level had reached US $2940.

Brazil's industrial boom was in some measure coincident with the world-wide upsurge in industry and trade. Both the dramatic rise in oil prices and the resulting world economic depression which terminated the good years, seriously affected Brazil. Once again she resorted to import-substitution where it was possible, but she had, of course, acquired a capacity geared to the external dimensions of the market. To sustain her growth, Brazil needed to borrow on the external money market. The need for capital was all the greater because of the size of her industrial sector. By 1979 Brazil had the highest level of indebtedness to commercial banks of any country in the Developing World. Her commitment in terms of public borrowing in 1981 was second only to that of Mexico which, unlike Brazil, is a major oil producer. The issue of debt servicing, referred to in Chapter 9, has, in Brazil's case, come to the fore requiring an expansion of exports, a cut-back in domestic expenditure and a greater and exacting efficiency in the management of her resources. In 1980 Brazil's total debt service as a percentage of all exports of goods and services was the equivalent of 63.1 per cent of the value of her exports (World Bank 1993). President Figuerido (1980–84) came to office as head of the military government in that same year. The subsequent decade was to see Brazil's inflation average 327.7 per cent per year and Brazil's forward momentum brought almost to a halt.

THE SPATIAL DIVIDE: BRAZIL THE COLONIZER

Prior to the late 1950s, the governments of Brazil paid little attention to where the economic activities which they were encouraging or suppressing were taking place. Developments which took place before 1914 had tended to be concentrated in the central and southern coastal areas and the nearer interior valleys, setting up a regional bias which has persisted. The successive governments in the republic pursued a centralized form of administration in part, perhaps, in an attempt to stress national unity in such an extensive but unintegrated territory. It may well have been a policy running counter to the maximization of economic development. Though it is debatable, greater regional autonomy could, possibly, have more effectively promoted the development of still unused resources. The risk was, of course, political separatism. As it is, in the last years of the twentieth century Brazil is still divided spatially between the economic core on the geographical periphery and the economic periphery in the geographical core. The focus of population and economic activity still leaves much of Brazil little integrated into the body of the political economy.

President Vargas had been disturbed by the neglect of the greater part of Brazil and had attempted, by reducing the powers of regional governments, to bring the nation into a more cohesive and integrated state. During the periods of office of his successors, planning bodies for the main regions were established. The Superintendency for the Development of the North East, SUDENE, was set up in 1959 and that of Amazonia, SUDAM, in 1966. A similar body, SUDESUL, for the densely-populated and economically vigorous core area in the south-east was also established. The presidency of Kubitschek (1956–61) also saw two much-publicized developments which could be taken as tokens of governmental recognition of the issues of regional development. Brasilia, the new purpose-built capital referred to in Chapter 10, was inaugurated in 1960 with a design size of 500,000. Located in an area of low population density, over 900 kilometres from Rio, it represented a step into the interior by national government and national interests. Though still peripheral to much of the country in many respects, it none the less represents a significant frontier gesture and in its appearance symbolizes the emergence of the new modern Brazil. In a more practical sense its impact on Brazil's economic periphery is limited. In Katsman's view, Brasilia will not contribute much to the industrialization of the state of Goias, though it will stimulate agricultural developments by the very presence of a large urban market. The interior areas have long suffered from their remoteness from the markets along the coast (Katzman 1977). The second development was the commencement of the Brasilia–Belem Highway as part of a plan to connect the new capital to the rest of the country. Begun in 1958, it was completed as an all-weather road in 1973. Katzman calculates that between 1960 and 1970 the Brasilia–Belem Highway, by stimulating settlement, accounted for between one-third and one-half of the growth of the population in Goias (Katzman 1977). A more ambitious project has been the construction of the Transamazonica Highway reaching westwards from the arid lands of north-eastern Brazil across the rain forest of Amazonia to the Peruvian border. Sufficient time has not yet elapsed for the full impact of the work of the regional development agencies to be felt nor yet that of Brasilia and the Trans-Amazonica Highway, but the development of most of Brazil's interior is still in its most early stages. There is little indication of either a transport network or hierarchy of urban centres which would integrate this vast territory into a cohesive space economy. Brazil indeed raises all the issues discussed in Chapter 10.

Ivo Babarović in his penetrating statistical analysis of the regionalization of Brazil's economy makes the point that while regional problems exist within regions themselves, they raise issues which can only be addressed at national level (Babarović 1978). He states that regional imbalances, as was indicated in Chapter 10, are common in developing countries that have reached the stage in which Brazil finds herself and stresses that the resulting undesirable effects cannot be left to some speculative restoration of spatial equilibrium by natural processes. He lists in Brazil's case the catalogue of adverse consequences so often associated with spatial skewness and its linked social and economic inequalities. First, the urban agglomerations exceed optimum size and produce diseconomies which retard natural growth. Second, regional concentration of growth, the mushrooming of the core, means the periphery is neglected, which is so clearly the case in Brazil. This neglect in turn leads to an impoverished group whose market stimulus to the economy is, as has been seen, thus enfeebled. Third, Babarović refers to the moral and social injustices which are inherent in spatial inequalities in welfare; and fourth, he points out that too great a discrepancy in regional levels of economic activity, social and income benefits and participation in the nation's progress, can lead to social and political conflict and put at risk both the political and territorial integrity of Brazil. This is no small risk in a country so large as Brazil in which her distant borders are so tenuously connected with her centre of economic and political gravity. Brazil shares her border with no fewer than ten other countries and with most of it more than 1600 kilometres from Brasilia. Babarović holds to the view that the near or 'frontier' regions of Brazil's periphery should be settled and encompassed into the national economy, while regions further out, but none the less settled, should also be linked into the national economy so as to enhance the domestic market for industrial growth. Conversely, he feels some attempt must be made to inhibit the growth of the big urban complexes. His solution is a greater degree of spatial deconcentration of economic activities, a theme which has been referred to in many of the case studies in this book.

POOR GOVERNMENT AND POOR PEOPLE

In the Brazilian metamorphosis that has been described and which raised Brazil into the higher middle-income group of the world's countries and made her a significant industrial nation, little mention has been made of governmental development plans. Prior to the Second World War, they did not exist. Governments in Brazil have played a major role in economic development but it has had a distinctive flavour imparted to it by the patrician attitudes of its leaders, by the paternalistic views of the ruling aristocracy and, in recent years, by an abhorrence of socialism and with it the concept of a planned command economy. Plans there have been, with the first series running through until 1960, with subsequent plans covering the 1960s, 1970s and into the first half of the 1980s. These National Plans have become more comprehensive and sophisticated as they have evolved, but they have been formulated in an atmosphere of capitalism and free enterprise even though some enterprises are totally state-owned while in others state participation is considerable. Planning has been pragmatic and consultative rather than rigid and ideological. Consultations have been essentially between industrialists and investors, both Brazilian and foreign, and the departments of Government. Government policies which, as has been seen, have significantly influenced Brazil's development, have often taken the form of import controls, of tariff and exchange rate manipulation and of investment subsidies, all of which have often determined the allocation of resources. Herein lies the cause of Brazil's uncertain progress in the past, her massive indebtedness, her inability to react positively to world recession and the existence of one of the most skewed distribution of incomes in the world. Government intervention is on a large scale and despite the successes of the 1970s, its effectiveness, by both military and civilian governments, can be called into question. Brazilian governments of the rich élite, supported by the élite, have attempted to promote the economy by investing in manufacturing. To do so, funds have been borrowed from abroad on an increasing scale; investment has been made in large capital-intensive plant when jobs were urgently needed; in agricultural developments large-scale agro-businesses have been favoured, leaving the plight of the peasant untouched. The tax structure, in need of urgent reform, achieves no significant redistribution of incomes and indeed in the 1980s and 1990s disparities increased. All this is taking place in a country where a large proportion of the population live in poverty. State industries have commonly proved to be inefficient; subsidies have distorted the operation of the market. Attempts have been made to solve problems by large capital injections but so often the funds have been inappropriately allocated; for example a large, and necessary, investment in education was so apportioned that the major part went to universities attended by the middle-class, leaving little for the whole of the school population. Interventionism in Brazil, as so often elsewhere, has been a failure because it has invoked inappropriate policies and neglected major issues.

Old political structures, entrenched attitudes and vested interests are not easily changed and certainly not by fleeting presidencies. The civil service of Brazil, which in so many ways resembles that of India, is the embodiment of government attitudes and a repository of state regulations. It has fed on the vast minutia of licences, permits, subsidies and other regulations created by Brazil governments over many years. Bureaucracy has flourished. In such a situation even the most simple social and economic transactions can involve many documents and much time; more complex affairs can involve hundreds of forms and several ministries. Such is the obstruction of this proliferation that a profession, the *despachante*, has evolved which for a fee applies its expertise to the penetration of the maze and the saving of precious time. Attempts have been made to excise red tape and to cut delays. In the early 1980s many procedures and documents were abolished yielding a saving estimated at US $3 billion (World Bank 1987).

The need is seen by many, both outside and inside Brazil, for radical reforms which will liberalize the economy, remove trade-barriers and bring about deregulation of the economy on a large scale. Many Latin American countries have already begun this process ahead of Brazil. Of prime concern is the need for Brazil to bring inflation under control and for the Government strictly to control its finances. The World Bank, as so often with other countries, advocates a restructuring of the economy in which the state divests itself of ownership of farms and factories. Unless Brazil keeps its house in order, foreign investment will be less forthcoming and debts more difficult to re-schedule.

Brazil's great development which has raised her into the higher middle-income group of the world's nations, ranked by per capita GNP, has made her a significant industrial nation. In terms of total value-added, her manufacturing exceeds that of any other nation in the Developing World, save China. In part this is a function of her giant size, in part it is the result of the measures taken towards industrialization by successive governments. That her economy is far from secure is indicated by the state of her indebtedness; that her society is as yet immature is shown in the inequitable distribution of incomes. Brazil possesses a society and economy which are clearly dualistic. A small affluent community participates in, and has access to the benefits of, her sophisticated industrial and service sector. At the same time the majority of the population is poorly paid. Thirty per cent still work in agriculture and this group is the most separate and deprived. The equity which is so noticeably present both in socialist China and capitalist Taiwan is equally noticeable by its absence in Brazil. The work of Kuznets, Paukert and Aluwalia, referred to in the discussion of Taiwan in Chapter 19, would seem to be more applicable in the case of Brazil (Kuznets 1955; Paukert 1973; Aluwalia 1974). It remains to be seen whether the present stage of Brazil's development will be followed by one of more equal opportunity and spread of benefits. If the status of a major world power is to be achieved, these socio-economic goals must be achieved – and achieved quickly.

18 Peru

'From the time of the Spanish Conquest, foreigners have displayed energy,
skill and daring in their attempts to open up the country and carry away its
riches, and in rendering up its wealth, first to the enterprise of capital
investors, Peru has suffered a fate analogous to that of the oyster, which it,
too, is 'opened up' for its riches'

R J Owens
Peru 1963

Peru shares with Brazil many of her characteristic Latin American features,
though in Peru they are more dramatically drawn and neither Peru's present
level of attainment nor her future prospects compare favourably with those of
Brazil. Peru's people are poor. In 1991 the average per capita GNP was US $1160.
Furthermore, this average conceals, as it does in Brazil, a markedly skewed
distribution of income. In 1985, based on expenditure patterns, the poorest 20
per cent of Peruvian families accounted for only 4.4 per cent of the national total
of family expenditure; indeed the bottom 60 per cent of household income
groups represented only 26.6 per cent. Half the population of Peru in 1990 had
incomes of only US $116 (Psacharopoulos 1990). Peru's development history has
produced a most inequitable distribution of the material benefits of
development and the situation is not improving. The average per capita GNP
has experienced an average decline of 0.2 per cent per year from 1965 to 1990
with the poorest groups experiencing the biggest fall in incomes (Thumm *et al.*
1981; Glewwe and Hall 1992). Poverty is thus extreme and widespread. This is
the country of the Incas. Its present situation has its beginnings in the destruction
of their empire.

THE SPANISH CONQUEST

Between 1530 and 1533 the Spaniard Francisco Pizarro destroyed the political
and military power of the Incas and disrupted the socio-economic system they
had established. In 1544 the Spanish Crown established the Vice-royalty of Peru,
covering a territory much larger than that of the present Peru, and attempted to
exert an absolute control over its lands and people. Though some negro slaves
were imported from Africa, their numbers were very small, partly because of
Peru's location and partly because of the greater size of the indigenous Indian
population compared with the situation in Brazil. Instead the people of the Inca
Empire were made virtual slaves in their own country. The new colony was
administered in systems which gave the appointed Spaniards, the *encomenderos,*

total control over the local people in their charge. Their labour could be directed at will. Land was taken from Indian use and redistributed to the Spanish officials or sold to Spaniards.

The extent to which this occupation took place varied within the contrasted physique of this large territory. Along Peru's entire coast is a long stretch of desert lowland (Fig. 32). Inland, foothills rise into the cordillera of the high Andes where peaks reach to 6000 metres in a series of ranges and high plateaux separated by deep valleys. To the east of the Andes the land drops down to the hot wet lands of Amazonia, the Selva. At the time of the conquest the main focus of Peruvian population was in the valleys of the Sierra and it was in this area that Spanish interest was focused since it possessed resources of gold, silver and mercury.

For the next two centuries Spain extracted large amounts of silver and significant amounts of gold from the Andean Sierra of Peru. Neither the deserts of the Costa nor the rain forests of the Selva were of much interest to them. The colonial economy was extractive and the colonists, made rich by their exports,

Figure 32 Peru

had little cause to develop agriculture and manufacturing of goods which could be imported from the home country. Gradually, however, the descendants of the original Spanish settlers began increasingly to organize their estates, the *haciendas,* and produce a range of crops both for local consumption and, in the case of sugar, cotton and tobacco, for export. Likewise manufacturing for local needs developed despite Spanish laws to the contrary.

By the beginning of the nineteenth century, Peru, along with other Spanish colonies in Latin America, had grown to be largely independent of Spain. The revolutions led by Simon Bolivar and Jose de San Martin, and beginning in 1810, resulted eventually in many countries achieving their independence. The process in Peru's case was an involved one and included participation in a confederation of Peru and Bolivia, which was dissolved in 1839. Peru had grown out of armed revolution, her final independence involved the warring of generals, and her first civilian government did not come into existence until 1872, fifty-one years after San Martin had taken Lima and declared Peru free of Spain. It was not an auspicious beginning, nor was the subsequent history of military rule, punctuated by revolutions, a period conducive to an ordered economic and social development let alone development planning. The foundations of a long period of a prodigal misuse of resources, of inefficiency and injustices were laid.

SILVER AND SUGAR, COTTON, COPPER AND BIRD DROPPINGS

Independent Peru was little different from the colony out of which it emerged. A series of laws were enacted which would, in theory, have given back rights to land to the Indians who had never known individual land ownership, as well as distributing land more widely among the descendants of the settlers. In practice this legislation facilitated the assemblage of large land-ownership units on the better land upon which the Indians worked mainly as share-croppers. These *haciendas* were located in the broader valleys of the Sierra and on the coastal plain, the Costa, where land could be reclaimed by irrigation. In the more difficult high Sierras and in the Selva, land was farmed in systems little different from those of pre-Conquest days. In mining the Indians continued to provide labour for the dangerous and difficult work of silver and mercury extraction. The Indian remained as the bottom layer of society in all these activities and the Peruvian economy, like that of Brazil, was characterized by an abundance of cheap labour.

In 1841 the discovery that the guano, the bird droppings which accumulated in the huge colonies of sea birds along the arid coastline of Peru, was a valuable source of nitrogen, invested Peru with a new and valuable resource. The guano could be, and was, used as a rich nitrogenous fertilizer on Peru's cotton and sugar estates. The extractive mentality in which the exploitation of Peru's mineral resources had first developed soon identified guano as a potentially valuable export. Using Indian labour, it was mined on a massive scale in an uncontrolled exploitation. Bird droppings came to exceed the value of silver exports, guano mining was declared a state monopoly and its export revenue did much to sustain the Peruvian economy. In the southern-most coastal areas of Peru, the arid climate had given rise to another valuable mineral fertilizer, the nitrate deposits at Tarapaca. These too were exploited as a state monopoly. The contractors associated with the extraction of guano and nitrate deposits came to be a rich and politically powerful group known as the *guaneros*.

Peru's economy was seriously affected by her participation in two wars in the second half of the nineteenth century. In the first, Peru and her neighbours fought off an attempt by Spain to re-exert her authority over her former colonies. In the second, Peru was herself in conflict with both Bolivia and Chile over the nitrate deposits in the desert lands on her border with Chile. Peru was defeated and lost her valuable nitrate fields to Chile. Peru entered the twentieth century with an insecure, disorganized economy. The capital for her infrastructural developments and the exploitation of her mineral resources had come largely from Europe and North America. Political power lay in the hands of a small group of landowners and mineral developers. In the periods between the succession of military coups when elected governments were returned, the Indian majority in the country were effectively disenfranchised since literacy was a prerequisite. Exports of sugar, cotton, guano, silver, copper and other minerals had sustained Peru in the nineteenth century. Both exports and imports were taxed and such manufacturing as there was produced only for the small home market.

Peru was essentially an exporter of primary products and, though not a political colony, bore all the marks of a colonial form of economy. She is fortunate in that the mineral resources she possesses are not only large but varied. The orogenic belt of the Andes yields substantial amounts of zinc, lead and copper and in addition manganese, antimony, silver, tungsten, bismuth, cadmium, tin and gold are also mined. Even today the mineral exploration of the Sierras is far from complete. As the exploitation of these resources progressed, the small-scale mining companies were replaced by large and mainly foreign companies which could raise the capital and provide the technology. Mining became increasingly capital-intensive so that while mineral exports formed a large part of material exports, mining sustained only a small labour force. Sugar and cotton were both important export items. The increasing concentration of land-ownership in the late-nineteenth century which has been mentioned, meant that the production of these two leading agricultural exports was in the hands of a few. The result of these developments was that the growth in Peru's economy associated with increased exports spilled over little to the benefit of the majority of the population. The greater part of Peru's population and economy was detached from these export developments. There was growth, therefore, without development in its true sense. As has been discussed in earlier chapters, this colonial type of export economy was characterized by infrastructural developments specific to it, by the import of technology and equipment for its particular use, and by its relative insulation from other industries. The bulk of Peru's population because they did not benefit remained poor. This limited the size of the domestic market and so inhibited the development of import-substitution industries financed by export earnings. Too little of these earnings was retained in Peru and those which were entered too few pockets. Since labour was abundant in Peru, the wages of even those participating in the export economy remained low.

While the variety of her exports gave Peru a degree of protection against the market fluctuations in any one particular commodity, her mineral exports in particular were geared to the needs of the industrialized countries of Europe and North America, so that when the economy of these countries collapsed in the great recession of the late 1920s and early 1930s, so too did the markets for most of Peru's exports. The capital for her mining was from these same areas and it too diminished with the markets. In Peru this world-wide recession did not result in a vigorous drive for industrialization since much of the economy was untouched by it. Only gradually did Peru begin to develop some import-substitution industries and even in this activity was disadvantaged by the world trends referred to in Chapter 9, in which the relative value and demand for manufactured goods, including production machinery, compared with primary products changed to the detriment of the latter. Peru's politically and financially powerful minority had long been associated with land-owning and mining rather than industrial expertise and entrepreneurship. Her population had very little experience of factory production. As a result, when the world economic crisis faded away Peru maintained her primary export economy and this persisted through and after the Second World War. In the 1950s, when many former colonies achieved their independence and were characterized by colonially initiated primary export economies, Peru, which had achieved independence some 130 years earlier, was still in that same category. Changes were to come but they came slowly.

THE ECONOMIC SITUATION IN THE MID-TWENTIETH CENTURY

Peru entered the second half of the twentieth century with a government and business community which still viewed the development of the economy as a process of expanding the export of raw materials both in volume and variety. The important foreign investors obviously saw their business as one of extracting primary products for use by the industrial nations. The developments which took place after 1950 continued to reflect these attitudes so that while Peru experienced a period of rapid growth in production it was, in the main, concerned with this characteristically insulated export activity.

Oil had been discovered in Peru in the nineteenth century and output from the Brea–Parinas field on the coast near Ecuador's border reached two million barrels per day by 1914. In the 1950s, however, oil exploration was actively encouraged. New oil leases were granted in 1952 with the Peruvian Government participating in 50 per cent of the profits. Exploration was extended east of the Andes and in 1962 the Mobil Oil Company tapped natural gas in the Aguayita region. Oil is now produced near Ganso Azul on the Pachitca River. By the 1970s

Peru, with contracts arranged with Japan, had become an oil exporter. The output of copper, long a significant Peruvian export, was greatly increased when the newly-discovered deposit in the coastal ranges of the Andes at Toquepala was brought into production. The first exports from this field took place in 1959 through the former fishing port of Ilo developed specifically to export the ore and carried to it by a railway built to connect the mines. The ore body is large, some 400 million tonnes, and its development made copper Peru's most important export in 1960, overtaking cotton which had long held the premier position. In this copper development the companies involved were American. Iron ore, which had not figured prominently in Peruvian mining, took on a new significance when in 1955 the Marcona Mine, 560 kilometres south of Lima, began producing a very rich iron ore which, with a 60 per cent iron content, was readily exportable. The high grade component in the ore body was estimated at 70 million tonnes with the lower grade at eight times that amount. By 1960 the output from Marcona reached 3 million tonnes, most of it being shipped to North America (Owens 1963). Other iron ore bodies were developed in the 1950s by American companies, as in the case of Marcona.

In the agricultural sector the export of primary products was further increased throughout the 1950s as the acreage under cotton and sugar was enlarged and yield increased. Production was on large estates and both Peruvian and foreign capital was involved in both the agricultural and manufacturing aspects. Coffee production also grew sufficiently to allow it to meet home demand and become an export commodity.

Fisheries resources were developed for the first time in Peru during this period. For centuries fish had been regarded as of little importance. When the guano deposits came to be worked it was realized that inshore fish shoals were supporting the dense bird colonies which produced the guano. Because of the great importance of guano as an export in the nineteenth century, the influential guaneros were fearful, with some justification, that fishing might deplete the bird population and with it the source of guano. The importance of guano mining declined as alternative fertilizers became available and in the 1940s a small fishing industry, canning bonitos, was established. Peru came gradually to realize that she possessed one of the world's richest fishing grounds off her shores. Expansion in the 1950s was rapid and involved not only the canning and freezing of the big fish, notably bonito and tuna, but also the harvesting of the huge shoals of oil-rich anchovies which became, by far, the most important component in Peru's fishing industry. Pressed for oil in coastal oil-extraction plants, the residue of the anchovies becomes fish meal which is a very high-protein livestock feedstuff. By 1963 there were 138 fishmeal factories along Peru's coast. Fish exports, which had totalled 21,500 tonnes in 1950, reached 592,000 tonnes by 1960; by 1970 the amount was 2.25 million tonnes and formed an important new export commodity. The industry began as an essentially Peruvian venture with 70 per cent of the invested capital in 1960 being Peruvian (Owens 1963). Despite an awareness of the risk of over-fishing, this important new development began to run into difficulties as the inshore shoals were depleted, resulting in an increase in fishing costs. At this stage foreign companies began to take over more of the industry since they could supply the needed capital (Furtado 1976).

The pattern of development in the 1950s and 1960s was thus clear. It was very much the situation as before. Much of it came from external stimulus and much involved foreign capital and expertise, and it was largely a development in which the value added to Peru's raw materials took place outside Peru. It is true, of course, that the smelters for refining the ores such as copper, zinc and lead were set up in Peru by the mining companies, but their products were for export. Expansion did, however, take place in Peru's iron- and steel-making capacity with the inauguration in 1957 of the integrated iron and steel works at Chimbote, powered by a hydro-electric station and with an associated steel rolling mill (Fig. 32). A government-owned concern, the small plant proved a loss-maker even though it was supplied with iron ore from the Marcona mines at a low cost. In 1960 the Government initiated plans for a considerable enlargement of the plant with a Swiss firm responsible for its operation. It was a move into the phase of import-substitution. The plant continued to lose money while its products were protected by high tariff barriers. The Chimbote plant produces solely for the home market but cannot meet its total demand, while its products cost more than imports. Utilizing Chimbote's output together with imported semi-processed products, metal-working and engineering industries grew significantly in the 1950s and 1960s. The most important manufacturing industries were, however, those concerned with processing agricultural products, and the food and textile industries characteristic of early developments in industrialization.

The result of this ensemble of developments was a marked increase in Peru's GDP. Between 1950 and 1960 the annual growth rate of GDP averaged 5.3 per cent. Much of this growth can be attributed to foreign investments and most, though not all, was concerned with the export of primary products. The multiplier effect of these developments was limited and a large part of the population was not involved. In consequence, the ability of Peru's economy to absorb and make use of the retained earnings of her export activities was much diminished. The essential corollary of absorptive home developments, particularly in the manufacturing sector but also in essential infrastructural developments, was missing. Export-led growth was not leading to a full and self-sustaining development. As has been seen, the situation was not entirely negative. Industrialization was beginning to accelerate, social and physical infrastructural investments were taking place and even schemes to ameliorate the worst features of the traditional land-holding structure were attempted. All these were changes which had been unduly delayed and all were insufficient in both the scale and rigour of their implementation to lift Peru's population to any significant extent up the ladder of social and economic development. In 1960 primary products still accounted for 99 per cent of exports. Of the work-force, 53 per cent were employed in agriculture and 19 per cent in mining and manufacturing (World Bank 1980a). Significantly, Peru's import coefficient, that is the value of imports as a percentage of the GDP, grew throughout the 1950s from 11.9 per cent in 1950 to 21.1 per cent in 1960 (Furtado 1976).

Peru's population was growing rapidly. In the 1950s annual growth appreciably exceeded 2.5 per cent, and between 1960 and 1970 average annual growth was 2.8 per cent. Her backward agricultural systems farming difficult terrain were unable to meet the demand for food. It should be remembered that only 25 per cent of Peru is suitable for agricultural use without major modifications and much of this area is devoted to grazing. A very large proportion of the country is mountain in the Sierra, much of the Costa is desert, unusable without investment in irrigation, and only the Selva possesses much untapped land potential though beyond its fringes it presents all the problems discussed in Chapter 5. The demand for non-durable consumer goods by Peru's poor population was small and met largely by home production. Peru's export activities did, however, require imports, and so too did her small affluent sector, while the beginnings of industrialization further increased the import coefficient. Peru was earning revenue from export but her problem, and in some respects it is not dissimilar from that of Nigeria, was in developing her internal economy in such a way as to be able effectively to broaden her economic base, modify her economic structure and provide employment for her growing population. As in Brazil, Peru's dual economy was not only inequitable but was retarding development. The average annual growth in GDP which had reached 6.1 per cent between 1960 and 1965 fell to 3.4 per cent in the next five years. The import coefficient rose to 26.1 in 1965 and was still 25.6 in 1970.

The nature of land-holding both impeded agricultural efficiency and denied many rural dwellers the opportunity of sharing the benefits of economic growth. The small size and immature nature of the manufacturing sector, much of it located in and around Lima, was a further index of Peru's unbalanced development. Both were features which required urgent attention. Restricted opportunities in the countryside stimulated an increasing rural to urban migration to towns ill-prepared to receive the influx. Between 1960 and 1980 Peru's urban population rose from 46 per cent of the total population to 67 per cent. Action had been taken by successive governments in an attempt to address these problems but with little enthusiasm. The situation changed in 1968 when the army seized power and set about the implementation of a radical reform of Peruvian society and economy. It was a policy socialist in its approach, with a greater equity in society as its goal and nationalization of wide sections of the economy as its method. In retrospect it would seem to have been more concerned with the reorganization of ownership and control rather than with improving productivity and efficiency and its actions transferred more power to the state than to the people. The Military Government's initial policies were incorporated in the Peruvian Agrarian Reform Law of 1969. Here the new government faced up squarely to the seemingly intractable issue of land ownership and initiated radical changes. In the Costa region much of the land was farmed in large estates growing chiefly sugar and cotton with the associated processing industries located on them; most were foreign owned.

These estates and factories were nationalized and the process began of converting them into co-operatives. Second, the Military Government addressed the problem of the *minifundios*, the tiny holdings which accounted in 1960 for 88 per cent of all farms but only 7.4 per cent of farmed land. To make more viable units the Government sought to amalgamate groups of small farmers into collectives organized and operated by their members in their common interests.

The hold of the *latifundio* owners was to be weakened by placing an upper limit on the size of holding which an individual could own. The precise limits varied according to the region of the country and to whether the land was irrigated and, indeed, according to the level of wages paid to farm workers. *Latifundios* had covered 82.4 per cent of agricultural land while accounting for only 1.1 per cent of farm units. This process of land distribution was carefully and methodically implemented throughout the country. In practice most of the land was reallocated to various types of co-operative or collective holdings and very little to individuals. The socialization of agriculture appeared to be the intent. Furtado, writing in 1973, believed that this drastic agrarian reform did not appear to have resulted in a fall in agricultural production as is so often the case (Furtado 1976). Later evidence, however, as will be seen, shows Peru to have been no exception and agricultural output fell considerably. Agrarian reform was essential. The form it took appears not to have achieved either the equity or improvement it presumably intended. The military government of Peru in the 1970s could have learned much from either China or Taiwan.

The Government also took decisive action in the fishing industry and in 1970 created state marketing authorities for all fish products, three years later setting up PESCAPERU, a government organization which took complete charge of all fishing and the fish processing industries. Public utilities, transport, communications and energy production likewise were taken into the control of the State. In manufacturing, state control was to some extent extended and private enterprise hedged around with restrictions, but more significantly the developments in industry were characterized by an increasing move towards capital-intensive activities and the state borrowed extensively to fund developments. This phase of import-substitution development suffered from all the problems which Griffin, Enos and others have identified and which were discussed in Chapter 7. The measures enacted by the military government failed to improve the quality of life of the majority of Peru's population, did not put the country's economy on a secure footing and in 1977–78 brought the country near to economic collapse. In 1980 the military government gave way to elected civilian rule and a new democratic government under Fernando Belaúnde came to power.

PERU IN 1980: THE REPORT OF THE ECONOMIC MISSION

In 1980 during the period of transition from military to civilian rule, an economic mission led by Ulrich Thumm visited Peru and made a report on its economy for the World Bank (Thumm *et al.* 1981). It is a report which reveals the many weaknesses in the Peruvian economy but puts forward recommendations for development action.

In the case of agriculture, the mission points out that while the military government had focused upon tenurial reform and other institutional changes, it failed to raise agricultural production. It had devoted most of the trained personnel of the Agricultural Ministry to the agrarian reform and had totally neglected its advisory services. These were of vital importance to the small farmers who suffered in consequence. In the establishment of the co-operatives the same problems referred to in Chapter 6 were identified by the mission. They mention that the newly erected co-operative enterprises required to be operated in a co-operative spirit. This was something new to the farmers of Peru and it was not cultivated as an essential component of radical reform as it had been in China. The same lack of adequately trained staff which bedevilled Zambia's ventures with state farms impeded progress in Peru. On the sugar estates disputes between the new co-operative members and the management and technical staff led to resignations of the latter, leaving the estates without the leadership and technical skills they required. The state could not engender the requisite capital or provide the necessary skills to operate the agricultural economy as a branch of government. The marketing system was, according to the mission:

heavily burdened by a large number of intermediaries and excessive government regulation as well as direct government involvement. During the past decade, the Government relied heavily upon controls (price controls, subsidies on food imports, prohibition of farm storage, control of inter-regional movements of agricultural products) to achieve its goal of reasonable prices for both producers and consumers. Virtually nothing was done, however, to make marketing through private channels more efficient and less costly. In practice, all government policies had a strong bias towards the urban consumer and penalized the producer through inadequate prices, thus contributing to the stagnation of agricultural production

(Thumm *et al.* 1981).

The mission is equally critical of the Military Government's industrial policy, stating:

Weak performance and present structure of industry are the results of the strategy begun in the early 1960s and vigorously pursued during 1969–75. Abundant credit at subsidised interest rates, cheap imports of capital and intermediate goods through an over-valued exchange rate and low tariffs, high protection for finished products, subsidised power rates and fuel prices, and relatively high labour costs created a highly capital-intensive, inefficient and inward-looking modern industrial sector. Although the traditional industries (food, beverages, textiles, garments) continued to be important, industrialization during the 1970s moved strongly towards the processing of natural resources and was heavily concentrated in chemicals, basic metals, metal products, and non-metal minerals . . . industries that are all relatively capital-intensive and, therefore, contributed to an increase in the sectors, overall capital intensity . . . by at least 50 per cent in real terms during 1971–78. Increasing capital intensity has not been a matter of 'appropriate technology', but rather a matter of shifting the industrial structure towards industries that are not best adapted to the country's resource endowment which, besides minerals and some agricultural raw materials, fundamentally consists of abundant labour

(Thumm *et al.* 1981).

The mission was similarly critical of the inadequacies of investment in the physical infrastructure, particularly transport, in the development of energy resources, notably hydro-electric power, in public utilities such as water supplies and sewerage facilities and comments on the low levels of efficiency manifest over a whole range of activities. The mission's analysis presents a dismal picture resulting from years of mismanagement. It lists the problems which it regards as acute such as the persistent high inflation (60 per cent in 1980), the high public sector deficits, and the high levels of unemployment and under-employment which reach 50 per cent in the urban labour force. It refers to stagnating agricultural production 'as a result of ill-conceived agricultural policies during the past decade and exacerbated by drought' and speaks of the further deterioration in income distributions, of increased poverty and malnutrition, especially in the towns. It points out that Peru's primary goods export pattern makes the balance of payments situation a precarious one. In speaking of deeper-seated and longer-term problems the mission's report points to widespread poverty, to the problems created by the rapid increase in population placing as it does burdens on the social and economic infrastructure, and to the increasing magnitude of unemployment. These features all contribute to the high incidence of disease and infant mortality rates, 130 per thousand, which characterize Peru to this day.

The full integration of the country in a regional or spatial sense is still far from complete as the transport network is quite inadequate and, as a result, its resources are neither fully nor rationally utilized. The mission's recommendations were designed to ameliorate these problems while recognizing that tackling short-term problems must not be regarded as solving the fundamental basic problems. Central to them is the need to increase production. This is basic to the provision of jobs and so must be achieved by methods which make full use of the resource of labour. It is also the means whereby measures to improve health and nutrition can be implemented. In both agricultural and industrial sectors the mission recommended a decontrol of activities and a liberalization of trade and marketing, and the decentralizing of the management of private enterprises away from the state. It is a proposal spelt out in some detail but in general terms it favours private enterprise and the creation of incentives which will allow all to participate in the developments. It is the pursuit of a middle-road between the capitalist exploitive and *laissez-faire* approach of decades of Peruvian governments and the extreme, negative and enforcing regime of the military government of the 1970s which, in purporting to put people first, froze them into a mould of regulations set firm by inefficiency.

DEBTS AND DETERIORATION

Peru's elected governments have proved to be governments of false hopes. The new government of President Belaúnde came to power when improvements in the nation's economy had emerged but Peru, along with other Latin American nations, was becoming increasingly indebted. The International Monitary Fund, IMF, together with foreign investors had urged upon Peru economic stabilization policies. The balance of payments had markedly improved, though as Crabtree points out, this was as much the result of the coming into production of the new Cuajone copper mine and the completion of the oil pipeline in northern Peru as of any economic policies (Crabtree 1992). In this situation Belaúnde embarked upon a policy of economic liberalization. Lowder points out that Belaúnde's policy was one of urban-based import-substitution industries established to expand urban employment and act as multipliers. Agricultural investment tended to be in capital-intensive infrastructural and large-scale commercial projects. The peasants were ignored (Lowder 1982). The IMF pressed Peru for a reduction in government expenditure. Belaúnde's policies resulted in high inflation, minimal growth and a fall in living standards. Peru's international debts increased. In the election of 1985 Alan Garcia became President at the head of the Aliaza Popular Revolucionaria Americana party. IMF demanded adoption of its restructuring policy in return for support in renegotiating Peru's debt repayments. Garcia refused and chose to restrict debt repayment; his policies failed. Whereas there had been modest increases in all sectors between 1965 and 1980, the output of all, save agriculture, fell between 1980 and 1990. The growth in energy production, essentially oil, began to slow down and then declined. Exports remained almost static and terms of trade deteriorated by 22 per cent. Glewwe and Hall, of the World Bank, are critical of Peru's abandonment of orthodox restructuring policies (Glewwe and Hall 1992). The World Bank had recommended the removal of government interventions which had produced price and market distortions. Peru had introduced price controls, job creation programmes, investments in health and education together with subsidies to help the poor. All were understandable, socially desirable interventions but they were not adequately funded by revenue receipts. The Government diverted funds from debt repayments to fund both these developments and consumption. A very large fiscal debt emerged, prices increased and inflation rose to dramatic levels (2776 per cent in 1989). Garcia's policies resulted in a fall in per capita production of 25 per cent and a reduction in the real minimum wage of 60 per cent between 1987 and 1990.

The nation was close to bankruptcy (Glewwe and Hall 1992). The fall in living standards among the already poor was severe in its impact.

Selowsky's critique and remedy for both Brazil and Peru is that both still require to undergo the most basic readjustment of their economies. A major reduction in their fiscal deficits must be achieved by cutting expenditure rather than funding the deficit by borrowing. Public expenditure cuts are difficult to implement in a poor nation and Selowsky advocates that this be achieved by a more judicial allocation of assistance involving the focusing of help upon the needy and withdrawing it from the better off (Selowsky 1990). A greater efficiency in the operation of public services, by privatization of some functions and the introduction of competition together with a more efficient raising of revenue, is urged by Selowsky as good and essential housekeeping. These actions he recommends be then accompanied by the positive promotion of productive enterprises encouraged by carefully conceived incentives within a liberated economy.

World recession has adversely affected Peru's development prospects but it cannot explain the nation's inability to develop a resilient economy. Peru had not been able to invest borrowed money effectively in the diversification of her productive base. The development of a capacity to absorb investment is crucial in attracting foreign capital. In the closing years of the twentieth century no less than 84 per cent of Peru's merchandise exports are made up of primary products. This is one measure of both her lack of economic diversity and vulnerability. Another measure of her economic and social failure is the deep poverty in which so large a proportion of her population live. The peasants of the Sierra have been virtually excluded from any improvements and in their thousands have flocked to the shanty towns around Peru's cities to seek an existence in the informal economy. In such a situation of deep poverty and few hopes it is small wonder that coca growing and the trade in cocaine flourishes in the Peruvian countryside and may well constitute Peru's major export. President Fujimori, elected in July 1990, faces a daunting task. Decades of mismanagement have blighted Peru's development prospects. In Furtado's words '. . . to a greater extent in Latin America than in any other important area, obstacles to development are mainly of an institutional nature' (Furtado 1976).

19 East Asian NICs

'For other Third world countries, the developing market economies of East Asia may be seen as providing a powerful 'demonstration effect' through the outward orientation of their economic policies, their reliance on market signals, the stability and predictability of their macroeconomic policies, and their avoidance of confrontation between government and business'

Parvez Hasan
Adjustment to external shocks 1984

If the countries of the world form a continuum of development in the way in which Berry's analysis has suggested and if their relative positions are not fixed, some countries may at present occupy that indeterminate borderland between less developed and more developed while others may have achieved development goals and status indistinguishable from the industrialized nations of Europe. Hollis Chenery believes industrialization to be the engine which powers economic development and writes, 'successful development in virtually all countries has been characterised by an increase in the share of manufacturing in total output. This structural change is both a cause and effect of rising incomes' (Chenery *et al.* 1979). Four countries, all in East Asia, are recognized as having achieved this transition based on industrialization. Called the Newly Industrialized Countries, NICs, they are the two city states of Hong Kong and Singapore and the two former Japanese colonies, South Korea and Taiwan. Not only are these four countries industrialized, they are also among the world's leading exporters of manufactured goods. It is of particular interest to examine why these countries should have been so successful when so many others have not. Are there common characteristics which might afford some explanation of their success or is each distinct, following its own route to industrial affluence? The development history of each will be examined beginning first with Taiwan, then the city states of Hong Kong and Singapore, and concluding with South Korea, the least affluent of the quartet and a World Bank upper middle-income nation.

TAIWAN

In the light of the evidence presented in the case of China it might be thought that Maoist-socialism is the only way to secure a sound and equitable development. However, one has only to look across the straits which separate the People's Republic from the island called Taiwan to see another. There in a free-enterprise, albeit authoritarian, society, equally impressive achievements in the equitable development of a formerly rural and colonial society are to be seen. There is an alternative way. As the People's Republic has much to teach other countries in the Developing World, so too has Taiwan. Her development path provides a marked contrast with those of other former colonial territories which have been discussed such as Sri Lanka and Zambia. The work of a number of economists, notably J. C. H. Fei, S. P. S. Ho, A. Y. C. Koo, S. W. Y. Kuo, S. Kuznets, G. Ranis and E. Thorbecke, has made it possible not only to trace and evaluate Taiwan's approach to development but also to relate it to the basic issues of development found throughout the Developing World.

Taiwan is an island of some 35,988 square kilometres. In the mountains which cover half the island, altitude and steepness of slope inhibit agricultural development though the forests which clothe them are inhabited by aboriginal tribes who practise shifting cultivation. The environment in the lowlands favours agriculture. A wide range of minerals occur but few are of any significance and there is no iron. Small amounts of oil and natural gas have been raised but coal is of greater importance though even this resource is modest; the recoverable reserves are put as low as 210 million tonnes (Rhynsburger 1956; Chen 1963). Clearly richness in natural endowment is not the explanation of Taiwan's success in the development stakes.

FROM ABORIGINAL ISLAND TO COLONY

The island now known as Taiwan, lying off the coast of China but originally neither part of it nor inhabited by Chinese, was referred to by the Chinese as Pakkau-tao or Hsiao Liu-ch'iu. When Europeans first saw the island in the sixteenth century they called it Ilha Formosa, the beautiful island. According to Kerr, it was at this time a 'formidable wilderness' inhabited mainly by aborigines who did not speak Chinese, did not regard their island as part of a China they had never seen and did not trade with China (Kerr 1974). There had been, however, a long history of sporadic Chinese settlements and when the Dutch set up a trading base on Pakkau-tao in 1624 there was already a small Chinese settlement. The first Dutch fort was built on an islet called T'aiwan, a name the Dutch subsequently applied to the whole island, and from it they began to conquer and organize the tribes of the western lowlands and to engage in farming. During the period of the Dutch settlement substantial numbers of Chinese crossed from the mainland to settle in Taiwan and by 1650 numbered at least 100,000 (Ho 1978). Dutch rule came to an end in 1662, when they were defeated by a Chinese army led by the Ming general Koxsinga. His army seized all Dutch lands and, together with the thousands of Chinese civilians which accompanied it, settled and farmed the accessible lowlands. In 1683, when the ancient Ming Dynasty of China was replaced by the Manchu Ch'ing Dynasty, Taiwan became a prefecture of Fukien Province and so, for the first time, was formally incorporated into the administration of Imperial China (Ho 1978). Further substantial migrations of Chinese enlarged Taiwan's population, which reached two million by 1827. In 1887 Taiwan was raised to the status of a Chinese province administered by its own governor. Within a few years, however, following her defeat in the first Sino-Japanese war, China had ceded Taiwan to the Japanese and the island, reverting to its earlier non-Chinese name of Formosa, became a colony of Japan and was developed as such from 1895 until 1945.

As Ho points out, in the two centuries of its government of Taiwan, the Ch'ing Dynasty of China had done little to enhance the economic development of the island (Ho 1978). The Japanese attitude was in marked contrast; Taiwan, as a colony, was to be actively developed to supply Japan with agricultural produce, notably rice and sugar. The subsistence agricultural community was transformed into a society of commercial farmers and as such became a market for Japanese manufacturers. It was a classic example of the colonial model. Japan saw her own interests best served by taking control of the Taiwanese economy and making its agricultural sector as efficient and productive as possible. In order to secure her claim to the new colony, she deliberately kept Taiwan and the Taiwanese isolated from China. In consequence during the fifty years of colonial rule, the Taiwanese people, though remaining Chinese in culture, came to have a distinctive identity of their own. On the other hand their economy, as part of a colonial system, was changed from one discrete and internally-focused into one linked to an external source of capital and initiative and to an external market. This transformation served as a great stimulus to agricultural output as capital, new technology, and new economic systems were applied to Taiwan's resources of land and labour.

The Japanese participated directly and actively in the economic development of Taiwan. As Ho reports, 'Because of both its economic size and autocratic nature, the colonial government was the paramount force in the economy' (Ho 1978). Unlike many other colonial governments who confined their investments to infrastructural developments and their administrative functions, over 40 per cent of expenditure by the Japanese colonial government was commonly devoted to productive investment in agriculture. It was for Japan a cheap way of securing a food supply for her own large population without having to buy it with industrial exports. Taiwan became, in effect, an extension of Japan's agricultural area. The value of food imports from Taiwan greatly exceeded that of industrial exports to the island. Infrastructural investments were designed to support agricultural production. A road network was established in the agricultural lowlands and a railway was built to connect the northern port of Keelung to the southern port of Kaosiung. The Government was directly responsible for a number of major irrigation schemes and in addition financed many others promoted by the Irrigation Association, itself a body whose initiatives required its approval. Between 1900 and 1940 the area under irrigation was doubled and by 1942 accounted for 64 per cent of all cultivated land. Research stations and advisory services were established to modernize agriculture further while the quality of the work-force was improved by investment in education and health facilities. The package of investment, research endeavour and advisory services had the effect of greatly increasing agricultural production while at the same time providing additional employment on the land for the rapidly growing population (Ho 1978; Thorbecke 1979). The dissemination of new ideas, techniques, seed varieties and other innovations brought in and demonstrated by Japanese agronomists, was greatly facilitated by the formation of the *nokai*, the Farmers' Associations which had their beginnings in 1900 near Taipei. The original initiative came from a group of wealthy farmers and landowners but the Japanese seized upon it not only as a vehicle for the dissemination of agricultural innovations but also as a means of controlling agricultural development. Farmers' Associations were, therefore, set up throughout Taiwan but with Japanese officials and scientists in control. Membership was compulsory. In effect the Farmers' Associations became a means of managing the agricultural economy in a disaggregated way but with a central overall control. In this respect, despite many other differences, it was not dissimilar to the commune system of Mao Tse Tung. Unlike the commune model, that of the *nokai* was not one of self-help. Experts, both scientists and administrators, were brought in from Japan and speeded up the transformation of agriculture from a largely subsistence peasant economy into a reorganized, commercial system capable of continually generating new developments. As the Green Revolution in India many years later was to show, one innovation or technical development requires the support of other changes if it is to be effective. The Japanese package ensured the changes taking place in concert. Neither the distribution of land ownership nor the tenurial system was altered in any radical way.

The development of a modern manufacturing sector in the colonial period was much slower. Most industrial units were small, scattered throughout the countryside and concerned with processing agricultural products or in servicing agriculture with fertilizer and tools. In 1930 food processing accounted for three-quarters by value of manufacturing production (Ho 1978). By far the most important industry was sugar refining, followed by the chemical industries, chiefly producing fertilizer, and the non-metallic manufacturing industries. Factories averaged only ten workers each and six large firms, five of them sugar refiners, accounted for 80 per cent of capital investment in industry. Many industrial products were imported and, though in her preparations for war Japan sought to reduce her industrial exports to Taiwan by diversifying industrial activity on the island, by 1941 Taiwan's industrial sector was still small and overwhelmingly concerned with agriculture. Japan throughout her colonial rule exerted an almost complete control over manufacturing. Over 90 per cent of the invested capital was Japanese and many Japanese immigrants, who settled in Taiwan between 1920 and 1940, took up employment in manufacturing as well as in mining and transport operations. In addition to this Japanese-dominated sector of corporate industries, many long-established, small-scale craft industries owned and operated by Taiwanese still persisted. In all respects they differed from those in the modern sector (Ho 1978).

THE POST-WAR TRANSITION

When the Second World War came to an end the United States ensured that Taiwan was returned to China once again to become one of her provinces. This move was complicated by the continuation of the civil war in China. When the Communists emerged victorious the Nationalist Government fled to Taiwan and established an alternative government with the intention of using the island as a base for a reconquest of the mainland. They were joined by many thousands of refugees. By 1950 a million mainland Chinese had settled in Taiwan and its population had increased from 5.87 million in 1940 to 7.54 million in 1950. The people of Taiwan who had regarded themselves as Taiwanese rather than Chinese were now ruled by yet another government which came from outside rather than within. The colonial power had gone and with it the whole *raison d'être* of the colonial economy. Taiwan had to face an economic future destined to be entirely different from the preceding half century.

The new Taiwan, which called itself the Republic of China, was largely agricultural, had a rapidly growing population augmented by a million refugees from the mainland and was cut off from both China and Japan. It had suffered extensively from war damage and possessed an economy unable, at first, to absorb the influx of people but it acquired assets in two ways. First, the immigrants from the mainland included many capitalists, entrepreneurs and professionals possessing both skills and resources. Even some industrial plant was transferred to the island. As in Hong Kong which similarly received refugees, the injection of this human resource proved to be important. In Taiwan's case it was especially significant since the industrial sector had been almost entirely in the hands of the now vanquished Japanese. Secondly, for political reasons, the Nationalist Government of Taiwan began to receive development aid from the United States. These two elements were catalytic in Taiwan's development process rather than its sole cause. In other respects Taiwan's position was like that of so many newly independent and former colonial territories. She was faced with an entirely new situation and an uncertain future, with her former markets gone, a larger population than she could effectively employ and few significant natural resources. Her agriculture was, however, better organized, more commercial and more scientifically based than in many developing countries of the day and it was upon this resource that the restructuring of the Taiwanese economy began. In this enterprise the Joint Commission on Rural Reconstruction (JCRR), an American–Taiwanese organization, played an important role (Ho 1978; Thorbecke 1979). Prior to 1945 almost all the economically viable land in Taiwan had been brought under cultivation. The greatly increased population therefore increased pressure upon a limited resource and future development could only be ensured by increasing agricultural output per capita and per unit area and by developing alternative employment. The new government, in part no doubt because it was in one sense alien and with no political ties to local invested-interests, introduced early and radical reforms in land holding. First, agricultural rents were reduced by law. Second, state-owned or public land was sold, not to speculative landowners, but to practising farmers, and third, privately owned land was redistributed by compulsory sale to farmers. These successive actions, completed by 1953, represented a significant redistribution of land holdings, some 25 per cent of all cultivated land being affected. A land-owning small-farmer class was securely established. By 1953 owner-occupiers accounted for 55 per cent of all farm holdings in Taiwan and the proportion continued to increase so that by 1970 nearly 80 per cent of farmers owned the land they farmed (Ho 1978; Tang and Hsieh 1961; Lin 1981; Wei 1981). These early government actions did much to ensure a wide measure of equity in the agricultural sector in terms of both income and ownership of the major factor of production. Upon this secure base increased agricultural productivity resulting from a greater intensity of use of land and labour took place. It took place in what can be described as a labour-surplus economy, a type commented upon in the context of both India and China, in which it is essential for developments to occur which enable the more effective use of labour either by increasing the marginal returns upon it in agriculture or by providing alternative employment or by doing both (Fei and Ranis 1975; Ho 1978; Fei *et al.* 1979). Taiwan both reorganized her agriculture and developed a manufacturing sector initially supported by agriculture.

Two organizational structures assisted in this post-war development of Taiwan's agriculture. One was the American sponsored JCRR which administered American aid. Not only did it handle monetary aid, it also provided much scientific and administrative help, was responsible for many innovations and identified new overseas markets for Taiwanese crops. It replaced the former Japanese role with one which was even more supportive and free, of course, of the colonial framework. The other organization was created by the Taiwanese Government out of the former Farmers' Associations. They were converted into true farmers' institutions representing and serving the common interests of their members. They gave co-operative strength to the multitude of small-holders and further strengthened the agricultural sector and the rural economy. The results were impressive. Agricultural production rose by 42.8 per cent and output of crop production per worker by 96.1 per cent between 1950 and 1960. This alone was not enough to accommodate the problem of population pressure, nor was the redistribution of land in itself sufficient to promote growth and retain equity. The agricultural population rose by 33 per cent between 1952 and 1964 and pressure on the land increased accordingly (Fei *et al.* 1979).

People left the countryside for the cities but in the rural areas themselves three developments were significant. One related to changes in the crops grown as higher value crops began increasingly to replace longer established staples. Asparagus and mushroom-growing are examples of innovations which helped retain higher incomes on smaller holdings. A second related to the technologies used in farming. As Ranis has pointed out, technological innovations in agriculture can be put into two groups: the engineering which involve the greater use of machines and which are more appropriate on large-scale enterprises where labour surplus is not a problem; and the biological which relate to improved varieties of seed and livestock, improved techniques of husbandry, and more effective use of fertilizers (Ranis 1979). This second type is able to use more effectively increases in supplies of labour and is, therefore, more suited to labour-surplus conditions such as those in Taiwan. This use of appropriate technologies enabled Taiwan to absorb more people on the land and at the same time increase productivity per worker. Together with the activities of the Farmers' Association, the provision of rural credit and the work of the JCRR, this application of appropriate technology to the extensive and well-engineered irrigation systems established in colonial times, meant that agricultural income levels remained buoyant and helped to reduce the differential between farm and industrial wages. The third development was that of the location of industries in the countryside. In part this was an inheritance of the widespread Japanese agriculturally-related industries. They provided alternative full-time or part-time employment in rural areas and their expansion was necessary to absorb labour and to prevent the rural economy becoming a mere adjunct to the growing cities as in so many countries. The sum total of these developments was the maintenance of an equitable distribution of incomes and a diffusion of growth points in the economy in this first phase of independent development.

Developments in the industrial sector were likewise significant. There was no in-built industrial power group in Taiwanese society since manufacturing in terms of both capital investment and managerial skills had been in the hands of the Japanese. The influx of refugees from the mainland provided entrepreneurial and technical skills, some capital and industrial labour and, of course, it provided the government. This new government identified the industrial gap and set about building up this sector. It did so, not by separating industry from agriculture, but by working the two in harness. Agricultural exports provided revenue for industrial investment, and the processing of agricultural products provided a basis for industry. The strategy adopted was that of import-substitution, the policy followed by so many developing nations. Taiwan's industries developed rapidly as both consumer goods and intermediate products, previously imported, were produced at home. As in agriculture, in Taiwan's labour-surplus society this expansion was based upon labour intensive methods allowing a greater take-up of the unemployed and under-employed. Import-substitution, as has been discussed in Chapter 7, can be a self-containing policy, particularly in small countries if it persists too long. In Taiwan it was a phase that was used effectively. The traditional agricultural exports together with some new higher-value ones had earned exchange revenue, while import-substitution had reduced its subsequent loss. More savings were available for investment and much of this went into capital goods. A larger and experienced industrial labour force was created. By the end of the 1950s this phase had run its course. It was important not to become entrapped by it but to build upon it.

EXPORT-LED INDUSTRIALIZATION

In the early 1960s the Taiwanese Government moved the economy into a second phase of development. Import-substitution of consumer goods was to be replaced by the development of export-led manufacturing. Agricultural exports were not to be diminished but instead of being almost the sole export earners they were to be augmented and then surpassed in a relative sense by industrial export. The over-valued Taiwanese dollar which had limited export opportunities and made quota-restriction against imports necessary, was gradually devalued in the late 1950s and early 1960s. Incentives were given to foreign investors and industrialists to set up factories in Taiwan and a number of Export Processing Zones were established, the first being in the southern port of Kaohsiung in 1966. The protection of home consumer industries by quota restrictions on imports was gradually removed, the tax system was modified to encourage exports and a number of further measures were introduced by the Government to stimulate the growth of export-orientated industries. All these initiatives were taking place at the beginning of a decade in which the world economy was booming and the world market was reaching dimensions never before experienced. This second phase of export-led industrialization proved to be singularly successful. Industrial output rose by an annual average of 20.1 per cent between 1963 and 1973 and, using labour-intensive methods, Taiwanese exports of manufactures were able to secure a place in the international market. By the early 1970s Taiwan was experiencing trade surpluses. This phase of a dramatic industrialization of the economy made full use of the labour surplus in Taiwan and changed the balance of employment to one in which, by 1971, equal numbers were employed in agriculture (35.1 per cent) and industry (35.6 per cent) including mining, construction, energy and transport (Galenson 1979). So great was the capacity to absorb labour of these new economic developments that by 1970 the labour surplus had disappeared and labour shortages emerged. Unemployment levels which were 4.3 per cent of the labour force in 1964 were down to 1.7 per cent in 1970 and stood at 1.3 per cent in 1979 (Wu 1981). By the 1970s Taiwan's manufacturing sector had overtaken agriculture as the major contributor to exports and its needs accounted for an increasing proportion of imports.

THE ISSUES OF EQUITY

Taiwan has emerged as a newly industrialized country. Unlike the two city states she has a substantial and efficient agricultural sector and, unlike South Korea, used it to fuel her initial economic acceleration. What is remarkable about her development is the equitable distribution of incomes which she has been able to maintain throughout this process. It is an achievement which runs counter to the experience of almost every other country, where increasing disparities in incomes throughout the population have been so commonplace as to suggest to many scholars that this is an inevitable phase in development. This inevitability, it is argued, explains why incomes are more markedly unequally distributed in developing countries than in the mature-economies of the industrialized world. Many economists have held to the view that some sectors of the economy must inevitably develop more quickly than others, and with them the incomes of their participants, in order to maximize growth opportunities. It will be recalled that these issues are implicit in the discussions of balanced and unbalanced growth strategies. To attempt to achieve equity, it was felt, might dramatically retard the total growth of the economy. Some would argue that the People's Republic of China could have advanced its total industrial and agricultural productivity more quickly if less attention had been paid to equity. The tendency has been to stress increase in production as the first goal, with a fairer distribution of rewards taking place later when there was something substantial to redistribute. In this model inequality would increase as production went up, reach a maximum at some stage in the modernization process, level off and then decline as society matured and the nation became more affluent (Kuznets 1955; Paukert 1973; Aluwalia 1974). In Taiwan this has not happened and it is this feature which has attracted the attention of John Fei, Shirley Kuo and Gustav Ranis (Fei and Ranis 1975; Fei *et al.* 1979; Kuo *et al.* 1981a).

As in all case studies it could be argued that Taiwan's situation is unique and in some respects this must be so. On the other hand, her former colonial status, her agricultural based economy, her rapidly expanding population and low levels of income placed her very much in the Developing World. The peculiar historical circumstances which left her as an independent territory in 1949 also burdened her with a need to maintain a large army. Her mineral resources were poorer than those of many less successful developing nations. She did possess advantages deriving from the particular nature of the Japanese occupation and, as it transpired, from acquiring an alien government. Crucially important was the way in which the Taiwanese Government identified and used to good purpose the advantages, and recognized and overcame the disadvantages. Government policies did play a significant part in the achievement of equity and may, therefore, have a wider applicability outside Taiwan. Kuo, Ranis and Fei argue that equity in levels of income can be achieved by the kind of economic growth that is created. In Taiwan's creation of the appropriate kind of growth several significant factors can be identified. First, there was a well-established agricultural infrastructure. This consisted of the irrigation works set up by the Japanese and the road and rail networks they constructed. Second, the rent control and land re-allocation legislation of the Nationalist Taiwan Government distributed the resources of agriculture production widely and essentially equitably throughout the countryside. Thirdly, initial manufacturing was associated with agriculture and was again dispersed throughout the countryside. Subsequent manufacturing enterprises were likewise widely distributed throughout the villages, rural towns and regional centres as well as in the cities. Urban primacy, a characteristic legacy of colonization, was not present in Taiwan. The urban hierarchy is much more rank-size in its character and its distribution favours the spatial dissemination of economic activities (Pannell 1973).

In 1951 Taipei and its semi-urban prefectures accounted for 22.1 per cent of all industrial establishments; in 1971 the proportion was 26.1 per cent. For the same years the proportions in rural areas were 21.5 per cent and 23.9 per cent. The proportion of industrial establishments in the five largest cities remained essentially unchanged throughout the period of rapid industrialization and the proportion of persons employed in them actually declined from 43 to 37 per cent between 1956 and 1966 (Ranis 1979). This dispersion of non-farm activities enabled the rural areas to participate in the growth and prosperity of industrialization. In a labour surplus economy, industrial wages did not rise dramatically above agricultural wages, particularly since labour-intensive techniques were used in all factories both large and small. Likewise on the land, labour-intensive techniques were associated with the more intensive use of the land characterized by double-cropping. The returns to labour did not diminish and the growing of high-value crops sustained agricultural income levels. The net result of all these factors was a close association of manufacturing and agriculture in both a spatial and, at least initially, in an industrial-linkage sense. Importantly, distinctions between an urban-industrial sector and a rural-agricultural sector were far less clear and significant than in most countries. Ranis quotes average annual growth rates in real wages for agricultural labour as 2.5 per cent between 1952 and 1960 and 5.1 per cent from 1960 to 1969. For the same two periods the industrial wage rates rose by 5.6 and 5.9 per cent respectively (Ranis 1974). Although agricultural exports were used to fuel the economy and finance industrialization in its early stages, agriculture was not drained of resources for changes and development or neglected as a backward underpaid sector. In the Developing World this is exceptional. The re-allocation of land and the improvements to agricultural infrastructures were indications of government support and gave strength to the sector. Crucially the import-substitution phase was not allowed to persist overlong and lead to economic stagnation.

The export-led phase of industrialization was also characterized by features which promoted an equitable distribution of incomes. Taiwan's comparative advantage lay in her abundant cheap labour. Using this but also drawing upon American capital for the bigger industrial projects, the industrial upsurge of this phase took up even more surplus labour and allowed agriculture to stave off further the point of diminishing returns on labour and a fall in wage rates. Labour-intensive techniques in both agriculture and industry were, of course, associated with lower levels of investment in capital equipment. This linked the returns on labour much more closely to increases in production so that labour benefited directly from economic growth. Also manufacturing in Taiwan has been characterized by a large number of small- to medium-sized businesses rather than a few giant firms. This has meant that a greater number of entrepreneurial, supervisory and senior skilled posts have been available; the opportunities for advancement are more widely available and this in itself promotes a more equitable distribution of income. Furthermore, the indigenous ownership of firms has meant that industrial wage levels have not been distorted by international companies, and foreign enclaves with foreign wage levels do not exist. Taiwanese society has emerged as one characterized by economic and social mobility with more freedom and opportunity to move between income classes than in almost any other developing country. The wide availability of uniformly high-standard educational facilities has done much to promote this mobility and economic and cultural equity.

In the period since the 1960s Taiwan's labour surplus has disappeared and increasing equity has been the product of higher wage levels in a situation where differentials in wage rates between skilled and unskilled workers have been diminished due to labour shortages. This is a phenomenon experienced in mature industrialized nations. Prior to 1970, however, Taiwan's history of growth and equity is exceptional. The transition from a land-based agricultural society into a labour-based industrial society has been achieved in a free-enterprise context. Comparative wage rates have been manipulated neither by trade unions nor by the Government.

Having built up an industrial expertise in the manufacture of cheap consumer durables, clothing and synthetic fibres, Taiwan underwent a programme of backward integration whereby the industries which made the raw materials and components for Taiwan's factories were installed. Steel plants, petro-chemical and plastic work and other basic industries widened the industrial spectrum. Shipbuilding grew in a highly competitive market while motor vehicle assembly plants produced cars for the home market and export. As the world economic recession deepened throughout the 1980s, Taiwan's multitude of small highly adaptable firms continued to prosper. The GNP grew by 8.9 per cent in the 1970s and 8.2 per cent in the 1980s. The average annual growth rate of members of the OECD was 2.4 per cent between 1965 and 1990. Having joined the industrial world, Taiwan is having to face the ever-changing technological and economic conditions of the international community. Her labour, with unemployment near zero, is no longer as cheap as it was. Increasingly Taiwan, like the other NICs, is adjusting to the evolving situation as ever more sophisticated technologies come to dominate the lead sectors in manufacturing and other countries enter the phase through which Taiwan has recently passed. Her success in the purposeful planning of an integrated economy under the supervision of an authoritarian government gives much food for thought. As her society and economy begins to liberalize in the 1990s her accumulated wealth is seeking out investments in offshore opportunities. Taiwan is becoming an important agent for integration in the growing economics of the western Pacific rim.

HONG KONG

The colony of Hong Kong extends over a mere 1062 square kilometres. It comprises the small island of Hong Kong, taken by the British in 1841 and 75 square kilometres in area, the Kowloon peninsula on the Chinese mainland ceded to the British in 1860, and a much larger area of over 900 square kilometres made up of islands in the Hong Kong archipelago and a portion of the Chinese mainland which were leased to the British as from July 1898 for 99 years. This leased area is known as the New Territories. The distinctions are significant. Without renewal of the lease, the New Territories were due to be returned to China in 1997. The Treaty of Nanking which ceded Hong Kong island and the Convention of Peking of 1860 which ceded the Kowloon peninsula to Britain were treaties in which Britain insisted on diplomatic equality and ensured it by force. The Chinese regarded the treaties as 'unequal' and this view, maintained by all successive Chinese governments, has meant that they have seen little distinction between the territory leased in perpetuity and that leased for 99 years whereas the British Government has held a different view. It could be argued that this difference of opinion and the uncertainty it created has had a bearing upon economic development in this colonial territory. As the title of one book on Hong Kong (Hughes 1970) has it, *'Borrowed Place, Borrowed Time'*.

A sliver of China and a sprinkle of islands Hong Kong may be, but it has a very distinctive character of its own which is a product of its separate existence. Between Hong Kong island, Lantau island to the west and the mainland to the north, lies a deep and sheltered stretch of the sea which forms a large and secure natural harbour. Like other European powers Britain was seeking, by establishing a secure base, to further her trading opportunities along the Chinese coast. It was this harbour rather than the island itself, which Palmerston had described as 'a barren island with hardly a tree on it', that was a major asset. This barren island was to become the centre of the Colony of Hong Kong which some 140 years later contained a population of over 5 million and where land values in its central business district exceeded those of central London.

For over a century after the acquisition, Hong Kong functioned as a Far Eastern entrepôt port. As trade links widened, as colonialism brought more areas into international trade and as improvements in ships and shipping facilitated long distance movement, so the reach and complexity of Hong Kong's trading relationships grew. The original military settlement on Hong Kong island grew into the town of Victoria. Across the harbour, Kowloon was transformed from a small walled Chinese town into a growing commercial community. The population of the colony grew from 35,000 in 1851 to 841,000 in 1931. Much of this increase was due to immigration. With the collapse of the Manchu Dynasty many had fled from China into Hong Kong. As the Japanese, who invaded China in the 1930s, penetrated further into the country, refugees poured in their thousands into the colony. By 1941 when the Japanese invaded Hong Kong itself, the population had reached 1.6 million. Hong Kong had become what essentially she is today, a society composed of refugees and the descendants of refugees. During the Japanese occupation many fled Hong Kong seeking to avoid Japanese oppression in the parts of China less securely under their control. When the British returned in 1945 the population was some 600,000. The years immediately following witnessed a flood of refugees, the like of which Hong Kong had not experienced before. They came in from China both during the civil war between the Communist and Kuomintang armies and after the establishment of the People's Republic. In 1950 the population stood at 2.3 million. Until the early 1970s refugees from China continued to augment Hong Kong's natural increase in population. The population in 1993 was estimated at 6 million. This small territory with no material resources, an accented mountainous topography, a population which has increased by over 760 per cent in 36 years, a refugee society of town dwellers, where urban population densities are the highest in the world, has somehow transformed itself into a major industrial nation. Hong Kong has done this with but two assets, its people and its fine natural harbour in a strategic location. It is a development of interest and significance.

Hong Kong began as a trading centre; it was her prime function until the Second World War. She was geared to the handling of ships and their maintenance, to storage and transhipment of goods and to the financial transactions associated with trade. She was outward looking, since there was nothing to look in upon, and she was very much concerned with Chinese trade. Merchants of Chinese, European and American origin settled and traded there. Many of the great Hongs, or business houses, of Hong Kong today can trace their origins back into the nineteenth century. Hong Kong's position on the Chinese coast had been her beginning and *raison d'être*. The revolution which created the People's Republic in China did much to change this situation. The United Nations placed an embargo upon trade with the People's Republic, an embargo to which Britain was a party, and the long-established link was severed. Out of China had come more than a million refugees who needed accommodation and employment. Among them were Chinese entrepreneurs and shrewd businessmen from Shanghai and other industrial centres skilled in management, with knowledge of production techniques and experienced in commercial dealings with the industrial nations. Though Hong Kong's trade and entrepôt activities continued, they were restricted by the United Nations embargo and there was an urgent need to develop other aspects of the economy both to maintain it and to provide employment. Agricultural developments on any large scale, even though many of the refugees came from rural peasant backgrounds, were made impossible by Hong Kong's small size and the nature of her terrain. People were the resource, not land, and the harnessing of their abilities and productive capacity has been the colony's remarkable achievement and the basis of its transformation. It was a productive capacity which had to be used in a special way. The territory had no raw material or power resources of any significance. It had a work-force, a harbour with a potential for considerable development and a commercially strategic location, and it had a history of experience in international trade. The genius of Hong Kong was in bringing these attributes together in such a way as to mark out a meaningful path of development. The path was that of manufacturing for export, using imported raw materials and energy supplies. The ships, the port, the trading experience and connections were fully utilized to facilitate this inward and outward movement. The entrepôt trade had played a similar role in the movement of goods; now the significant difference was that Hong Kong was a manufacturing, value-added, intermediary so that which went out was quite different from that which came in. There was no question of import-substitution policies, none of indigenous growth, no issue of core and periphery imbalances except on a town-planning microscale. The policy was simple and clear, to manufacture goods to sell abroad. To generate this activity and sustain it, there was a need for capital and labour. Hong Kong's growing population proved to be a most efficient hard working and adaptable labour force and was seen as a positive asset which attracted capital. The capital came from the long-established Hongs, from Chinese capitalists who had left China, and importantly from Britain, other European countries and North America. It was an economy as vulnerable to the vicissitudes of world trade as any other yet it has remained a success. Success tends to breed success and Hong Kong's achievements have received an increasing flow of international capital. The keys to that success have been first, and importantly, Hong Kong's flexibility in enterprise and labour. With no subsidy or protection her industrialists have had to demonstrate an ability to identify changing conditions in both market and methods of production and adjust both what they made and how they made it to these changes. The industrial ear in Hong Kong has been finely tuned to detect such changes. Secondly, this industrial endeavour has not taken place in an isolated-outpost sense solely concerned with manufacturing. Rather it has developed within the commercial context in which it was born so that the one has nourished and stimulated the other and it has taken place alongside the continuing development of Hong Kong's port facilities and trading connections. Expertise in, and facilities for, financial transactions have so developed that Hong Kong has become the third largest banking and financial centre in the world. Her container terminal is also the world's third largest while her airport is one of the busiest in the Far East. The service sector accounts for 73 per cent of GDP. This ensemble of developments has both nurtured and been nurtured by Hong Kong's transformation into an industrial nation.

The approach to economic development could hardly provide a greater contrast with that of the equally successful People's Republic of China. On the one hand the world's giant is transforming its society while on the other the world's midget is transforming its economy. The one is experimenting with perhaps the most interesting and original planning programmes for economic development in the Developing World, the other not even contemplating economic planning. Hong Kong represents free enterprise capitalism in perhaps its most classic form. The port is a 'free port' with neither import nor export duties. No restriction is placed on the movement of the world's currencies. Manufacturing and commerce are wholly in the hands of private enterprise as are such public utilities as bus and ferry services and fuel and power supplies. The Colonial Government does play a major role in providing social services and exerts overall control over urban planning developments. It has built a large part of the accommodation needed to house the growing population though, despite its great achievements, the influx of people has meant that it has never been enough. There are, however, no state industrial or commercial activities, nor is there a state bank issuing currency. This is done by three private banks. The Colonial Government does not set out an economic development policy for the Colony. Its rôle is one of active non-participation.

Hong Kong's entrepreneurial resources applied to a labour force which is hardworking, adaptable, eager to acquire new skills and both psychologically and socially receptive of factory-machine techniques, has generated capital for further development and attracted it from overseas. But entrepreneurial skills and an excellent labour force cannot maximize their opportunities in a country the size of Hong Kong without an external dimension. Capital is one, markets are another. To sustain her level of activity Hong Kong requires access to world markets, an access sometimes impeded by tariffs, quotas and other devices. To enter these markets the sensitivity to changes in them, to which reference has been made, is vital; so too is competitiveness. Hong Kong's industrialists do not devote capital to the research and development of new techniques. Instead they take in the very latest in techniques and products from the outside world and use and produce them. Only by remaining at the forefront of development can they remain competitive. This is another aspect of Hong Kong's flexibility and clearly it can most effectively apply to the lighter high-technology industries rather than heavy basic manufactures. By utilizing the latest techniques in textile manufacturing, in engineering and electronics from Japan, the United States and Europe, she draws upon the development resources of the industrialized world. The factors involved thus become mutually reinforcing ones as indigenous enterprise, adaptable labour force, excellent infrastructure, international technology, international capital and international trade, interact. Hong Kong has striven to become, and is, a window where the world can look in, invest and buy. She sustains a large tourist traffic not only to derive revenue but also to facilitate contact and encourage international involvements. In the late 1970s, of foreign investments in Hong Kong, 46 per cent were American, 20 per cent Japanese and 7.5 per cent British. The remainder were a mix of Asian and European investors. These foreign investments were made in electronics (26 per cent of the total), textiles (15.8 per cent), chemical products (12 per cent), electrical goods (10 per cent), printing and publishing (7 per cent) and toys (3 per cent).

In an investigation carried out at the time by members of the Chinese University of Hong Kong, the most influential factors determining foreign investment in the colony were revealed (Lin *et al.* 1979). American investors ranked these factors in the following order: peaceful industrial relations; efficient labour force; the stability of the Hong Kong Government; freedom to repatriate earnings; freedom to transfer capital assets; low tax-ceiling; easy to serve export markets. The British order of factors was: low taxation; freedom of capital transfer; freedom to repatriate earnings; local market potential; peaceful industrial relations; efficient labour; easy to serve export markets; the existence of supporting industries. In Japan's case, the most important inducement by far was the low tax-ceiling. In all three cases lower wages were considered less significant than the factors listed. Though prevailing wages were much lower than those of Japan, the USA or Britain, those in Korea, Philippines, Taiwan and Singapore were even lower than in Hong Kong.

Production for export has meant two things. First, a vulnerability to market changes whether these were the result of recessions and taste changes, of free market forces or protective tariffs. The response to this has been the adaptability which has been mentioned. Secondly, it has meant that Hong Kong producers have continually to identify comparative advantage and minimize their opportunity costs. This in turn requires them to specialize in a narrow range of industrial activities and while this allows them to enjoy scale economies it does require a special alertness and an awareness of changing events.

The transformation of the Hong Kong economy has thus seen three phases. First the stage characterized by the entrepôt trade with associated activities such as shipbuilding and repairing, commercial developments and, as the years went by, the emergence of manufacturing concerned with textiles, engineering, electrical goods and sugar refining. The second phase emerged after 1945 with the China trade cut off and the influx of refugees encouraging a rapid restructuring of the manufacturing sector and massive increases in the production of textiles, clothing, plastics, toys, shoes, electrical goods and transistor radios. The emphasis was largely on cheap consumer goods produced by labour-intensive methods. During this period from the late 1940s to the late 1960s, Hong Kong became known throughout the world for these commodities. What is commonly not realized is that this phase has been replaced by another. Hong Kong's industrial growth has continued but its nature has changed precisely because of changing market conditions and because her comparative advantage with respect to labour costs has altered. In her main market areas, restrictions on the import of the goods Hong Kong was making have meant that she has switched to different levels of products within the same section of manufacturing. Lower labour costs among her competitors and import restriction of her markets have moved Hong Kong firms into the manufacture of higher-value and more sophisticated products. More sophisticated forms of electronic and mechanical engineering products are now characteristic. In the textile-clothing sector there is a much greater emphasis upon high-quality clothes rather than cheap cloth. The image of cheap low-quality plastic products is no longer accurate. Manufacturing accounts for some 40 per cent of the labour force and of this over a half are employed in the textile and clothing industries, electrical and electronic industries, plastic products, optical photographic and watch-making industries. These industries account for some three-quarters of Hong Kong's exports. In recent years a polystyrene plant, an outboard engine factory, a steel rolling mill and a machine tool plant have been examples of further additions to Hong Kong's industrial spectrum.

The physical infrastructure of Hong Kong has continued to be improved and augmented unceasingly at a rapid rate. There is a developing urban motorway system with a cross-harbour road tunnel. A mass transit system, which is an underground–overground railway, is capable of carrying a million passengers each day and places 40 per cent of homes and 50 per cent of work-places within ten minutes walk of its stations. The railway from Kowloon to the Chinese border has been double-tracked and a motorway to China, commenced in the early 1980s, has now been completed. All the paraphernalia of a large modern city exists in the heavily built-up areas which fringe the harbour. The growth in population has been so large, however, that the Hong Kong Government planned and built new towns out in the mainland New Territories, towns with populations of a million or so each. These infrastructural and economic developments raise, of course, the issue of Hong Kong's relationship with China.

The former insecure settlement on a barren island has become a large densely-populated city state. It has demonstrated a successful expanding economy which has weathered the post-war strains and sustained itself throughout the world economic recession. It is legally and politically a colony yet it is the most peculiar of colonies. It was never a populated area whose people were exploited in a colonialist fashion. Its population is made up of people who chose to come in. If conditions were not good, they were none the less preferred alternatives. The activities which it has practised have benefited local workers and manufacturers as well as foreign capitalists. The benefits have not been equitably distributed but the distribution of income is not as skewed as some would have it. The top 20 per cent of households accounted for 47 per cent of household incomes and the bottom 20 per cent for 5.4 per cent. The equivalent figures for France for example are, 40.8 per cent and 6.3 per cent. It is a territory whose economic development is totally free of any influence from the colonial power; even its currency is not related to the pound sterling. It is self-governing and being essentially a collection of contiguous cities this is carried out by elected city councils but its supreme government, the Colonial Government, is not elected. It has been described as experiencing freedom without democracy. In such circumstances, the relationship between the Colonial Government and the people of the colony is a complex and subtle one (Kuan 1979). It is one in which the role of China, the motherland of Hong Kong's population and the cradle of their culture, is of fundamental importance and it is one which turns upon the past and the future of Hong Kong. The British have regarded the Kowloon peninsula and Hong Kong island as their sovereign territory; in the development of Hong Kong, the leased New Territories and these sovereign areas have become inextricably linked. A severance of the two would be unthinkable in economic and social terms. The people of Hong Kong have increasingly expressed the natural wish for a greater say in their own affairs. On their part the British have not been averse to granting independence to their colonial possessions as the dismantling of the British Empire since 1945 has demonstrated. An independent Hong Kong would, however, have been unacceptable to the government of Chairman Mao. It would have meant another Taiwan on China's shores, for the Chinese, not accepting the 'unequal treaties' of the nineteenth century, have regarded the colony, in all its parts, as being sovereign Chinese territory. There was another dimension. Once the United Nations embargo on trade with China was lifted, trade with her resumed and cordial relationships prevailed between China and Britain. Far from being a capitalist sore on the socialist body, Hong Kong became China's window on the world and with her excellent harbour facilities, far superior to those of Canton on the shallow Pearl River, handled in the 1970s over 40 per cent of China's foreign trade. China invested in factories and department stores in Hong Kong, operated a dry dock and established a banking system. The harbour of Hong Kong received many Chinese ships and the improvements to transport connections with Canton were further examples of the growing communication between the two areas. As has been described, the establishment in post-Mao China of Industrial Development Zones across the border from Hong Kong which replicate many of the features of foreign capitalist investment in Hong Kong, further stressed inter-connection rather than separation. Prior to 1984 it was possible only to speculate on Hong Kong's future and with speculation came uncertainty as 1997 approached. The whole edifice of Hong Kong's commercial and industrial success was based, as it must be in a capitalist free-enterprise system, upon confidence. Once confidence in the future was gone, capital support would be withdrawn and the whole edifice would collapse leaving modern buildings and an elaborate infrastructure, inarticulate. This point was almost reached in early 1984. It was staved off by the agreement drawn up, after long negotiations, between the Chinese and British Governments over the future of Hong Kong. Its Chinese sovereignity has been recognized and a Chinese governmental presence will be established in Hong Kong. The administration of the territory was to be carried out by the Hong Kong people and though becoming a province of China it was to remain distinctive in form and character from the administrations in the People's Republic. Likewise the economy was to remain for fifty years from 1997 a free-enterprise capitalist economy. The future in reality is dependent upon developments in China. The disaster of the massacre on Tiananmen Square gave rise to anti-Chinese demonstrations in Hong Kong. China, seeing her future province as a possible focus of of opposition to Chinese political policies, embarked upon a series of meetings with British representatives as to the future governance of Hong Kong. Attempts were made to limit British freedom of action before 1997 and Britain was accused of breaking the agreements made in 1984. Fears have been expressed as to the impact this episode has had and is having upon Hong Kong's economy. China has, however, embarked upon an economy policy described as socialist capitalism which, as has been seen, is embracing foreign technology and investment. Hong Kong is already intimately connected with the People's Republic not only by the new transport links and the heavy Chinese investment in the colony but also by the involvement of Hong Kong's entrepreneurs and capital in the coastal Special Economic Zones, particularly those at Shuhai and Shenzen. At one time to cross the Lo Wu bridge from Hong Kong to China was to enter a different world: today the factories and high rise buildings of the zones are indistinguishable from those of Hong Kong.

Most of the entrepreneurs in Hong Kong's manufacturing sector are Hong Kong Chinese; other foreign investments tend to be in the servicing sector. The Hong Kong industrialists are at one with the Cantonese language and traditional Chinese culture. The economic fabric of these two areas is now so interwoven that Hong Kong's economic future is secure though its life styles may change. The Hong Kong–China relationship can be seen as another bond linking the economies of the western Pacific rim.

SINGAPORE

Singapore is a fascinating rejection of the long-held belief that the humid tropics, plagued by climate, ravaged by disease and inhabited by peoples stunted in intellectual and physical development by the environment, could never be part of the modern world. Singapore, near the equator, experiencing high rainfalls with continuously high temperatures, with a natural vegetation of tropical rain forest and mangrove swamps on her coasts, has been transformed into a large modern city which, to an even greater extent than Hong Kong, lays claim, in appearance at least, to be more a part of the modern world than Manchester, Marseilles or Milan.

Singapore's history is longer that that of Hong Kong. Singapore Island at the tip of the Malay peninsula has always held a strategic position in a zone where the great cultures of the Indian sub-continent and China have mingled in Malaya and the East Indies. In the late-eighteenth century the British identified its strategic importance to their East Indian and Chinese trade but it was the Dutch who came to dominate the East Indian trade and to control the Straits of Malacca and the Sunda Straits. Under the aegis of the Governor General of India, Sir Stamford Raffles explored the area for possible bases for British trade and ships. He landed on Singapore Island in January 1819 and, after treaties were signed with the Sultan who owned the island, Britain secured a trading base and began to develop her commercial interests. Subsequently these interests spread to the peninsula where Britain displaced the Dutch from Malacca, and Singapore became the seat of an administration which ruled Malacca, Penang and Singapore as the Straits Settlement. This administration was managed by the British East India Company until 1867 when the Colonial Office took over. After the Second World War the Straits Settlement was abolished; Singapore became a separate colony in 1946 with the rest of the Settlement joining the Malayan Union. In 1959 Singapore became an independent country. However, in 1961 it was proposed by Malaya that a new nation be created composed of Malaya, Sarawak, North Borneo, Brunei and Singapore. This new state, named Malaysia, was created in 1963, though Brunei chose not to join it. After some difficulties within Malaysia, Singapore herself became a separate country in 1965.

Independent Singapore is a small country made up of one island, 616 square kilometres in extent, and a few islets measured in hectares rather than square kilometres. Her population in 1991 was 2.8 million of whom 76 per cent were Chinese, 15 per cent Malay and 7 per cent Indian. The population density is over

4800 per square kilometre (compared with 5460 per square kilometre in Hong Kong).

Like Hong Kong, Singapore has no natural resources of consequence. Her population is well educated with schools at all levels, polytechnics and a university. It is a population which grew at an annual rate of 1.5 per cent between 1965 and 1980, a rate controlled by family planning services, fiscal measures to encourage small families and by immigration restrictions. More recently shortages in the labour market have led the Government to favour slightly larger families and the rate of increases reached an average of 2.2 per cent in the decade to 1990. Its strategic trading location and its well-trained and disciplined labour force again suggest comparisons with Hong Kong but there are differences. First, Singapore is independent and, second, in its foreseeable future it is not envisaged that it will be absorbed into some other nation. Third, both its economy and society are much more strictly controlled than those of colonial Hong Kong. Indeed it is this organized, detailed and ultra-efficient planning of so many aspects of life in Singapore which its critics have focused upon. Yet the achievements have been substantial and impressive (Ooi and Chang 1969; Grice and Drakakis Smith 1985; Dicken 1987).

From an economy still very much tied to its port activities at the time of its emergence as a separate nation in 1965, Singapore has become much more involved in manufacturing, though the port activities have not been neglected and facilities have been markedly expanded and modernized. Singapore today is one of the world's busiest ports with a substantial entrepôt trade in south-east Asia and a large modern container terminal. The British maintained a military base on Singapore Island which provided employment for 40,000 persons, a very large number in the Singapore context. When the base closed in 1968 these workers, including highly-trained and skilled personnel, were added to the already high number of unemployed and the economy of this new country appeared at risk. The decision was taken to embark upon a vigorous policy of industrialization; there was indeed little alternative. To be successful it required not only labour but investments, a high level of export activity and the establishment of a range of diverse export markets. The political party in power, the People's Action Party, saw its role in this economic endeavour as an involved, positive-planning role giving not only political stability but direction, with its plans to be carried out with a clinical efficiency. The contrast with the *laissez-faire* approach to economic developments in Hong Kong could hardly have been greater. Singapore is a democracy with its government elected in free elections, yet no party other than the People's Action Party has been returned to power and opposition parties are virtually non-existent. In practice, Singapore has become a one-party state. Led by Prime Minister Lee Kwan Yew from 1965 to 1990, the Government has become a vigorous, planning and enforcing administration reaching out into all aspects of life. It has proved to be efficient and successful and Singaporeans enjoy the benefits of this success, though critics of the regime would say that equity has not been to the forefront of the Government's goals.

In 1968 an industrial estate was set up at Jurong, which is now run, together with twenty other smaller estates covering 6300 hectares, by the Jurong Town Corporation (JTC) (Chiang 1969). Some 100,000 workers in 1099 factories are employed on the main Jurong estate, with a further 98,000 on the smaller estates. These estates have planned development schedules extending into the 1990s which take into account population projections and the overall distribution of population in the development of the industrial estates and their supporting industries. The Jurong Town Corporation designs and builds factory buildings as well as providing factory sites. It operates a port which handles raw materials for the estate and it has a marine base which serves as a central distribution point for south-east Asia for a range of oilfield equipment and services, including exploration. It operates repair yards and provides a spare parts service. JTC is an example of the integrated, planned approach to industrial development which consolidates foreign enterprises within the country's plans. The Singapore Government has signed investment guarantees with Canada, France, Germany, the Netherlands, Switzerland, the United States, the United Kingdom, Belgium, Luxemburg and Sri Lanka. This foreign investment is largely characterized by multinational companies which account for over half of Singapore's industrial employment. The external dimension is very much present in Singapore's economy, as it is in Hong Kong.

To retain an ability to react quickly to changing technological demands and market circumstances, a Skills Development Fund has been established to promote training and retraining in skills to facilitate the restructuring of the economy. An Economic Development Board, a government agency, is active in attracting high-technology, skill-intensive industries to the country, of which microwave links, pneumatic controls, precision optical instruments and computers are examples.

Between 1960 and 1970 the average annual growth in manufacturing was 13 per cent. Between 1970 and 1981, despite the world recession, it was 9.7 per cent, and in 1981–82 reached 12 per cent and employment went up by 7 per cent. Labour has been brought in from other countries, such as Sri Lanka for example, to work in the industrial sector. Manufacturing contributed 29 per cent to the GDP in 1991 compared with 12 per cent in 1960 which had included work in the military base. Not all industries have made progress but, as in Hong Kong, the electronics and electrical sectors have made particular advances. Singapore now has the world's third largest oil refinery complex and the functions of this investment are clearly related to the interaction between oil prices and the level of world industrial activity. Tourism, with the attractions of modern hotels and duty free goods as in Hong Kong, is buoyant and brings the world to see Singapore's achievements, for despite the recession of the 1970s and 1980s Singapore demonstrated health, strength and flexibility.

The great city supported by this economy is modern, well-planned and growing. Great attention is paid to preserving open space and greenery and to creating amenities in what is a very densely populated area. Over 70 per cent of the population live in government-built housing, mostly high-rise flats. They represent a great improvement in material living conditions though there are those who say that Singapore's urban renewal is designed to remove the core of the radical proleteriat. There are also those who point out, quite accurately, that the years which saw the transformation of Singapore's economy were the boom years of the 1960s and early 1970s when world production, trade and demand were at a level never exceeded before and when the contribution of the new Singapore could only be welcomed. There are those who say that in a world recession Singapore's position is revealed as weak and vulnerable. She has, of course, to compete in a diminishing market and her small size, like that of Hong Kong, means she has to depend upon external markets whether they be for tourism or manufactured products. Her port functions and her financial services all depend upon the level of trade. All this is true yet the evidence shows how well Singapore, like Hong Kong, has survived these adverse circumstances and in real terms both GDP and per capita incomes have grown significantly. With a GDP still growing at over 6 per cent per annum and manufacturing output at a similar rate, Singapore's economic achievements compare favourably with most nations of the world. As in the other NICs a flexible response to market changes has allowed her to remain competitive (Geiger 1976; Goh 1977; Wang and Tan 1981).

An evaluation of the society which has been created as part of Singapore's development process depends upon priorities, values and, to some extent, political stance. It is very much a state-run society. The main physical and social infrastructure is in government hands. Contributions to the Central Provident Fund at the rate of 45 per cent of earned income, are taken from all individuals and used to support national economic development projects. The rate of saving for investment, so central to early development theories, is determined by the Government. The Government's record is one of sound investment and as yet it has not had to borrow from overseas to fund its own industrial development projects, which involve shipyards, shipping, airlines and smaller companies. The foreign companies operating in Singapore tend to be wholly foreign-owned and capitalized, and joint schemes and licensing agreements are few. Despite its achievements the interventionist government of Singapore, the economic basis of the island's progress, has been called into question (Young 1992). In an interesting comparison of Singapore and Hong Kong the work of Alwyn Young reveals that in Singapore's case growth has been based on very large investments in the economy; the Central Funds resources are ploughed back into the state-owned companies; foreign investments are subsidized by the Government. The rate of investment as a proportion of GDP exceeds 40 per cent compared with less than half that in Hong Kong whose output per worker has gone up slightly more than in Singapore. Hong Kong has been using capital, labour and technology more efficiently than Singapore whose ICOR is twice that of Hong Kong. There is a limit to the amount of capital Singapore can extract from its population; if Young's analysis is correct this will put a ceiling on Singapore's growth through investments. Lee Kuan Yew has already indicated that he feels Singaporean graduates are insufficiently entrepreneurial and need to explore the outside world for further business.

The model republic created by Lee Kuan Yew remains as yet in good heart. With Hong Kong, Taiwan and now South Korea (to be examined next), it has moved along the development conveyor belt, if indeed such a mechanism exists, into the league of industrial nations. It has achieved the development goals which it set itself.

SOUTH KOREA

South Korea with the lowest per capita GDP of the four East Asian NICs, less than half that of Hong Kong and Singapore, is still in the transition stage from developing country to fully industrialized nation. For over a thousand years Korea, both North and South, was an independent and civilized nation until in 1910 she was subjugated by the Japanese and like Taiwan became a colony. A similar pattern of colonial development was initiated and resulted in a more efficient agriculture able to supply Japan with food. To a larger extent than in Taiwan, the Japanese installed industrial plant in Korea, mainly in the north where coal and iron deposits occur. Ownership and management remained in Japanese hands but Koreans gained experience in factory work both in Korea and in Japan where some two million were working during the Second World War. They experienced a society different from that of a still medieval Korea but were denied the right to determine the destiny of their own country. After the defeat of Japan in 1945, the United States and the Soviet Union divided Korea into ideological spheres of influence along the 38th parallel of latitude. The division took on a permanency and fostered a mutual antagonism. In 1950 North Korea invaded the South and the resulting three-year war devastated the economy and infrastructure of the whole country and imparted a politico-economic shock to South Korea which has influenced her subsequent development. For purely ideological reasons the United States, once the war was over, participated actively in establishing an economy strong enough to ensure the South a degree of security and independence. American technological and financial aid played a major part in laying the foundations for post-war growth.

The government system which emerged was one so authoritarian as to be tantamount to dictatorship. A strict control of the economy was established with a focus upon import-substitution protected by quotas, tariffs and exchange controls. Manufacturing came to be dominated at first by light industries and small firms set up and run by Korean entrepreneurs. In the countryside Japanese-owned land was redistributed among Korean peasants on American insistence, but Korean land ownership remained unreformed and large estates worked by a poor peasantry remained the norm in a stagnant agricultural scene. This situation, so contrasted with Taiwan's experience, led to a major movement of workers from the countryside to the towns where, fortunately, industries using labour-intensive methods were able to absorb the influx. The switch from an agrarian to an industrial state was beginning but agriculture was not used as the springboard.

By the late 1950s the import-substitution policy had run its course and every effort was made by the Government to encourage exports. Export subsidies were granted, access to credit provided, and tax advantages and other fiscal measures installed to promote an export-led economy (Mason *et al.* 1980; Hamilton 1986). Such help could only be justified if Korean exporters could demonstrate a real comparative advantage and not an artificial one created by government help. The view now held is that Korea's rapid development, although it was initiated within an incentive structure, was none the less very efficient and competitive in its export of labour-intensive manufactures in exchange for capital-intensive imports (Mason *et al.* 1980). In the domestic sector of the economy equally significant development had taken place. The society and economy were changing; a new class of entrepreneur emerged where before all had been Japanese; a new work-force, educated, well-trained and industrious, was created; capital and financial facilities were set up.

The bulk of the economy was in private hands but private enterprise worked in cooperation with the Government. Five-year development plans were formulated and the authoritarian Government ensured that they were strictly followed. The new class of entrepreneurs is entirely composed of Koreans, unlike Taiwan and Hong Kong where immigrants from China have been to the fore. They come largely from the aristocratic and landed classes and from the professions, and while many own small firms, the Korean industrial sector has come to be dominated by large conglomerations known as *chachol* which operate on a very large scale covering a whole range of manufactures. The state owns a number of industrial enterprises particularly among the heavy basic industries such as heavy chemicals, fertilizers and iron and steel. All Korea's governments have been interventionist, giving overall direction to the way in which the economy operates, though not on ideological grounds.

Agriculture, at first neglected, has now been given attention as the exodus of labour to the towns exacerbated rural problems. Agriculture has become more mechanized and productive. By the mid-1960s the rural infrastructure was being improved and education, and to some extent health-care, provided but parity with the towns is nothing like the level attained in Taiwan.

Like the other NICs, South Korea has had to tune her production to changing markets and technological developments. Increasingly she is producing technologically more sophisticated goods but she has also further extended her manufacturing in heavy engineering and ship building. Her economic growth is truly remarkable averaging in terms of per capita GDP 7.1 per cent per year from 1965 to 1990, a rate exceeded in the world only by Taiwan and Botswana, the latter growing from a very low base. The education of the whole population has been promoted to improve the quality and adaptibility of both work-force and management. Greater affluence, government pressure and the provision of family-planning services have all served to reduce population growth. All has been achieved under a succession of totalitarian regimes which have done much to restrict individual expression and to promote the concept of the state before the individual. It has been more readily accepted by a nation so fearful of its northern neighbour and a population experiencing a steady improvement in material well-being. The Korean economy has had many of the features of a command economy and yet it is based upon private enterprise, and governmental intervention has been pragmatic rather than ideological. In the early 1990s the economy has begun to slow down and President Kim Young Sam, the first civilian president for thirty years, has signalled his intent to introduce a more liberal regime encouraging a freer economy while at the same time strictly scrutinizing business dealings.

The success of this East Asian quartet raises the question as to whether they share common characteristics and have adopted similar approaches to the development of their economies. All are former colonies though the experience of Taiwan and Korea under Japanese rule was markedly different from that of British-ruled Hong Kong and Singapore. All suffered from the disruption of the Japanese occupation during the Second World War and had to rebuild themselves anew. All experienced subsequent politico-economic shocks and significantly all four countries reacted in a positive and constructive way. South Korea experienced the Korean War of 1950–53; Taiwan and Hong Kong were both flooded with Chinese immigrants and Hong Kong suffered from the embargo on trade with China imposed after the Korean War; Singapore experienced the withdrawal of the British naval base, her major employment facility, in 1968. With the exception of Hong Kong, all have been ruled by strong autocratic governments which have not only given political stability but have ensured that governmental plans have been carried through completely and efficiently. Again with Hong Kong's exception, these governments have been interventionist but they have not dictated artificial costs, prices and production levels as in a socialist command economy; instead they have set economic frameworks within which they have encouraged both individual and corporate enterprise to flourish. In three of the four countries the overwhelming majority of the population is Chinese, a population imbued with Confucian beliefs and values which stress the virtues of hard work, self-discipline, the need for self-improvement and the value of education. The Korean people hold similar values. This resource has been nourished in all four NICs. Education facilities have been provided at all levels including the tertiary (Taiwan, for example, has 117 polytechnics and universities) with instruction linked to the technical and economic needs of the country. The people of the East Asian quartet have proved to be adaptable, compliant, tolerant of regimentation and quick to learn. It is perhaps these qualities of the work-force that have enabled the first newly industrialized nations of the Developing World to emerge in East Asia. It is not the only factor but it is an important one. Labour has been cheap, lending a comparative advantage to the economies. It is becoming better paid and more affluent as industries shift into higher value-added production and this in countries which have stressed the equitable distribution of the benefits of development, has helped to make the work-force more tolerant of authoritarian imposition. Most importantly the governments of all four countries have maintained an international stance and outward-looking economies which have attracted foreign investments notably from Japan, the power house of East Asia, and the United States.

As their economies have evolved in the manner described, so too have their international linkages become even greater and more complex. All are seeking investment and manufacturing opportunities overseas and are strengthening or creating new political relationships to achieve this. China is the biggest recipient. South Korea restored diplomatic relations with China in 1992 and so recognized the People's Republic; Lee Kuan Yew, the arch anti-communist, has made diplomatic visits to China. Hong Kong's entrepreneurs are busy in China, so too are Taiwan's operating through Hong Kong.

The successful package appears to be to produce cheaply and efficiently goods of world class; keep continuously aware of changing market opportunities and exploit them; create a hard working, skilled and contented work-force; achieve an economic openness which encourages inward investment and at the same time seek out new investment opportunities both within and without the nation; maintain the confidence of the economy by political stability. Can this package by transferred to other areas? Allowing for the degree of diversity that is presently exhibited, it is a package which appears transferable to Malaysia, Thailand and Indonesia (at least within Java). It could be applicable to the Philippines if that nation could rid itself of internal dissension. Vietnam is another possible candidate if it can relinquish communism or at least can replicate the economic revolution taking place in China. The model has much to teach and can be viewed, understood and perhaps applauded world-wide, but it is the Far East which confirms itself as the most likely focus of new industrialization achieved in this way.

20 Retrospective and prospective views

'Ah! What avails the classic bent
And what the cultured word,
Against the undoctored incident
That actually occurred?'

Rudyard Kipling (1865–1936)
The Benefactor

'The Golden Rule is that there are no golden rules.'

Bernard Shaw (1856–1950)
Maxims for Revolutionaries

Forty years of experience of formulation of development theories and implementing development plans has allowed their evaluation and called into question their efficacy. It is now possible to identify those actions which seem to have done most to promote development and whose neglect has increased the risk of failure. There are still large variations in achievement among developing nations and in that sense there are still worlds within the Developing World.

The framework of ideas in the orthodox model for the promotion of development is set out in Fig. 33. It demonstrates the derivation of capital, the input of technology and the application of both to the increase in production. In turn this creates a series of beneficial effects which reinforce the momentum of growth in inter-connected and consequential changes. The outcome is depicted as one of greater competitiveness, continuing growth and an improvement in standards of living for all. Such a schema has underpinned almost all development programmes in the belief that society and economy could be engineered and managed as smoothly as a machine. The techniques used and the importance given to the various sectors of the economy has varied from country to country and so too has the outcome. Marxist societies chose the total management of the command economy. Other ideologies, while favouring the market economy, none the less thought that development plans could be designed which would embrace all significant aspects of the development process and ensure its speedy progress. Rostow, it will be recalled, wrote of a period of two to three decades (Rostow 1956). At all stages, however, difficulties have in practice appeared. In Fig. 33 the commonly experienced difficulties and failings are indicated in italics. The models have proved to be too abstract from reality, too clean of people.

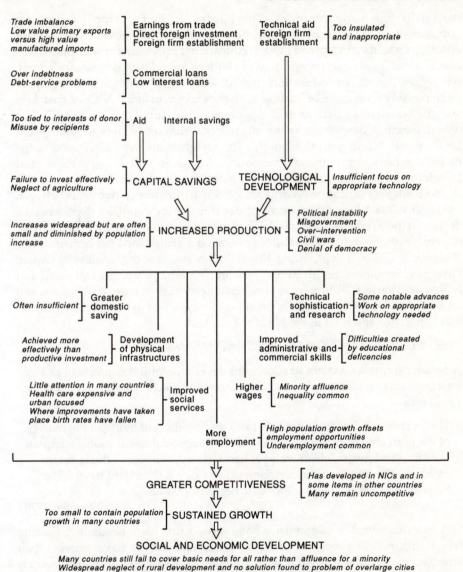

Figure 33 Schematic presentation of a development model

In the real world complete success is not possible and in development there is no ultimate level of attainment – it is a matter of progress. Over the past half century progress has been made. Many nations are moving through the demographic transition though few have, as yet, completed it and some, notably in tropical Africa, have scarcely begun. World health has improved and contributed to falling death rates though tragically high infant mortality rates remain in too many countries. The burden created by rapidly growing populations is still present and will persist well into the next century. Adult illiteracy remains widespread though educational provisions now in place have ensured that a very high proportion of children attend at least primary school. At higher levels of education the developing countries still compare unfavourably with those of Europe, with the exception of the NICs of East Asia and one or two other nations. If development is measured by per capita GNP, a useful measure despite its many drawbacks, marked differences still exist between the 'North' and the 'South'. The weighted average GNP among the nations which form the Organisation for Economic Cooperation and Development (OECD) is almost ninety times that in the low-income group of the World Bank's classification (World Bank 1993). In the twenty-five years between 1964 and 1990 no fewer than twenty-three developing countries experienced a decline in the average annual rate of change in their per capita GNP; at the other extreme, thirteen developing nations averaged a growth of 4 per cent and over (Botswana, China, Egypt, Hong Kong, Indonesia, Lesotho, Malaysia, Oman, Paraguay, Singapore, South Korea, Taiwan, Thailand). By almost all social and economic criteria, the differences between the established industrial nations and those of the Developing World remain the most marked in the countries of sub-Saharan Africa and least in the NICs of East Asia.

THE RECONSIDERATION OF ENDS AND MEANS

A nation cannot lay a claim to successful development if it neglects to provide for a large section of its population. Paul Streeten based his definition upon such a provision:

> The purpose of development is to raise the standard of living of the masses of the poor countries as rapidly as feasible, to provide secure jobs, adequate nutrition and health, clean water at hand, cheap transport, education for children. The test of any (development) strategy is the extent to which it helps or hinders in meeting the basic needs of people

(Streeten 1973).

A growing number of countries have passed this basic needs test. Policies which ensure the provision of these needs and which have proved their effectiveness are now available; with a billion people living in a state of poverty there is an urgent need for their widespread adoption (Walton 1990). Development strategies which set out programmes for an efficient and increasingly productive economy, as in Fig. 33, are by themselves not enough since they are incomplete. They must be accompanied by programmes which specifically cater for the lagging sectors of economies, for backward areas, for those whom the 'trickle down' never reaches, the poor, the forgotten people. Such policies must recognize that differing stages of development exist not only between countries but also *within* countries. Without forgoing or neglecting opportunities for structural change, industrialization and modernization, it is necessary and possible to make special provision. The free operation of market forces does not necessarily cater for basic needs, nor does the command economy with its social engineering as the opening of the closed world of the former Soviet bloc has revealed. Special provision has to be made. As Joan Robinson has written, 'it is a very remarkable fact that there is no discussion at all in orthodox economics of what form of investment is desirable from the point of view of society (Robinson J 1979). Successful policies have shown the need to put people, all of them and not some, first.

The present panorama of success, partial success, stagnation and failure which makes up the Developing World identifies the policy considerations most likely to ensure vigorous economic growth and to provide for basic needs. They are first to invest in people, second to ensure a greater degree of equity both in participation and benefits from development initiatives, and third to provide employment. On the evidence available a necessary adjunct to the attainment of these goals is the creation of a thriving rural economy. All these considerations are interlinked, the achievement of one both facilitating and deriving from achievements in the others.

The successful nations such as those of East Asia have invested in their inhabitants. Education has been made widely available at all levels and there has been little discrimination against women. Social service facilities have been created and health care provided for the whole population. These investments, particularly in the education of women and in health care, feed back into decisions on family size and speed progression through the demographic transition; this in turn reduces the pressure on employment and social service provision by lowering the population growth rate. Educated people are more adaptable, more readily acquire skills and are likely to be more innovative. Education if widely available contributes to the attainment of equity, the national work-force becomes more productive, increases its value as a resource and gives bigger returns on investments. The educated work-forces of Hong Kong, Malaysia, Singapore, South Korea and Taiwan have been a major factor in the success of their nations.

For successful progress the developing countries need to provide employment; it is an essential component in the creation of equity and the abolition of poverty. Greater industrialization does not necessarily maximize employment opportunities unless accompanied by deliberate policies of employment creation. The over-manning of trivial service jobs or the enlargement of the petty bureaucracies is no adequate solution. Productive jobs created by the deliberate selection of labour-intensive means of production are a necessary beginning. Such means are available as an alternative to capital-intensive processes over a wide range of manufactures. They are an example of appropriate technology and what is appropriate will change as the economy grows and matures and in the process creates further jobs. It does not mean the retention of inferior methods of production. Taiwan is an outstanding example of such a development in both farming and manufacturing sectors (Chapter 19).

Manufacturing has proved to be a vital element in economic progress but its promotion has often led to the neglect of both farming and the rural areas, and in doing so has helped to perpetuate rural poverty. The China of Chairman Mao with all its defects was none the less aware of the need to 'walk on two legs' to achieve a balance between town and country and industry and agriculture. In rural areas in many, perhaps most, developing countries, under-investment is the norm and unemployment and underemployment high; in consequence while poverty is not confined to the countryside, the rural poor exist in their millions. The need is for investment not solely in agriculture but in the creation of an *integrated rural economy*. Almost all analysts have argued this case and emphasized its potential for the reduction in rural poverty.

DECENTRALIZED DEVELOPMENT

In a situation where technological developments are emphasizing central places, agriculture the lavish user of space, whose work-force is the very fabric of rural society and often the keeper of traditional values, requires a positive development stimulus not in an isolated, singular way, but in conjunction with other developments in the countryside. The rural sector cannot be left stranded to accommodate change and development as best it can. This has happened in Nigeria and many other countries. It did not happen in Taiwan partly because of historical accident and partly because of intent. In China it was avoided by the planning of the rural economy. It is the main cause of the failure to attain the development goals which Streeten cites. It plays a major role in the rural–urban migration and with it the many consequent problems which have been discussed. The need is for an integrated rural economy which embraces not only a healthy and maturing agriculture but also industrial centres, energy generation, servicing centres with health and education facilities, and a marketing arrangement which further welds rural regions into cohesive space economies. Manufacturing can range not only over the obvious processing of agricultural products and the provision of agricultural needs such as fertilizer and implements, but also include more sophisticated manufactures not sensitive to scale restrictions or dispersed locations. Industrial development is a means of increasing both the level of employment and the marginal productivity of agricultural labour.

What is envisaged is the articulation of an urban hierarchy in which rural towns function as centres of development interrelated with the surrounding rural areas in an economic and social sense. As has been discussed, the urban hierarchy in most developing countries is imperfect and attempts to complete and activate it have often failed because the towns remained outposts of the metropolitan centres rather than becoming embedded in the surrounding economy. The case is, therefore, for truly integrated functional cities. It is a scheme incorporated into the development plans in Fiji and is the substance of Babarović's proposals for Brazil. Its need is widely evident in countries of the Developing World. Only in this way can the poverty that is the product of sectoral lag be eliminated and the drainage of the periphery restrained. It could do much to create an integrated space economy. It puts the fabric over the frame of the umbrella.

Clearly, in many countries such developments will require a radical reorganization of land tenure in order to allow changes in the agrarian economy. The land tenure situation present in most Latin American countries is, for example, incompatible with the developments suggested. The reorganization of land can take many forms. In China, though the State acquired the ownership of land, the genius of the commune concept lay in the degree of local management and involvement it allowed and in the disaggregation of development. The answer would not appear to lie in the establishment of state farms. All the evidence runs counter to this. In Taiwan the key to rural success was the redistribution of land, including state land, among the working farmers.

APPROPRIATE TECHNOLOGIES

To achieve the implementation of integrated rural development in conditions of labour surplus and capital scarcity requires the use of labour-intensive methods which are widely available as alternatives in most economic processes. The stages of development concept requires the use of appropriate technologies; technologies appropriate to the stage of development, to the skills and resources which can be used and appropriate to the particular sector – a rural manufacturing enterprise could use a technology different from that used contemporaneously in a large urban centre. The virtues of this approach have been demonstrated in China but it is widely applicable. Since countries vary in their stages of development, as sectors of economies likewise differ, appropriateness is the crucial concept. Without it terms like intermediate or elementary or primitive creep into use. It is not a concept which sees virtues in the home-made, the small-scale, the unsophisticated for their own sake, nor is it one rooted in some nostalgia for the pre-technical past. It is concerned with techniques which make possible a greater productivity from resources which are too small to justify large capital investments, from people who have not received a technical training, and for purposes which are meaningful to the communities which engender them. The last point is important. Properly conceived and implemented, the use of appropriate technologies draws upon the inventiveness and initiative of the people involved and produces things which they see as useful. The answers must come in the involvement of local populations from the beginning, with their identification of priorities and in the light of their awareness of social and economic implications. It is important to appreciate that changes in the nature of manufactured goods and the manner of their production have meant that a wide range of industrial products can be made economically in dispersed rural locations. In conjunction with the widening of educational opportunities, industries can succeed in the countryside (Robinson A 1979; Barnett *et al.* 1982; World Bank 1975, 1980a, 1980b, 1990, 1992; FAO 1992).

THE ROLE OF THE STATE

Variations in levels of attainment, speed of change and methods used have brought about both debate and re-assessment as to what constitutes good management and what rôle, if any, the state should play in economic affairs. The collapse of the Soviet Union, long seen by many developing nations as an examplar of successful socialist command economies, has further heightened the debate as to the rôle of the state. As a proselytizing Marxist nation, the Soviet Union had attempted to spread its ideology widely. It succeeded in Cuba, for a time influenced China, for decades provided a model for Indian industrialization, and in Africa it helped to produce a number of pseudo-Marxist regimes. In a detailed catalogue of African countries Bartlett sets out the disastrous effect of Marxism upon the continent and indicates the present widespread rejection of the doctrine (Bartlett 1990). In the Soviet bloc, Marxist command economies have not created the equity and well-being promised, while in Africa they have been associated with low levels of economic and social attainment.

In the least developed countries the state has frequently had to play a leading part in initiating developments in the absence of a significant local community of entrepreneurs; difficulties have resulted from the over-extensive nature and overlong retention of that rôle. In India, Nigeria and Zambia the over-involvement of the state, often in the most minor of economic matters, has been demonstrated in the case studies. With the exception of Hong Kong, the successful NICs of East Asia have also been characterized by considerable intervention of the state but within free-market economies. The debate thus becomes one of identifying the parts of the national life in which the government has an essential rôle to play and what form that intervention should take. The accumulated evidence suggests that the state should not take upon itself the management of economic matters which others can do better; in non-democratic societies the temptation to do so has often proved impossible to resist. Since social provision and equity are not an inevitable outcome of free-market economics, the state clearly has an important part to play in those matters which have been seen to be essential in the successful attainment of development. It would appear therefore that the *manner* of state involvement is of crucial significance. Success would appear to derive from a mutually supporting combination of the forces of the market and the participation of the state. The achievement of macro-economic stability is essential. The missing elements in the pursuance of successful development have been sound management and good government. This deficiency, compounded by communal strife, insurrection and civil war has, tragically, blighted the prospects of many developing countries.

Though their styles have differed, the successful developing countries have been characterized by stable governments, outward looking and pragmatic in their approach both to problems and opportunities. They have been flexible rather than doctrinaire and have not hesitated to change policies as circumstances demanded or when they have outlived their usefulness. All have participated forcefully in the world economy and have welcomed foreign investments including the operation of multinational companies. Development has been nurtured by a process of transfusion not only of capital but also of technology, business methods and information. The four NICs are the foremost examples of the success of good management but other nations are manifesting this involvement; Indonesia, Malaysia, Mexico and Thailand are further examples. The most interesting development of all is taking place within Marxist China where modern management techniques are being adopted, the latest technology is being set in place and involvement in the world economy is strongly emerging. It would appear that the key to success is the creation of a socio-economic environment which is attractive to international investment and is able to maximize its comparative advantages so as to participate successfully in the world economy. Exogenous forces interacting with endogenous developments may well set the pattern for the unification of the world economy in the twenty-first century and in doing so reduce the contrast between the rich and the poor nations.

Bibliography

Abrams, C. (1964) *Man's Struggle for Shelter in an Urbanizing World*, Cambridge, Mass., MIT Press.

Adams, W. M. (1986) 'Traditional agriculture and water use in the Sokoto valley Nigeria', *Geog. J.*, **152**, 30–43.

Adams, W. M. (1991) 'Large scale irrigation in northern Nigeria: performance and ideology', *Trans. Inst. Brit. Geog.*, **NS16**, 287–300.

Adelman, I. and **Morris, C. T.** (1973) *Economic Growth and Social Equity in Developing Countries*, Stanford, California, Stanford University Press.

Agnew, C. T. (1982) 'Water availability in the development of rainfed agriculture in south west Niger, West Africa,' *Trans. Inst. Brit. Geog.*, **7**, No. 4, 419–57.

Ahmed, M. and **Gooptu, S.** (1993) 'Portfolio investment flows to developing countries', *Finance and Development*, March 1993, 9–12.

Allan, W. (1965) *The African Husbandman*, London, Oliver and Boyd; New York, Barnes and Noble.

Allan, W. (1969) (2nd edn) *Studies in African Land Usage in Northern Rhodesia*, Rhodes-Livingstone Papers No. 15, Rhodesia, Rhodes-Livingstone Institute.

Alonso, W. (1968) 'Urban and regional imbalances in economic development', *Economic and Cultural Change*, **17**, 1–13.

Aluwalia, I. J. (1985) *Industrial Growth in India*, Delhi, Oxford University Press.

Aluwalia, M. S. (1974) 'Income inequality: some dimensions of the problem' in Chenery *et al.*, 1974, 3–31.

Amin, S. (1972) 'Underdevelopment and dependence in Black Africa', *J. of Modern African Studies*, **10**, 503–24.

Amin, S. (1974) *Accumulation on a World Scale: A Critique of the Theory of Underdevelopment*, 2 vols, New York, Monthly Review Press.

Amin, S. (ed.) (1974) *Modern Migrations in West Africa*, London, Oxford University Press.

Angier, F. R. *et al.* (1960) *The Making of the West Indies*, London, Longman.

Anthony, K. R. M. *et al.* (1979) *Agricultural Change in Tropical Africa*, New York, Cornell University Press.

Arnold, G. (1977) *Nigeria Today*, London, Longman.

Arnold, G. (1979) *Aid in Africa*, London, Kegan Paul.

Atkins, P. J. (1989) 'Operation Flood: dairy development in India', *Geog.*, **74**, 259–62.

Auer, P. (ed.) (1981) *Energy and the Developing Nations*, New York, Pergamon.

Aziz, S. (1978) *Rural Development: Learning from China*, London, Macmillan.

Babarović, I. (1978) 'Rural marginality and regional development policies in Brazil' in Kuklinski 1978, 189–319.

Bairoch, P. (1975) *The Economic Development of the Third World Since 1900*, London, Methuen.

Baklanoff, E. N. (ed.) (1969) *The Shaping of Modern Brazil*, Baton Rouge, Louisiana State University Press.

Baldwin, R. E. (1964) 'Patterns of development in newly settled regions' in Friedmann and Alonso 1964, 266–84.

Banister, J. (1987) *China's Changing Population*, Stanford, California, Stanford University Press.

Barber, W. J. (1961) *The Economy of British Central Africa*, London, Oxford University Press.

Barnett, A., Bell, M. and **Hoffman, K.** (1982) *Rural Energy and the Third World*, London, Pergamon.

Barraclough, S. L. (1976) 'Interactions between agrarian structures and public policies in Latin America' in Hunter, Bunting and Bottrall 1976, 93–105.

Barraclough, S. L. and **Domike, A.** (1966) 'Agrarian structure in seven Latin American countries', *Land Economics*, **42**, 391–442.

Bartlett, B. R. (1990) 'Capitalism in Africa', *J. Dev. Areas*, **24**, 327–50.

Bascom, J. B. (1990) 'Border pastoralism in eastern Sudan', *Geog. Rev.*, **80**, 416–30.

Basgupta, B. (1977) 'India's Green Revolution', *Econ. and Polit. Weekly*, Feb. 1977.

Bassett, T. J. (1986) 'Fulani herd movements', *Geog. Rev.*, **76**, 233–48.

Bayliss-Smith, T. and **Sundhir Wanmali** (1984) *Understanding the Green Revolution*, Cambridge, Cambridge University Press.

Becker, B. K. and **Egler, C. A. G.** (1992) *Brazil: A New Regional Power in the World Economy*, Cambridge, Cambridge University Press.

Beckford, G. L. (1969 and 1973) 'The economics of agricultural resource use and development in plantation economies', *Social and Econ. Studies*, **18**, 321–47. Reproduced in Bernstein 1973 as Ch. 6.

Beckford, G. L. (1972) *Persistent Poverty: Underdevelopment in Plantation Economies of the Third World*, Oxford, Oxford University Press.

Bedford, R. D. (1984) 'The Polynesian connection: migration and social change in New Zealand and the South Pacific', in Bedford (ed.) 1984, 113–41.

Bedford, R. D. (ed.) (1984) *Essays on Urbanisation in South-East Asia and the Pacific*, Proceedings of symposium, 15th Pacific Science Congress Dunedin, 1983, Dpt. Geog. U. of Canterbury, Christchurch, New Zealand.

Belshaw, C. S. (1964) *Under the Ivi Tree: Society and Economic Growth in Rural Fiji*, London, Routledge and Kegan Paul.

Benneh, G. (1972) 'The response of farmers in northern Ghana to the introduction of mixed farming: a case study', *Geogr. Annaler*, 95–103.

Bergsman, J. (1970) *Brazil: Industrialisation and Trade Policies*, New York, London, Toronto, Oxford University Press.

Bernstein, H. (ed.) (1973, reprinted 1978) *Underdevelopment and Development*, London, Penguin.

Berry, B. J. L. (1960) 'An inductive approach to the regionalisation of economic development' in Ginsburg 1960, 78–107.

Berry, B. J. L. (1961) 'Basic patterns of economic development', in Ginsburg 1961, 110–19.

Berry, B. J. L. (1972) 'Hierarchical diffusion: the basis of developmental filtering and spread in a system of growth centres' in Hansen 1972, 108–38.

Bertram, I. G. and **Watters, R. F.** (1985) 'The MIRAB economy in South Pacific microstates', *Pacific Viewpoint*, **26**, 498–519.

Bertram, I. G. and **Watters, R. F.** (1986) 'The MIRAB process: earlier analysis in context', *Pacific Viewpoint*, **27**, 47–9.

Best, L. (1968) 'Outlines of a model of pure plantation economy', *Social and Econ. Studies*, Sept. 1968, 283–326.

Bethell, L. (1970) *The Abolition of the Brazilian Slave Trade*, Cambridge, Cambridge University Press.

Bhagwati, J. N. and **Desai, P.** (1970) *India: Planning for Industrialisation*, London, New York, Bombay, Oxford University Press.

Blaikie, P. M. (1971) 'Spatial organisation of agriculture in some north Indian villages', *Trans Inst. Brit. Geog.*, Part I, **52**, 1–40, Part II, **53**, 15–30.

Bondstam, L. and **Bergstrom, S.** (eds) (1980) *Poverty and Population Control*, London, New York, Academic Press.

Bongaarts, J. and **Way, P.** (1989) *Geographic Variation in the HIV Epidemic and the Mortality Impact of AIDS in Africa*, Research Division Working Paper No. 1, New York, Population Council.

Boserup, E. (1965) *The Conditions of Agricultural Growth*, London, Allen and Unwin, Chicago, Aldine Publishing Co.

Boudeville, J. (1966) *Problems of Regional Economic Planning*, Edinburgh, University of Edinburgh Press.

Bradley, P. N. (1980) 'Agricultural development planning in the Senegal valley' in Simpson 1980, 41–58.

Bradley, P. N., Raynaut, C. and **Torrealba, J.** (1977) *The Guidimaka Region of Mauretania*, London, War on Want.

Brandt, W. (1980) *North–South: Report of the Independent Commission on International Development* (The Brandt Report), London, Pan Books.

Briggs, J. (1980) 'Rural policy in Tanzania since 1967: trends and issues' in Simpson 1980, 9–26.

Brookfield, H. C. (ed.) (1973) *The Pacific in Transition*, New York, St. Martins Press.

Brookfield, H. C. (1975) *Interdependent Development*, London, Methuen.

Brown, C. (1977) *China: 1949–76*, London, Heinemann Educational Books.

Brown, L. A. (1988) 'Reflections on Third World development: ground-level reality, exogenous forces and conventional paradigms', *Econ. Geog.*, **64**, 255–78.

Brown, L. R. (1970) *Seeds of Change: The Green Revolution and Development in the 1970s*, New York, Praeger.

Brunner, R. and **Brewer, G. D.** (eds) (1975) *Political Development and Change*, New York, London, Macmillan.

Bruton, H. J. (1965) *Principles of Economic Development*, New Jersey, Prentice-Hall.

Cairncross, S. and **Feacham, R.G.** (1983) *Environmental Health Engineering in the Tropics*, Chichester, Wiley.

Campbell, G. (1972) *Brazil Struggles for Development*, London, Charles Knight.

Cardozo, M. (1969) 'The modernisation of Brazil 1500–1808' in Baklanoff 1969, 3–18.

Cassen, R. H. (1978) *India: Population, Economy and Society*, London, Macmillan.

Ceylon Planning Secretariat (1955) *Six Year Programme of Investment 1954–1959/60*, Colombo, Ceylon Planning Secretariat.

Ceylon Planning Secretariat (1959) *The Ten Year Plan*, Colombo, Ceylon Planning Secretariat.

Chakravarti, A. K. (1973) 'Green revolution in India', *Annals Ass. Am. Geog.*, **63**, No. 3, 319–30.

Chang, J. H. (1968) 'The agricultural potential of the humid tropics', *Geog. Rev.*, **58**, 333–61.

Chang, Jung (1991) *Wild Swans*, London, Harper Collins.
Chen, C. S. (1963) *Taiwan: An Economic and Social Geography*, Taipei, Research Report No. 96, Fu-Min Geographical Inst. of Econ. Development.
Chenery, H. B. (1960) 'Patterns of industrial growth', *Americ. Econ. Rev.*, **50**, 624–54.
Chenery, H. B. and Syrquin, M. (1975) *Patterns of Development 1950–70*, London, Oxford University Press.
Chenery, H. B. and Taylor, L. J. (1968) 'Development patterns among countries and over time', *Rev. of Econ. and Stats.*, **50**, 391–416.
Chenery, H. B. *et al.* (1974) *Redistribution with Growth*, London, New York, Oxford University Press.
Chenery, H. B. *et al.* (1979) *Structural Change and Development Policy*, New York, London, Oxford University Press.
Chiang, T. C. (1969) *The Jurong Industrial Estate: Present Pattern and Future Prospect*, Singapore, Nanyang University.
China, Republic of (1953) *First Four Year Plan for Economic Development*, Economic Stabilisation Board, Taipei, Rep. of China.
Chisholm, M. (1980) 'The wealth of nations', *Trans. Inst. Brit. Geog.*, New Series, **5**, No. 3, 255–76.
Chisholm, M. (1982) *Modern World Development*, London, Hutchinson.
Chiu, S. K. (ed.) (1979) *Modernisation in China*, Centre of Asian Studies Occasional Papers and Monographs No. 35, Hong Kong, Hong Kong University.
Clark, J. and Allison, C. (1989) *Zambia: Debt and Poverty*, Oxford, Oxfam.
Coale, H. J. and Hoover, E. M. (1958) *Population Growth and Economic Development in Low Income Countries*, Princeton, New Jersey, Princeton University Press.
Cody, J., Hughes, H. and Wall, D. (1980) *Policies for Industrial Progress in Developing Countries*, London, New York, Oxford University Press.
Cohen, R. (1974) *Labour and Politics in Nigeria*, London, Heinemann.
Colson, E. and Gluckman, M. (eds) (1968) *Seven Tribes of British Central Africa*, London, New York, Oxford University Press.
Conklin, H. C. (1954) 'An ethnoecological approach to shifting cultivation', *Trans. New York Acad. Sci.*, **17**.
Conklin, H. C. (1961/2) 'The study of shifting cultivation', *Current Anthropology*, 27–58
Cook, E. K. (1953) *Ceylon: its Geography, its Resources and its People*, London, Macmillan.
Coppock, J. T. (1966) 'Agricultural developments in Nigeria', *Tropical Geog.*, **23**, 1–18.
Corbridge, S. (1986) *Capitalist World Development: A Critique of Radical Development Geography*, Totowa, New Jersey, Rowman and Littlefield.
Corbridge, S. (1988) 'The debt crisis and the crisis of global regulation', *Geoforum*, **19**, No. 1, 109–30.
Courtenay, P. P. (1980) (2nd edn) *Plantation Agriculture*, New York, Praeger; London, Bell.
Crabtree, J. (1992) *Peru under Garciá*, London, Macmillan.
Cuca, R. (1979) *Family planning programs: an evolution of experience*, World Bank Staff Working Paper No. 345, Washington DC, World Bank.
Curtin, P. D. (1969) *The Atlantic Slave Trade: A Census*, Madison, University of Wisconsin Press.

Dalby, D. and Harrison-Church, R. J. (eds) (1973) *Drought in Africa*, SOAS, London, University of London.

Darwent, D. F. (1969) 'Growth poles and growth centers in regional planning – a review', *Environment and Planning*, **1**, 5–32.

Dasgupta, A. K. (1974) *Economic Theory and the Developing Countries*, London, Macmillan.

Davies, H. R. J. (1966) 'Nomadism in the Sudan', *Tijdschrift fur Economische Geo.*, Sep / Oct.

Davy, E. G., Mattei, E. and **Solomon, J. I.** (1976) *An evaluation of climate and water resources for development of agriculture in the Sudano-Sahelian zone of West Africa.* Special environment Report 9, Geneva, W.M.O.

Dayal, E. (1983) 'Regional response to high yield varieties of rice in India', *Sing. J. Trop. Studies*, **4**, 2.

De, P. K. and **Mandal, L. N.** (1956) 'Fixation of nitrogen by algae in rice soils', *Soil Science*, **81**, 453–9.

De Silva, K. M. (ed.) (1977) *Sri Lanka: A Survey*, Institute of Asian Affairs, Hamburg, London, C. Hurst and Co.

Derrick, R. A. (1959) 'Fiji's darkest hour: an account of the measles epidemic of 1875', *Trans. and Proc. Fiji Society*, Suva, Fiji, **6**, 3–16.

Dicken, P. (1987) 'A tale of two NICs: Hong Kong and Singapore at the crossroads', *Geoforum*, **18**, No. 2, 151–64.

Dickenson, J. P. (1974) 'Imbalances in Brazil's industrialisation', in Hoyle 1974, 291–306.

Dickenson, J. P. (1982) *Brazil*, London, Longman.

Dixon, C. L. (1977) *The Agricultural Policy in Zambia 1964–71*, O'Dell Memorial Monograph No. 5, Dept. of Geog., Aberdeen, University of Aberdeen.

Domar, E. (1957) *Essays in the Theory of Economic Growth*, London, Oxford University Press.

Dos Santos, T. (1970) 'The structure of dependence', *Am. Econ. Rev.*, **60**, 289–90.

Dos Santos, T. (1978) 'Theory and the problem of dependence in Latin America' in Bernstein, 1978, 57–80.

Duckham, A. N. and **Masefield, G. B.** (1971) *Farming Systems of the World*, London, Chatto and Windus.

Dulles, J. W. F. (1969) 'The contribution of Getulio Vargas to the modernisation of Brazil' in Baklanoff 1969, 36–57.

Dumont, R. (1957) *Types of Rural Economy: Studies in World Agriculture*, trans. D. Magnin, London, Methuen.

Dunning, J. H. (ed.) (1971) *The Multinational Enterprise*, London, Allen and Unwin.

Eckstein, A. (1977) *China's Economic Revolution*, Cambridge, London, New York, Cambridge University Press.

Economist, The (1992) China Survey, Nov. 28, 1–22.

Eden, M. J. (1978) 'Ecology and land development: the case of Amazonian rain forest,' *Trans. Inst. Brit. Geog.*, **3**, No. 4, 444–63.

Edwards, C. (1988) 'The debt crisis and development: a comparison of major theories', *Geoforum*, **19**, No. 1, 3–28.

Eiches, C. K. and **Liedholme, C.** (1970) *Growth and Development of the Nigerian Economy*, Michigan, Michigan State University Press.

Emmanuel, A. (1972) *Unequal Exchange*, London, New Left Books.

Epstein, T. S. (1973) *South India: Yesterday, Today and Tomorrow*, London, Macmillan.

Evans, A. (1955) 'A study of crop production in relation to rainfall reliability', *East African Agric. J.*, April, 263–7.

Farmer, B. H. (1950) 'Agriculture in Ceylon', *Geog. Rev.*, **40**, 42–66.

Farmer, B. H. (1957) *Pioneer Peasant Colonization in Ceylon: A Study in Asian Agrarian Problems*, London, Oxford University Press.

Farmer, B. H. (1963a) 'Peasant and plantation in Ceylon', *Pacific Viewpoint*, **4**, 9–16.

Farmer, B. H. (1963b) *Ceylon: A Divided Nation*, London, New York and Bombay, Oxford University Press.

Farmer, B. H. (1981) 'The Green Revolution in South Asia', *Geog.*, **66**, 202–7.

Faucher, P. (1991) 'Public investment and the creation of manufacturing capacity in the power–equipment industry in Brazil', *J. Dev. Areas*, **25**, 231–60.

Fei, J. C. H. and **Ranis, G.** (1975) 'A model of growth and employment in the open dualistic economy', *J. of Develop. Studies* **2**, 32–63.

Fei, J. C. H., Ranis, G. and **Kuo, S. W. Y.** (1979) *Growth with Equity: the Taiwan Case*, London, New York, Oxford University Press.

Feuchtwang, S. and **Hussain, A.** (eds) (1983) *The Chinese Economic Reforms*, London, Croom Helm.

Fiji Education Commission (1970) *Report of the 1969 Fiji Education Commission*, Suva, Fiji, Govt. Printer.

Fiji Legislative Council (1966) *Fiji Development Plan 1966–70*, Development Plan Review, Legislative Council of Fiji, Council Paper No. 11 of 1966, Central Planning Office, Colony of Fiji, Suva, Fiji Govt. Printer.

Fiji, Parliament of (1970) *Fiji's Sixth Development Plan 1971–75*, Parliam. Paper No. 25 of 1970, Central Planning Office, Ministry of Finance, Suva, Fiji, Govt. Printer.

Fiji, Parliament of (1975) *Fiji's Seventh Development Plan 1976–80*, Parliamentary Papers 1975, Central Planning Office, Suva, Fiji, Govt. Printer.

Fiji, Parliament of (1980/81) *Fiji's Eighth Development Plan 1981–85*, Parliament of Fiji Papers 1980/81, 2 vols, Central Planning Office, Suva, Fiji, Govt. Printer.

Fleure, H. J. (1919) 'Human regions,' *Scot. Geog. Mag.*, **35**, 94–105.

Flinn, M. W. (1979) *British Population Growth 1700–1850*, London, Macmillan.

Food and Agricultural Organisation, FAO (1982) *World Report: State of Food and Agriculture 1981*, Rome, FAO.

Food and Agricultural Organisation, FAO (1985) *World Report: State of Food and Agriculture 1984*, Rome, FAO.

Food and Agricultural Organisation, FAO (1990) *World Report: State of Food and Agriculture 1989*, Rome, FAO.

Food and Agricultural Organisation, FAO (1991) *World Report: State of Food and Agriculture 1990*, Rome, FAO.

Food and Agricultural Organisation, FAO (1992) *World Report: State of Food and Agriculture 1991*, Rome, FAO.

Forde, D. (1934) *Habitat, Economy and Society*, London, Methuen.

France, P. (1969) *The Charter of the Land: Custom and Colonization in Fiji*, London, Wellington; New York, Oxford University Press.

Frank, A. G. (1967) *Capitalism and Underdevelopment in Latin America*, New York, Monthly Review Press. Historical Studies in Chile and Brazil Review, edition 1969.

Frank, A. G. (1969) *Latin America, Underdevelopment or Revolution*, New York, Modern Reader, Monthly Review Press.

Frankel, J. M. (1971) *India's Green Revolution: Economic Gains and Political Costs*, Princeton, New Jersey, Princeton University Press.

Frazer, F. R. (1952) 'Housing and planning in Singapore', *Town Planning Rev.*, **23**, 5–25.

Friedmann, J. P. (1966) *Regional Development Policy: A Case Study of Venezuela*, Cambridge, Mass., MIT Press.

Friedmann, J. P. (1968) 'The strategy of deliberate urbanization', *J. Am. Inst. of Planning*, **34**, 364–73.

Friedmann, J. P. (1972) 'A general theory of polarized development' in Hansen 1972, 82–107.

Friedmann, J. P. and **Alonso, W.** (eds) (1964) *Regional Development and Planning*, Cambridge, Mass., MIT Press.

Friedmann, J. P. and **Douglas, M.** (1978) 'Agropolitan Development: towards a new strategy for regional planning in Asia' in Lo and Salih 1978.

Friedmann, J. P. and **Weaver, W.** (1979) *Territory and Function*, London, Edward Arnold.

Friedmann, J. P. and **Wulff, R.** (1976) *The Urban Transition*, London, Edward Arnold.

Froehlick, W. (ed.) (1961) *Land Tenure, Industrialisation and Social Stability*, Milwaukee, Marquette University Press.

Furtado, C. (1959; trans. 1963) *The Economic Growth of Brazil*, Berkeley, CA, University of California Press.

Furtado, C. (1976) (2nd edn.) *Economic Development of Latin America*, Cambridge, Cambridge University Press.

Gaile, G. L. (1988) 'Choosing locations for small town development to enable market and employment expansion: the case of Kenya', *Econ. Geog.*, **64**, No. 3, 242–54.

Gaitskell, A. (1959) *Gezira: A Story of Development in the Sudan*, London, Faber and Faber.

Galenson, W. (ed.) (1979) *Economic Growth and Structural Change in Taiwan*, Ithaca, New Jersey, Cornell University Press.

Galloway, J. H. (1968) 'The sugar industry of Pernambuco in the nineteenth century', *Annals Assoc. Am. Geogs.*, **58**, 285–303.

Gann, L. H. (1958) *The Birth of a Plural Society: The Development of Northern Rhodesia under the British South Africa Company 1894–1914*, Manchester, Manchester University Press for the Rhodes-Livingstone Institute.

Gann, L. H. (1963) *A History of Northern Rhodesia*, London, Chatto and Windus.

Geertz, C. (1968) *Agricultural Involution*, Berkeley and Los Angeles, University of California Press.

Geiger, T. (1976) *The Development Progress of Hong Kong and Singapore*, Hong Kong, Macmillan.

Gilbert, A. G. (1975) 'A note on the incidence of development in the vicinity of a growth centre', *Regional Studies*, **9**, 325–33.

Gilbert, A. G. (ed.) (1976) *Development Planning and Spatial Structure*, New York, John Wiley.

Gilbert, A. G. and **Gugler, J.** (1981) *Cities, Poverty and Development*, London, Oxford University Press.

Ginsburg, N. S. (ed.) (1960) *Essays on Geography and Economic Development*, Chicago, University of Chicago Press.

Ginsburg, N. S. (1961) *Atlas of Economic Development*, Chicago, University of Chicago Press.

Glass, D. V. and **Eversley, D. E. C.** (eds) (1965) *Population in History: Essays in Historical Demography*, London, Edward Arnold, Chicago, Aldine Pub. Co.

Gleave, M. B. (1980) 'Some further thoughts on population density and agricultural systems in West Africa' in Simpson 1980, 59–74.

Gleave, M. B. (ed.) (1992) *Tropical African Development*, London, Longman.

Glewwe, P. and **Hall, G.** (1992) 'Unorthodox adjustment and poverty in Peru', *Finance and Development* December 10–13, Washington, World Bank.

Goddard, A. D. *et al.* (1974) 'Population movements and land shortages in the Sokoto close–settled zone' in Amin 1974.

Goddard, A. D., Fine and Norman (1971) *'A socio-economic study of three villages in Sokoto close-settled zone'*, Samaru Miscellaneous Papers No. 33, Samura, Nigeria, Ahmadu Bello University.

Goh, K. S. (1977) *The Practice of Economic Growth,* Singapore, Federal Publications.

Goldsmith, A. A. (1988) Policy dialogue, conditionality and agricultural development: implications of India's Green Revolution, *J. Dev. Areas,* 179–98.

Goldthorpe, C. L. (1987) A definition and typology of plantation agriculture, *Sing. J. Trop. Studies,* **8,** No. 1, 26–43.

Gottmann, J. (1983) 'Third World cities in perspective', *Area,* **15,** No. 4, 311–13.

Gould, P. R. (1970) 'Tanzania 1920–63: the spatial impress of the modernisation process', *World Politics,* **22,** 149–70.

Gourou, P. (1953) *The Tropical World,* London, Longman.

Gray, J. (1969) 'The economics of Maoism', *Bull. of Atomic Scientists,* **25,** No. 2, 42–51. Also in Bernstein 1973.

Gray, J. and **Gray, M.** (1983) 'China's new agricultural revolution' in Feuchtwang and Hussain 1983, 151–84.

Gray, R. H. (1974) 'The doctrine of mortality in Ceylon and the demographic effects of malaria control', *Population Studies,* **28,** 205–29.

Greaves, I. (1959) 'Plantations in the world economy' in Pan American Union 1959.

Greenwood, M. (1936) 'English death rates, past, present and future', *J. of Royal Stat. Soc.,* **99.**

Grice, K. and **Drakakis Smith, D.** (1985) 'The role of the state in shaping development: two decades of growth in Singapore', *Trans. Inst. Brit. Geog.,* **10,** 347–59.

Griffin, K. B. and **Enos, J. L.** (1970) *Planning Development,* Addison Wesley; see also Bernstein 1973, 141–53.

Grigg, D. (1982) 'Counting the hungry: world patterns of undernutrition', *Tijdschrift für Economische Geo.,* **73,** 66–79.

Grist, D. H. (1975) (5th edn) *Rice,* New York and London, Longman.

Grossman, L. (1981) The cultural ecology of economic development', *A.A.A.G,* 220–36.

Grove, A. T. (1967) *Africa South of the Sahara,* London, Oxford University Press.

Grove, A. T. (ed.) (1985) *The Niger and its Neighbours: Environmental History and Hydrobiology, Human Use and Health Hazards of the Major West African Rivers,* Rotterdam, A. A. Balkema.

Habakkuk, H. J. (1965) 'The economic history of modern Britain' in Glass and Eversley 1965, 147–58.

Habakkuk, H. J. (1971) *Population Growth and Development since 1750,* Leicester, Leicester University Press.

Hall, R. (1965) *Zambia,* London, Pall Mall Press.

Hamilton, C. (1986) *Capitalist Industrialisation in Korea,* Boulder Col and London, Westrew Press.

Hansen, N. M. (ed.) (1972) *Growth Centres in Regional Economic Development,* New York, The Free Press.

Harris, B. (1971) 'The Green Revolution in Ludhiana District India', *Geog.,* **56,** 243–46.

Harris, J. and **Harris, B.** (1979) 'Development studies', *Prog. Hum. Geog.,* **3,** 576–84.

Harrod, R. F. (1948) *Towards a Dynamic Economics*, London, Macmillan.

Hasan, P. (1984) 'Adjustment to external shocks', *Finance and Development*, **21**, No. 4, Dec., 14–17, World Bank.

Hatcher, J. (1977) *Plague, Population and the English Economy 1348–1530*, London, New York, Macmillan.

Hayami, Y. (1988a) 'Asian development: a view from the paddy fields', *Asian Development Review*, **6**, 50–63.

Hayami, Y. (1988b) 'Induced innovation, green revolution and income distribution', *Econ. Dev. and Cultural Change*, **30**, 169–76.

Heady, E. O. (1952) *Economics of Plantation Production and Resource Use*, New Jersey, Prentice-Hall.

Healey, D. T. (1972) 'Development policy: new thinking about interpretation', *J. Of Econ. Literature*, Sept. 1972.

Heathcote, D. (1972) 'Insight into a creative process: a rare collection of embroidery drawings from Kano', *Savanna*, Dec. 1972, **1**, 2, 165–74.

Hellen, J. A. (1968) *Rural Economic Development in Zambia 1890–1914*, Munchen, IFO-Inst. for Wintschaftsforschung, Munchen Afrika-Studienstelle.

Hendrix, W. E., Naive, J. J. and **Adams, W. E.** (1968) *Accelerating India's Food Grain Production 1967–68 to 1970–71*, Foreign Agricultural Economic Report No. 40, Dept of Agric., USA, Economic Research Service. Washington DC, US Dept of Agriculture.

Hirschman, A. O. (1958) *The Strategy of Economic Development*, New Haven, CT, Yale University Press.

Ho, R. (1967) *Farmers of Central Malaya*, Canberra, Australian National University Press.

Ho, S. P. S. (1978) *Economic Development of Taiwan 1860–1970*, New Haven and London, Yale University Press.

Hone, P. F. (1909) *Southern Rhodesia*, London, Bell.

Hoyle, B. S. (ed.) (1974) *Spatial Aspects of Development*, London, Sydney and Toronto, John Wiley.

Hsiung, J. C. (ed.) (1981) *The Taiwan Experience 1950–80*, New York, Praeger.

Hubback, E. M. (1947) *The Population of Britain*, London, Pelican.

Hughes, R. (1970) *Borrowed Place, Borrowed Time*, Hong Kong.

Hunter, G. (1969) *Modernizing Peasant Societies*, London, Oxford University Press.

Hunter, G., Bunting, A. H. and **Bottrall, A.** (eds) (1976) *Policy and Practice in Rural Development*, London, Croom Helm.

Hutchinson, H. H. (1961) 'The transformation of Brazilian plantation society', *J. Inter-American Studies*, **3**, 201–12.

Ilich, I. (1969) 'Outwitting the "developed" countries', *New York Review of Books*, Nov. 1969, 20–24. Reprinted in Bernstein 1978, 357–68.

Indian Planning Commission (1966) *Fourth Five Year Plan*, 10–11, New Delhi, Govt of India.

Isard, W. (1951) 'Regional and interregional input–output analysis: a model of a space economy', *Rev. of Econ. and Stats.*, **33**, 318–28.

Isard, W. (1956) *Location and Space Economy*, Boston, The Technology Press of MIT and John Wiley and Sons.

Isard, W. (1957) 'The value of the regional approach to some basic economic problems', *Regional Income, Studies in Income and Wealth, Vol. 2*, Princeton, Princeton University Press.

Isard, W. *et al.* (1960) *Methods of Regional Analysis: An Introduction to Regional Science,* London, John Wiley and Sons.

Islam, N. (1967) *Tariff Protection, Comparative Costs and Industrialisation in Pakistan,* Pakistan Institute of Development Economics, Research Report 57.

Jakubowski, M. (1977) 'The theory of demographic transition and studies in the spatial differentiation of population dynamics', *Geographia Polonica,* **35,** 73–89.

Jeyaratnan Wilson, J. (1977) 'Politics and political development since 1948' in De Silva 1977.

Johnson, C. (1981) 'The Taiwan model' in Hsiung 1981, 9–10.

Johnson, E. H. J. (1970) *The Organisation of Space in Developing Countries,* Cambridge, MA, Harvard University Press.

Johnson, H. G. (1964) 'Tariffs and economic development', *J. Dev. Studies,* Oct. 1964.

Johnson, L. L. (1967) 'Problems of import-substitution: the Chilean automobile industry', *Economic Development and Cultural Change,* Jan. 1967.

Johnson, S. (1972) *The Green Revolution,* London, Harper and Row.

Jones, B. G. (1986) 'Urban support for rural development in Kenya', *Econ. Geog.,* **62,** No. 3, 201–14.

Jones, W. O. (1968) 'Plantations' in Sills 1968.

Jorgenson, D. W. (1967) 'Surplus agricultural labour and the development of dual economies', *Oxford Econ. Papers,* Nov. 1967.

Kadt, E. de (1980) *Tourism – Passport to Development?* London, New York, Oxford University Press.

Kamarck, A. M. (1976) *The Tropics and Economic Development,* Baltimore and London, Johns Hopkins University Press.

Katzman, M. T. (1977) *Cities and Frontiers in Brazil,* Cambridge, MA, and London, Harvard University Press.

Kay, G. (1967) *A Social Geography of Zambia,* London, University of London Press.

Kenworthy, J. M. and **Glover, J.** (1958) 'The reliability of the main rains in Kenya', *East African Agric. J.,* **23,** 267–71.

Kerr, G. H. (1974) *Formosa 1895–1945,* Honolulu, University Press of Hawaii.

Kessel, J. F. (1961) 'The ecology of filariasis' in May 1961.

Kilby, P. (1969) *Industrialisation in an Open Economy: Nigeria 1945–66,* Cambridge, London, Cambridge University Press.

Killick, T. (1978) *Development Economics in Action: A Study of Economic Policies in Ghana,* London, Heinemann.

Kimmage, K. and **Adams, W. M.** (1990) 'Small-scale farmer-managed irrigation in Northern Nigeria', *Geoforum,* **21,** No. 4, 435–43.

Kloos, H. and **Thompson, K.** (1979) 'Schistosomiasis in Africa: an ecological perspective', *Journal Tropical Geography,* **48,** 1, 31–46.

Knight, C. G. and **Newman, J. L.** (eds) (1976) *Contemporary Africa: Geography and Change,* New Jersey, Prentice-Hall.

Koo, A. Y. C. (1968) *The Role of Land Reform in Economic Development: A Case Study of Taiwan,* New York, Praeger.

Kowal, J. J. and **Kassam, A. H.** (1978) *Agricultural Ecology of Savanna: A Study of West Africa,* London, Oxford University Press.

Krishna, R. (1980) 'The economic development of India' in *Economic Development,* Scientific American 1980, 78–87.

Kuan, H. C. (1979) 'Political stability and change' in Lin *et al.* 1979.

Kuklinski, A. (ed.) (1978) *Regional Policies in Nigeria, India and Brazil*, U.N. Research Inst. for Social and Economic Planning, No. 9, The Hague, Paris, New York, Mouton.

Kuo, S. W. Y., Ranis, G. and **Fei, J. C. H.** (1981a) *Rapid Growth with Improved Distribution in the Republic of China*, Boulder, CO, Westview Press and Praeger.

Kuo, S. W.Y., Ranis, G. and **Fei, J. C. H.** (1981b) *The Taiwan Success Story*, Boulder, CO, Westview Press.

Kuznets, S. (1955) 'Economic growth and income inequality', *Am. Econ. Rev.*, **45**, 1–28.

Kuznets, S. (1979) 'Growth and structural shifts' in Galenson 1979.

Lall, S. and **Streeten, P.** (1977) *Foreign Investment, Transnationals and Developing Countries*, New York, Macmillan.

Lange, O. (1960) *Essays on Economic Planning*, London, Asia Publishing House/Statistical Publishing Society.

Lappiere, D, (1986) *The City of Joy*, London, Century.

Larson, A. (1990) 'The social epidemiology of Africa's AIDS epidemic', *African Affairs*, **89**, 5–25.

Learmonth, A. T. A. (1957) 'Some contrasts in the regional geography of malaria in India and Pakistan', *Trans. Inst. Brit. Geog.*, **23**, 37–59.

Learmonth, A. T. A. (1978) *Patterns of Disease and Hunger*, London, Vancouver, David and Charles.

Learmonth, A. T. A. (1988) *Disease Ecology*, Oxford, Blackwell.

Leeming, F. (1985a) 'Chinese industry – management systems and regional structures', *Trans. Inst. Brit. Geog.*, New Series, **10**, 413–26.

Leeming, F. (1985b) *Rural China Today*, London, Blackwell.

Leeming, F. (1989) 'Rural change and agricultural development', *Geography*, **74**, 348–50.

Leff, N. H. (1982) *Underdevelopment and Development in Brazil*, Vol. 1, Economic Structure and Change 1822–1947; Vol. 2, Reassessing the obstacles to economic growth, London, Boston, Sydney, Allen and Unwin.

Leibenstein, H. (1954) *A Theory of Economic–Demographic Development*, Princeton, New Jersey, Princeton University Press.

Leibenstein, H. (1957) *Economic Backwardness and Economic Growth*, New York, John Wiley.

Leung, C. K. (1979) 'Modernisation of railways' in Chiu 1979.

Leung, C. K. (1980) *China: Railway Patterns and National Goals*, Department of Geography, Research Papers No. 195, Chicago, University of Chicago Press.

Leung, C. K. and **Chiu, T. N.** (eds) (1983) *China in Readjustment*, Centre of Asian Studies, Hong Kong, University of Hong Kong.

Leung, C. K. and **Comtois, C.** (1983) 'Transport reorientation towards the Eighties' in Leung and Chiu 1983.

Lewis, W. A. (1954) 'Economic development with unlimited supplies of labour', Manchester School of Economics and Social Science, Vol. 22, 139–91.

Lewis, W. A. (1955) *The Theory of Economic Growth*, London, Allen and Unwin.

Lin, C. Y. (1981) 'Agricultural and land reform' in Hsiung 1981, 140–3.

Lin, T. B., Lee, R. P. L. and **Simonis, U. E.** (1979) *Hong Kong: Economic, Social and Political Studies in Development*, Institute of Asian Affairs Hamburg, M. E. Sharpe Inc. Dawson.

Little, I. M. D. (1979) 'An Economic Reconnaissance' in Galenson 1979, 448–508.

Little, I., Scitovsky, T. and **Scott, M.** (1970) *Industry and Trade in Some Developing Countries: A Comparative Study*, London, Oxford University Press.

Little, M. (1991) 'Colonial policy and subsistence in Tanganyika 1925–1945', *Geog. Rev.*, **81**, No. 4, 375–88.

Lo, F. C. and **Salih, K.** (eds) (1978) *Growth Pole Strategy and Regional Development Policy*, Oxford, Pergamon Press.

Logan, M. I. (1972) 'The spatial system and planning strategies in developing countries', *Geog. Rev.*, **62**, No. 2, 229–44.

Lowder, S. (1982) 'Agrarian production and development; agrarian reform military style in Peru', *Tijdschrift für Economische Geo.*, **73**, 173–85.

Lucas, R. E. B. and **Papanek, G. F.** (eds) (1988) *The Indian Economy*, Delhi, Oxford University Press.

Mabogunje, A. L. (1965) 'The economic implications of the pattern of urbanisation in Nigeria', *Nigeria J. of Econ. and Soc. Studies*, **7**, 9–30.

Mabogunje, A. L. (1968) *Urbanisation in Nigeria*, London University Press.

Mabogunje, A. L. (1978) 'Growth poles and growth centres in the regional development of Nigeria' in Kuklinski 1978.

Mabogunje, A. L. (1980a) *The Development Process*, London, Hutchinson.

Mabogunje, A. L. (1980b) 'The dynamics of centre–periphery relations: the need for a new geography of resource development', *Trans. Inst. Brit. Geog.*, **5**, 3 (new series), 277–97.

Mackinder, H. J. (1902) *Britain and the British Seas*, London, Heinemann.

Malthus, T. (1798) *Essay on the Principle of Population*, London.

Mansfield, J. E. *et al.* (1974) *Land Resources of the Northern and Luapula Provinces Zambia*, Surbiton, England, Land Resources Division.

Manshard, W. (1968) *Tropical Agriculture*, London, Longman.

Mason, E. S. *et al.* (1980) *The Economic and Social Modernisation of the Republic of Korea*, Cambridge University Press and London, Harvard Press.

Mathur, A. (1966) 'Balanced versus unbalanced growth: a view', *Oxford Econ. Papers*, **18.**

Matzke, G. (1983) 'A reassessment of the expected consequences of tsetse control efforts in Africa', *Social Science and Medicine*, **17**, No. 9, 531–8.

Maude, A. (1973) 'Land shortage and population pressure' in Brookfield 1973, 163–85.

May, J. M. (1958) *The Ecology of Human Disease*, New York, M.D. Publications.

May, J. M. (ed.) (1961) *Studies in Disease Ecology*, New York, Hefner.

McDonald, W. H. (1963) 'Disease in Fiji', *Trans. and Proc. Fiji Society*, **7**, 63–72, Suva, Fiji.

McGee, T. G. (1967) *The Southeast Asian City*, London, Bell.

McGee, T. G. (1971) *The Urbanisation Process in the Third World: Explorations in Search of a Theory*, London, Bell.

McKeown, T. and **Brown, R. G.** (1965) 'Medical evidence related to English population changes in the Eighteenth Century' in Glass and Eversley 1965, 285–307.

Mehureta Assefa (1986) 'Towards a framework for spatial revolution of structural polarity in African development', *Econ. Geog.*, **62**, No, 1, 30–51.

Mikesell, R. F. (1968) *The Economics of Foreign Aid*, Chicago, Aldine.

Miracle, M. P. (1957) *Agriculture in the Congo Basin*, Madison, Milwaukee, London, University of Wisconsin Press.

Mitchison, R. (1977) *British Population Change since 1860*, New York, London, Macmillan.

Mohammad, A. (1978) *The Situation of Agriculture Food and Nutrition in Rural India*.

Molyneux, D. H. and **Ashford, R. W.** (1983) *The Biology of Trypanosoma and Leishmania, Parasites of Man and Domestic Animals*, London, Taylor and Francis.

Morgan, W. B. (1977) *Agriculture in the Third World: A Spatial Analysis*, London, Bell.

Morgan, W. T. W. (1963) 'The "White Highlands" of Kenya', *Geog. J. London*, **129**, 140–55.

Mortimore, M. J. (1967) *'Land and population pressure in the Kano close-settled zone*, northern Nigeria', *Adv. of Sci.*, **23**, 677–86.

Mortimore, M. J. and **Wilson, J.** (1965) *Land and people in the Kano close-settled zone*, Department of Geography, Occasional Paper No. 1, March 1965, Ahmadu Bello University, Zaria, Nigeria.

Morton, K. and **Tulloch, P.** (1977) *Trade and Developing Countries*, London, Croom Helm.

Moses, L. (1955) 'Interregional input–output analysis', *Amer. Econ. Rev.*, **45**, Dec. 1955, 803–32.

Mountjoy, A. B. (ed.) (1971) *Developing the Underdeveloped Countries*, London, Macmillan.

Myint, H. (1964) *The Economics of the Developing Countries*, London, Hutchinson.

Myrdal, G. (1957) *Economic Theory and Underdeveloped Regions*, London, Duckworth.

Myrdal, G. (1968) *Asian Drama*, 3 vols., New York, Twenty Century Fund.

Needleman, L. (ed.) (1968) *Regional Analysis*, London, Penguin.

Nelson, R. R. (1956) 'A theory of the low-level equilibrium trap', *Americ. Econ. Rev.*, **46**, 894–908.

Newman, P. (1970) 'Malaria control and population growth', *J. of Dev. Studies*, **6**, 133–58.

Nicholls, W. H. (1971) 'Agriculture and the economic development of Brazil' in Saunders 1971, 215–56.

Nigeria, Federation of (1962) *National Development Plan 1962–68*, Lagos, Nigerian Government .

Nigeria, Federal Republic of (1970) *Second National Development Plan 1970–74*, Lagos, Nigerian Government.

Nigeria, Federal Republic of (1974) *Third National Development Plan 1974–80*, Lagos, Nigerian Government.

Nigeria, Federal Republic of (1981) *Fourth National Development Plan 1981–85*, Lagos, Nigerian Government.

Nkrumah, K. (1965) *Neo-colonialism: The Last Stage of Imperialism*, New York, International Publishers.

Nongaarts, J. and **Way, P.** (1989) *Geographic variation in the HIV epidemic and the mortality impact of AIDS in Africa*, Research Division Working Paper No. 1, New York, Population Council.

Norman, D. W. (1967 and 1972) *An Economic Study of Three Villages in Zaria Province*, Vol. 1 Land and labour relations, Vol. 2 Input–output Study, Rural Economy Research Unit, Ahmadu Bello University, Samaru, Ahmadu Bellow University.

Norman, D. W. *et al.* (1976) *A Socio-Economic Study of Three Villages in the Sokoto Close-settled Zone*, No. 64, Samaru Miscellaneous Papers, Samaru Nigeria, Ahmadu Bello University.

North, D. C. (1964) 'Location theory and economic growth' in Friedmann and Alonso 1964, 240–55.

Nulty, L. E. (1972) *The Green Revolution in West Pakistan*, New York, Praeger.

Nurkse, R. (1953) *Problems of Capital Formation in Underdeveloped Countries*, Oxford, Blackwell.

Nye, P. H. and **Greenland, D. J.** (1960) *The Soil Under Shifting Cultivation*, Bureau of Soils, Reading, Reading Commonwealth Bureau of Soils.

Odell, P. R. and **Preston, D. A.** (1973) *Economies and Societies in Latin America*, London, New York, Sydney, Toronto, John Wiley.

Odingo, R. S. (1971) *The Kenya Highlands: Land Use andAgricultural Development*, Nairobi, East Africa Publishing House.

Okafor, F. C. (1985) River-basin management and food crisis in Nigeria, *Geoforum*, **16**, No. 4, 413–21.

Okri, B. (1991) *The Famished Road*, Jonathan Cape.

Ooi, J. B. and **Chang, H. D.** (eds) (1969) *Modern Singapore*, Singapore University Press.

Ousmane, S. (1970) *God's Bits of Wood*, London, Heinemann.

Owens, R. J. (1963) *Peru*, London, Oxford University Press.

Pacione, M. (ed.) (1981) *Problems and Planning in Third World Cities*, London, Croom Helm.

Paige, J. M. (1975) *Agrarian Revolution*, New York, The Free Press, London, Macmillan.

Palma, G. (1978) 'Dependency: a formal theory of underdevelopment or a methodology for the analyses of concrete situations of underdevelopment?', *World Development*, **6**.

Pan American Union (1959) *Plantation Systems of the World*, Washington Social Science Monographs No. VII, Washington, DC, Pan American Union.

Pannell, C. W. (1973) *T'ai-Chung, T'ai-Wan: Structure and Function*, Research paper No. 144, Dept of Geography, Chicago, Chicago University Press.

PANOS Dossier (1986) *AIDS and the Third World*, PANOS in association with the Norwegian Red Cross, London.

Parr, J. B. (1973) 'Growth poles, regional development and central place theory', *Regional Sc. Assoc., Pages and Proceedings*, **31**, 173–212.

Paukert, F. (1973) 'Income distribution at different levels of development', *Internat. Labour Rev.*, **108**, 97–124.

Paus, E. (1989a) 'Direct foreign investment and economic development in Latin America: perspective for the future', *J. Latin Americ. Studies*, **21**, 221–39.

Paus, E. (1989b) 'The political economy of manufactured export growths: Argentina and Brazil in the 1970s', *J. Dev. Areas*, **23**, 221–39.

Peet, R. (1983) Introduction: the global geography of contemporary capitalism', *Econ. Geog.*, **59**, 103–11.

Perloff, H. and **Wingo, L.** (1964) 'Natural resource endowment and regional economic growth' in Friedmann and Alonso 1964, 215–39.

Perroux, F. (1950) 'Economic space: theory and applications', *Quarterly J. of Econ.*, **64**, 89–104. Also in Friedmann and Alonso 1964, 21–36.

Peters, D. U. (1950) *Land usage in Serenji district*, Rhodes-Livingstone Paper No. 19, Rhodesia, Rhodes Livingstone Institute.

Phillips, J. (1961 and 1966) *The Development of Agriculture and Forestry in the Tropics*, London, Faber and Faber.

Population Reports (1986) *AIDS and public health crisis*, Series L, 6 Population Information Program, Baltimore and Washington, DC, Johns Hopkins University.

Population Today (1988) 'UN raises its global projections', *Population Today*, **16**, No. 12, 3.

Prebisch, R. (1959) 'The role of commercial policies in underdeveloped countries', *Am. Econ. Rev., Papers and Proceedings*, May 1959.

Preston, D. A. (1965) 'Changes in the economic geography of banana production in Ecuador', *Trans. Inst. Brit. Geog.*, **37**, 77–90.

Prothero, R. M. (1957) 'Land use at Soba: Zaria province Northern Nigeria', *Econ. Geog.*, **33**, 72–86.

Prothero, R. M. (1965) *Migrants and Malaria*, London, Longman.

Prothero, R. M., Zelinsky, W. and **Kosinski, L. A.** (eds) (1970) *Geography and a Crowding World,* New York, Oxford University Press.

Psacharopoulos, G. (1990) 'Poverty alleviation in Latin America', *Finance and Development,* March, 17–19, Washington, World Bank.

Radice, H. (1975) *International Firms and Modern Imperialism,* London, Harmondsworth, Penguin.

Ranis, G. (1974) 'Taiwan' in Chenery *et al.* 1974, 285–90.

Ranis, G. (1979) 'Appropriate technology in the dual economy: reflections on Philippine and Taiwanese experience' in Robinson 1979, 140–59.

Ranis, P. (1971) *Five Latin American Nations,* London, Macmillan.

Rawski, T. G. (1979) *Economic Growth and Employment in China,* London, Oxford University Press.

Reuber, G. L. (1973) *Private, Foreign Investment in Development,* London, Oxford University Press.

Reutlinger, S. (1985) 'Food security and poverty in LDCs', *Finance and Development,* **22,** No. 4, 7–11, New York and London, World Bank.

Rhynsburger, W. (1956) *Area and Resources Survey: Taiwan,* Taipei, International Co-operation, Administration and Mutual Security Mission to China.

Richards, A. I. (1958) 'A changing pattern of agriculture in East Africa: the Bemba of northern Rhodesia, *Geog. J.,* **124,** No. 3, 302–14.

Richards, A. I. (1961) *Land, Labour and Diet in Northern Rhodesia,* International African Institute, London, Oxford, Oxford University Press.

Richards, A. I. (1968) 'The Bemba of north-eastern Rhodesia' in Colson and Gluckman 1968.

Riddell, J. B. (1970) *The Spatial Dynamics of Modernization in Sierra Leone,* Evanston, North-western University Press. Also in Knight and Newman 1976, 393–407.

Rigg, J. (1989) 'The Green Revolution and equity: who adopts the new rice varieties and why?' *Geog.,* **74,** 144–50.

Rimmer, D. (1981) 'Basic Needs and the origins of the development ethos', *J. Dev. Areas,* **15,** 215–38.

Rios, J. A. (1971) 'The growth of cities and urban development' in Saunders 1971, 269–88.

Riskin, C. (1979) 'Intermediate technology in China's rural industries' in Robinson 1979, 52–74.

Robequin, C. (1954) *Malaya, Indonesia, Borneo and the Philippines,* London, Longman.

Robinson, A. (ed.) (1979) *Appropriate Technologies for Third World Development,* London, New York, Macmillan.

Robinson, D. A. (1964, reprinted 1971) *Peru in Four Divisions,* Lima, American Studies, Detroit, Blain-Ethridge Books.

Robinson, G. and **Salih, K. B.** (1971) 'The spread of development around Kuala Lumpur', *Regional Studies,* **5,** 303–14.

Robinson, J. (1979) *Aspects of Development and Underdevelopment,* Cambridge, Cambridge University Press.

Rosen, G. (1975) *Peasant Society in a Changing Economy: Comparative Development in South East Asia and London,* Urban, University of Illinois Press.

Ross, R. S. J. (1983) 'Facing Leviathan and public policy and global capitalism', *Econ. Geog.,* **59,** 144–60.

Rostow, W. W. (1952) *The Process of Economic Growth*, New York, London, Oxford, Oxford University Press.

Rostow, W. W. (1956) 'The take-off into self-sustained growth', *Econ. J.*, Mar. 1956, 25–48.

Rostow, W. W. (1960) *The Stages of Economic Growth*, Cambridge, Cambridge University Press.

Rostow, W. W. (ed.) (1963) *The Economics of Take-off into Sustained Growth*, London, Macmillan.

Rudolph, J. D. (1992) *Peru: the Evolution of a Crisis*, Westport, CT, London, Praeger.

Russett, B. M. (1967) *International Regions and the International System*, Chicago, Rand McNally.

Ruthenberg, H. (1971) *Farming Systems in the Tropics*, Oxford, Clarendon Press.

Saith, A. (ed.) (1987) *The Re-emergence of the Chinese Peasantry*, London, Croom Helm.

Santos, M. (1979) *The Shared Space: The Two Circuits of the Urban Economy in Underdeveloped Countries*, London, New York, Methuen.

Saunders, J. (ed.) (1971) *Modern Brazil: New Patterns and Development*, Gainsville, Florida, University of Florida Press.

Scammell, W. M. (1980) *The International Economy since 1945*, London, Macmillan.

Schlippe, P. de (1956) *Shifting Cultivation in Africa*, London, Routledge.

Schram, S. R. (ed.) (1974) *Chairman Mao Talks to the People*, New York.

Schultz, J. (1974) *Explanatory Study to the Land Use Map of Zambia*, Munich, Ministry of Rural Development, Govt of Zambia.

Schultz, T. W. (1953) *The Economic Organization of Agriculture*, New York.

Schumpeter, J. A. (1911) (trans. R. Opie 1934) *The Theory of Economic Development: An Enquiry*, Cambridge, MA, Harvard University Press.

Scientific American (1980) *Economic Development*, San Francisco, Oxford, Freeman and Co.

Seavoy, R. E. (1987) 'Hoe shifting cultivation in East African subsistance agriculture', *Sing. J. Trop. Stud.*, **81**, No. 1, 60–71.

Seers, D. (1969) 'The meaning of development', *Internat. Dev. Rev.*, **3**.

Selowsky, M. (1990) 'Stages in the recovery of Latin America's growth', *Finance and Development*, June, 28–31, Washington, World Bank.

Shahid Alam, M. (1989) 'The South Korean Miracle', *J. Dev. Areas*, 233–58.

Shand, R. T. (1969) *Agricultural Development in Asia*, Canberra, Australian National University.

Shand, R. T. (ed.) (1979) *The New Strategy in India: Agriculture in the First Decade*, New Delhi.

Shonfield, A. (1960) *The Attack on World Poverty*, London.

Siddayao, C. M. (1978) *ASEAN and the Multinational Corporations*.

Sills, D. L. (ed.) (1968) *International Encyclopedia of the Social Sciences*, 12, UNESCO.

Simmons, A. B., Diaz-Briquets, S. and **Laguian, A. A.** (1977) *Social Change and Internal Migration: A Review of Research Findings from Africa, Asia and Latin America*, Ottawa, International Research Centre.

Simon, J. L. (1977) *The Economics of Population Growth*, Princeton, NJ, Princeton University Press.

Simpson, E. S. (1969) 'Electricity production in Nigeria', *Econ. Geog.*, **45**, No. 3, 239–57.

Simpson, E. S. (1971a) 'Education, universities and geography in developing countries', Proceedings Sixth New Zealand Geog. Conference, Vol. 2, 118–24, New Zealand Geog. Soc. Conference Series No. 64, Christchurch, New Zealand, New Zealand Geog. Soc.

Simpson, E. S. (1971b) 'Northern Nigeria plans for peace', *Geog. Mag.*, **43**, No. 6, 3.

Simpson, E. S. (1972) 'Economic development in Fiji', *Perspective*, No. 9, 1–7.

Simpson, E. S. (ed.) (1980) *The Rural-Agricultural Sector*, Newcastle, Developing Areas Study Group, Institute of British Geographers.

Simpson, E. S. (1992) 'Energy Resources' in Gleave 1992, 122–52.

Singer, H. (1950) 'The distribution of gains between investing and borrowing countries', *Am. Econ. Rev.*, Papers and Proceedings, May 1950.

Singh, J. S. (1979) *World Population Policies*, UN Fund for Population Activities, New York, Praeger.

Smil, V. and Knowland, W. E. (eds) (1980) *Energy in the Developing World*, Oxford, Oxford University Press.

Smith, A. (ed.) (1987) *The Re-emergence of the Chinese Peasantry*, London, Croom Helm.

Soja, E. W. (1968) *The Geography of Modernisation in Kenya*, Syracuse, University of Syracuse Press.

Soja, E. W. and Tobin. R. J. (1975) 'The geography of modernisation: paths, patterns and processes of spatial change in developing countries' in Brunner and Brewer 1975.

Soligo, R. and Stern, J. J. (1965) 'Tariff protection, import substitution and investment efficiency', *Pakistan Devel_opt. Rev.*, Summer 1965.

Spate, O. H. K. (1959) *The Fijian People:Economic Prospects and Problems*, Council Paper No. 13, Legislation Council of Fiji, Suva, Fiji, Government Press Fiji.

Spate, O. H. K. and Learmonth, A. T. A. (1967) *India and Pakistan*, London, Methuen.

Spencer, J. E. (1966) *Shifting Cultivation in South-eastern Asia*, Berkeley and Los Angeles, University of California Press.

Stenning, D. J. (1959) *Savanna Nomads*, London, Oxford University Press.

Stevens, C. (1979) *Food Aid and the Developing World*, London, Croom Helm.

Stouse, P. A. D. (1970) 'Instability of tropical agriculture: the Atlantic lowlands of Costa Rica', *Econ. Geog.*, **60**, 78–97.

Streeten, P. (1972) *The Frontiers of Development Studies*, London, Macmillan.

Streeten, P. (ed.) (1973) *Trade Strategies for Development*, London, Macmillan.

Susman, P. and Schutz, E. (1983) 'Monopoly and competitive firm relations and regional development in global capitalism', *Econ. Geog.*, **59**, 161–77.

Swami, K. (1973) *Moisture Conditions in the Savanna Region of West Africa*, Montreal Dept of Geog., Savanna Research Centre, McGill University.

Taaffe, E. J., Morill, R. L. and Gould, P. R. (1963) 'Transport expansion in underdeveloped countries', *Geog. Rev.*, **53**, Oct. 1963, 503–29.

Tallroth, N. B. (1987) 'Structural adjustment in Nigeria', *Finance and Development*, Sept, 20–2.

Tang, H. S. and Hsieh, S. C. (1961) 'Land reform and agricultural development in Taiwan' in Froehlick 1961, 114–42.

Tarrant, J. R. (1980) 'The geography of food aid', *Trans. Inst. Brit. Geog.*, **5**, 2, 125–40.

Thom, D. J. and Wells, J. C. (1987) 'Farming systems in the Niger inland delta, Mali', *Geog. Rev.*, **77**, 3, 328–42.

Thorbecke, E. (1979) 'Agricultural development' in Galenson 1979, 263–307.

Thorp, E. (1956) *The Ladder of Bones: The Birth of Modern Nigeria from 1853 to Independence*, London, Jonathan Cape. Also 1966, Fontana Books.

Thumm, U. *et al* (1981) *Peru: Major Development Policy Issues and Recommendations*, Latin American and the Caribbean Regional Office, Washington, DC, World Bank.

Tiebout, C. M. (1956) 'Exports and regional economic growth', *J. of Polit. Econ.*, **64**, April 1956, 160–5.

Tiebout, C. M. (1968) 'Regional and inter-regional input–output models: an appraisal' in Needleman 1968, Chapter 3.

Todaro, M. P. (1985) (3rd edn) *Economic Development in the Third World*, New York and London, Longman.

Tosi, J. A. and Voertmann, R. F. (1964) 'Some environmental factors in the economic development of the Tropics', *Econ. Geog.*, **40**, No. 3, 189–205.

Tranter, W. (1973) *Population Since the Industrial Revolution*, London, Croom Helm.

Tugendhat, C. (1973) *The Multinationals*, Harmondsworth, Penguin Books.

Tully, M. (1992) *No Full Stop in India*, Penguin.

Twyford, I. T. and Wright, A. C. S. (1965) *The Soil Resources and Fiji Islands*, Suva, Fiji, Government Printer.

United Nations Organisation (1949) *Relative Prices of Exports and Imports of Underdeveloped Countries*, New York, UNO.

United Nations Organisation (1973) *Multinational Corporations in World Development*, New York, UNO.

United Nations Organisation (1978) *Transcontinental Corporations in World Development*, New York, UNO.

United Nations Organisation (1989) *World Population Prospects: 1988*, New York, UN Population Division.

Unwin, T. (1983) 'Perspectives on "development"', *Geoforum*, **14**, No. 3, 235–41.

Venezian, E. L. and Gamble, W. (1969) *The Agricultural Development of Mexico; Its Structure and Growth Since 1950*, New York, Praeger.

Vermeer, D. E. (1981) 'Collision of climate, cattle and culture in Mauritania during the 1970s', *Geog. Rev.*, **71**, 281–97.

Vernon, R. (ed.) (1968) *Multinational Enterprise in the 1960s*, London.

Vernon, R. (1977) *Storm over Multinationals: The Real Issues*, New York, Vail-Ballow Press.

Vianna, H. (1961/62) *Historia do Brasil*, 2 vols, Sao Paulo, Melhoramentos de Sao Paulo.

Vieira, D. T. (1971) 'Industrial development in Brazil' in Saunders 1971, 156–89.

Wade, R. (1992) *Governing the Market: Economic Theory and the Role of Government in East Asian Industrialization*, Princeton, NJ, Princeton University Press.

Wallach, B. (1988) 'Irrigation in Sudan since independence', *Geog. Rev.*, **78**, 417–34.

Wallach, B. (1989) 'Improving traditional grassland agriculture in Sudan', *Geog. Rev.*, **79**, No. 2, 143–60.

Walsh, F. (1985) 'Onchoceriasis: river blindness', in Grove 1985, Chapter 10.

Walton, M. (1990) 'Combating poverty: experience and prospects', *Finance & Development*, Sept, Washington, World Bank.

Wang, L. H. and Tan T. H. (1981) 'Singapore' in Pacione 1981, 218–49.

Ward, R. G. (1965) *Land Use and Population in Fiji*, London, HMSO.

Watts, J. and Basset, T. J. (1986) 'Politics, the state and agrarian development: a comparative study of Nigeria and the Ivory Coast', *Polit. Geog. Quat.*, **5**, No. 2, 103–25.

Webster, C. C. and Wilson, P. N. (1966 and 1980) *Agriculture in the Tropics*, London, Longman.

Weeks, J. R. (1988) 'The demography of Islamic nations', *Pop. Bull.*, **43**, No. 4, 3–53, Washington DC.

Wei, Y. (1981) 'Land reform and ownerships' in Hsiung 1981, 144.

Weil, C. and **Kvale, K. M.** (1985) 'Current research on geographical aspects of schistosomiasis', *Geog. Rev.*, **75**, 186–216.

Weinand, H. C. (1973) 'Some spatial aspects of development in Nigeria', *J. of Developing Areas*, **7**, 247–63.

Weitz, R. (1987) 'Rural development: the Rehovet approach', *Geoforum*, **18**, No. 1, 21–36.

Wickizer, V. D. (1958a) 'The plantation system in the development of tropical economies', *J. Farm Econ.*, Feb. 1958.

Wickizer, V. D. (1958b) 'Plantation crops in tropical agriculture', *Tropic. Agric.*, July 1958.

Wickremeratne, A. L. (1977) 'Planning and economic development' in De Silva 1977, 144–71 .

Wiesner, E. (1985) 'Domestic and external causes of the Latin American debt crisis', *Finance and Development*, March, 24–6, Washington, World Bank.

Wilber, C. K. (ed.) (1988) (4th edn) *The Political Economy of Development and Underdevelopment*, New York, Random House.

Wilber, C. K. and **Jameson, K. P.** (1988) 'The paradigms of economic development and beyond', in Wilber 1988.

Wilkie, R. (1971) *San Miguel: A Mexican Collective Ejido,* Stanford, CA, Stanford University Press.

Wilkinson, R. C. (1983) 'Migration in Lesotho: some comparative aspects with particular reference to the role of women', *Geog.*, **6**, 208–24.

Williams, G. J. (1977) 'Zambia balances Zambezi power', *Geog. Mag.*, **49**, 10 July, 611–14.

Windstrand, C. (ed.) (1975) *Multinational Firms in Africa.*

Wisseman, C. L. and **Sweet, B. J.** (1961) 'The ecology of dengue' in May 1961, 15–44.

Wolf, E. R. and **Mintz, S. W.** (1957) 'Haciendas and plantations in middle America and the Antilles', *Social and Econ. Studies*, Sept. 1957.

Wolf-Phillips, L. (1979) 'Why Third World?', *Third World Quarterly*, **1**, No. 1, 105–16.

Wood, A. P. and **Smith, W.** (1984) 'Zambia up for grabs? The regionalisation of agricultural aid', *Area*, **16**, No. 1, 3–8.

World Bank, The International Bank for Reconstruction and Development (1953) *The Economic Development of Ceylon,* Baltimore, World Bank.

World Bank, The International Bank for Reconstruction and Development (1955) *The Economic Development of Nigeria,* Baltimore, World Bank.

World Bank, The International Bank for Reconstruction and Development (1974a) *Population Policies and Economic Development,* Baltimore and Washington, DC, Johns Hopkins University Press.

World Bank, The International Bank for Reconstruction and Development (1974b) *Country Economy Report: Nigeria Options for Long-Term Development,* Washington, DC, World Bank.

World Bank, The International Bank for Reconstruction and Development (1975) *Rural Electrification,* Washington, DC, World Bank.

World Bank, The International Bank for Reconstruction and Development (1979a) *World Development Report 1979,* New York, London, Oxford University Press.

World Bank, The International Bank for Reconstruction and Development (1979b) *Family Planning Programs: An Evaluation of Experience,* World Bank, Staff Working Paper No. 345, July 1979, Washington, DC, World Bank.

World Bank, The International Bank for Reconstruction and Development (1980a) *World Bank Development Report 1980,* New York, London, Oxford University Press.

World Bank, The International Bank for Reconstruction and Development (1980b) *Energy in the Developing Countries*, Washington, DC, World Bank.

World Bank, The International Bank for Reconstruction and Development (1981) *World Bank Development Report 1981*, New York, London, Oxford University Press.

World Bank, The International Bank for Reconstruction and Development (1982) *World Bank Development Report 1982*, New York, London, Oxford University Press.

World Bank, The International Bank for Reconstruction and Development (1983) *World Bank Development Report 1983*, New York, London, Oxford University Press.

World Bank, The International Bank for Reconstruction and Development (1984) *World Bank Development Report 1984*, New York, London, Oxford University Press.

World Bank, The International Bank for Reconstruction and Development (1987) *World Development Report 1987*, New York, London, Oxford University Press.

World Bank, The International Bank for Reconstruction and Development (1988) *World Development Report 1988*, New York, London, Oxford University Press.

World Bank, The International Bank for Reconstruction and Development (1989) *World Development Report 1989*, New York, London, Oxford University Press.

World Bank, The International Bank for Reconstruction and Development (1990) *World Development Report 1990*, New York, London, Oxford University Press.

World Bank, The International Bank for Reconstruction and Development (1991) *World Development Report 1991*, New York, London, Oxford University Press.

World Bank, The International Bank for Reconstruction and Development (1992) *World Development Report 1992*, New York, London, Oxford University Press.

World Bank, The International Bank for Reconstruction and Development (1993) *World Development Report 1993*, New York, London, Oxford University Press.

World Health Organization (WHO) (1987) Expert Committee on Onchoceriasis, Third Report, *Technical Report 752*, Geneva, WHO.

Wu, Y. L. (1981) 'The Taiwan experience in economic development' in Hsiung 1891, 119–29.

Yang, T. M. (1970) *Socio-Economic Results of Land Reform in Taiwan*, Honolulu, Hawaii, East West Centre Publication.

Young, A. (1992) 'A tale of two cities: factor accumulation and technical change in Hong Kong and Singapore', *Microeconomics Annual*, National Bureau of Economic Research, Washington, USA.

Yuan Tien, H. (1989) 'Second thoughts on the second child', *Population Today*, **17**, No. 4, 6–8, Pop. Ref. Bureau.

Yuan Tien, H. (1991) 'The new census of China', *Population Today*, **19**, No. 1, 6–8, Pop. Ref. Bureau.

Zambia ,Republic (1966) *First National Development Plan 1966–70*, Lusaka, Zambia, Office of National Development and Planning.

Zambia, Republic (1972) *Second National Development Plan 1972–76*, Lusaka, Zambia, Office of National Development and Planning.

Zinyama, L. M. (1986) 'Agricultural development policies in the African farming areas of Zimbabwe', *Geog.*, **71**, 105–15.

Index